AF850287

Aging and Cell Structure

Volume 2

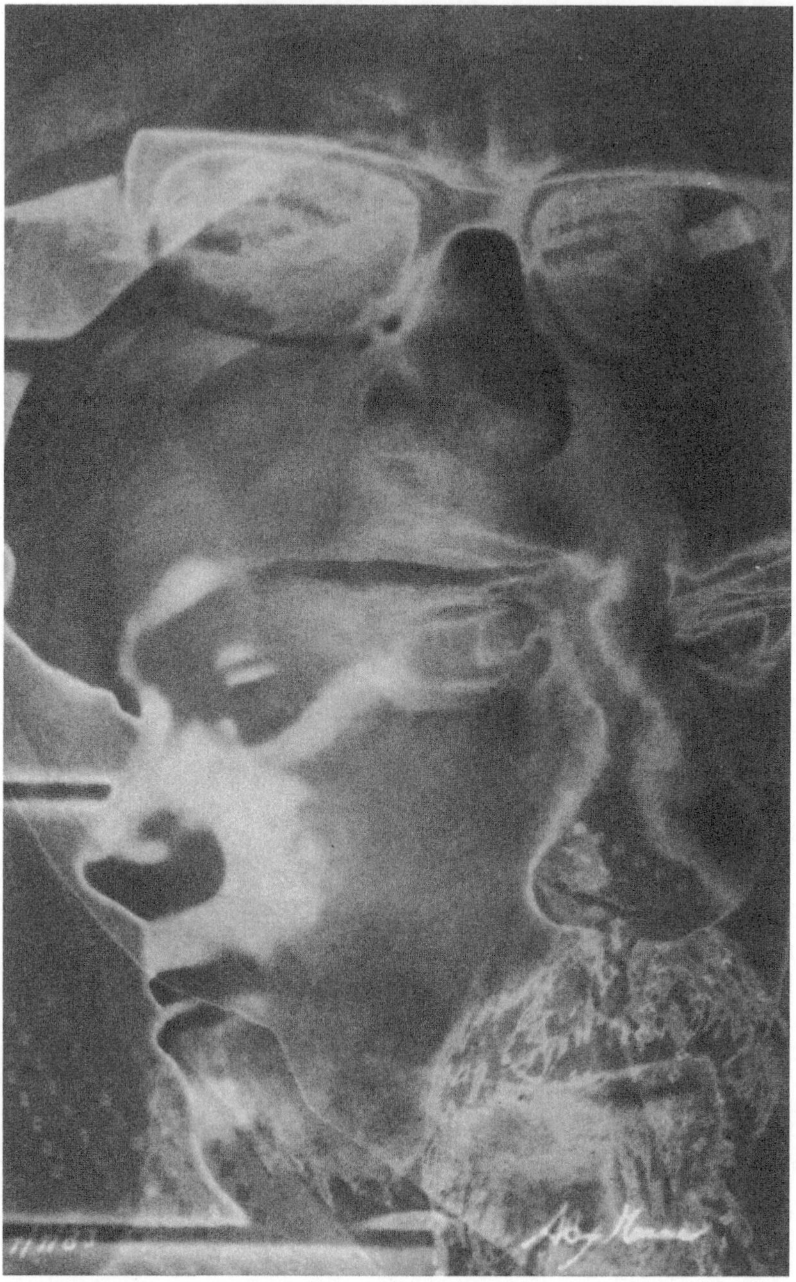

The frontispiece, a solarized photograph superimposing family members from three generations, is by Mr. Alex Marcus of New Paltz, New York, and was especially designed for this volume.

NORTH CAMPUS LIBRARY
Saddleback Community College District

Aging and Cell Structure

Volume 2

Edited by

John E. Johnson, Jr.

Department of Neurology
Johns Hopkins University School of Medicine
Baltimore, Maryland
and
National Institute on Aging, NIH
Baltimore City Hospital
Baltimore, Maryland

PLENUM PRESS • NEW YORK AND LONDON

Library of Congress Cataloging in Publication Data

Main entry under title:

Aging and cell structure.

Includes bibliographies and index.
1. Cells—Aging. I. Johnson, John E., 1945– . [DNLM: 1. Aging. 2. Cells. WT 104 A2664]
QH608.A37 574.87′6 81-17886
ISBN-13: 978-1-4684-4609-8 e-ISBN-13: 978-1-4684-4607-4
DOI: 10.1007/978-1-4684-4607-4

© 1984 Plenum Press, New York
Softcover reprint of the hardcover 1st edition 1984
A Division of Plenum Publishing Corporation
233 Spring Street, New York, N.Y. 10013

All rights reserved

No part of this book may be reproduced, stored in a retrieval system, or transmitted in any form or by any means, electronic, mechanical, photocopying, microfilming, recording, or otherwise, without written permission from the Publisher

"I was thirty. Before me stretched the portentous, menacing road of a new decade.... Thirty—the promise...of loneliness,...a thinning briefcase of enthusiasm, thinning hair.... As we passed over the dark bridge her wan face fell lazilv against my coat's shoulder and the formidable stroke of thirty died away with the reassuring pressure of her hand."

—F. SCOTT FITZGERALD
The Great Gatsby
1925

This volume is dedicated to my friends.

Contributors

SOL BERNICK Department of Anatomy and Cell Biology, University of Southern California School of Medicine, Los Angeles, California 90033

JOHN E. JOHNSON, JR. Department of Neurology, Johns Hopkins University School of Medicine, Baltimore, Maryland 21205; and National Institute on Aging, NIH, Baltimore City Hospital, Baltimore, Maryland 21224

ROBERT O. KELLEY Department of Anatomy, The University of New Mexico School of Medicine, Albuquerque, New Mexico 87131

JOHN D. METTER Department of Pathology, The University of Texas Health Science Center at San Antonio, San Antonio, Texas 78284

WENDELIN J. PAULE Department of Anatomy and Cell Biology, University of Southern California School of Medicine, Los Angeles, California 90033

THOMAS B. POOL Department of Anatomy, The University of Texas Health Science Center at San Antonio, San Antonio, Texas 78284

JULIE L. ROSENHEIMER Department of Physiology, University of Wisconsin, Madison, Wisconsin 53706

BERNARD G. SLAVIN Department of Anatomy and Cell Biology, University of Southern California School of Medicine, Los Angeles, California 90033

DEAN O. SMITH Department of Physiology, University of Wisconsin, Madison, Wisconsin 53706

KATHYRN G. VOGEL Department of Biology, The University of New Mexico College of Arts and Sciences, Albuquerque, New Mexico 87131

Preface

Although thousands of articles and hundreds of books on aging have been published, only a small percentage of this material has dealt with anatomy, particularly at the fine-structural level. It was with this in mind that *Aging and Cell Structure* was conceived.

Volume 1 of *Aging and Cell Structure* was published in 1981 and represented a current compilation of information, concentrating at the electron microscopic level, on morphological changes which occur in cells and tissues as they age. The present volume completes the two-volume set.

While Volume 1 highlighted structural changes occurring in the aging nervous system, Volume 2 centers its efforts on studies of *in vitro* aging. Chapters on other subjects are included as well. These include age-related changes seen in neuromuscular junctions, oral tissues, and the pancreas.

Although these two volumes represent a very small part of the published information on experimental gerontology, their approach is rather unique because they focus on anatomy, perhaps the most basic of all the biomedical sciences. Because many different tissue types are examined, we begin to see recurrent, definitive patterns in the aging cell which may not be fully apparent from studies taking one cell type at a time. This becomes even more evident in the present volume where changes seen in populations of cells grown in culture—isolated from hormones or nervous impulses from other body areas—are found to be similar to those changes found *in vivo.*

Aging and Cell Structure will be of vital interest to cell biologists, pathologists, physicians, anatomists, biochemists, physiologists, and all other scientists who wish to have up-to-date information which will provide an understanding of basic changes occurring in cells and tissues as they age.

John E. Johnson, Jr.

Baltimore

Contents

Chapter 5. Human Oral Tissues as a Model for the Study of Cellular Age Changes by Light Microscopy
Sol Bernick

Chapter 6. Morphological Changes in the Aging Mammalian Pancreas
Bernard G. Slavin, Wendelin J. Paule, and Sol Bernick

Aging and Cell Structure

Volume 2

1

The Aging Cell Surface

Structural and Biochemical Alterations Associated with Progressive Subcultivation of Human Diploid Fibroblasts

ROBERT O. KELLEY and KATHRYN G. VOGEL

1. INTRODUCTION

The cell membrane is a complex fluid mosaic of lipid and protein (Bretscher, 1973; Singer and Nicolson, 1972; Steck, 1974; Shinitzky and Henkart, 1979) between the cellular interior and the immediate extracellular environment. It provides the foundation for the structural and molecular mechanisms that govern cell-to-cell interaction, external signal reception, cell adhesion and motility, and establishment and maintenance of cell shape. Signal reception and intercellular communication play important roles in the control of cell cycle kinetics and proliferation (Clarkson and Baserga, 1974). Proteins which effect cell shape and motility also participate in mitotic events and the interphase stages of G_1, S, and G_2 (Baserga, 1968), in addition to establishing and maintaining contact with other cells and/or their normal or artificial substrates (Adelstein *et al.*, 1979; Weihing, 1979).

Numerous recent studies have discovered important differences in cell surfaces of normal, aging, and transformed cells. Indeed, many investigators have come to view this external boundary of the cell as a mediator of signal transduction and as a major component of the complex regulatory mechanisms which control growth. Hayflick (1965) and others (Mellman and Cristofalo, 1972; Macieira-Coelho, 1973; Wright and

---

ROBERT O. KELLEY • Department of Anatomy, The University of New Mexico School of Medicine, Albuquerque, New Mexico 87131. ***KATHRYN G. VOGEL*** • Department of Biology, The University of New Mexico College of Arts and Sciences, Albuquerque, New Mexico 87131.

Davison, 1980; Packer *et al.*, 1967; Hayflick and Moorhead, 1961; Martin *et al.*, 1970) have suggested that the functional losses which occur in cells prior to their loss of division potential produce age-related changes in animal cells before those cells have reached their maximum division limit. Hayflick and Moorhead (1961) have shown that this degeneration is not related in any simple, direct way to inadequate nutrition, pH variation, toxic metabolic products, or microcontaminants (however, see Bell *et al.*, 1978, for additional discussion). They have suggested that the limited life span of human diploid fibroblasts is probably programmed and reflects a cellular expression of aging. However, it is becoming increasingly clear that the biological processes which control and regulate senescence are exceedingly complex (Wright and Davison, 1980) and one must investigate the molecular mechanisms which underlie the aging process from levels which integrate several cellular compartments into functional units. In addition, it is far from clear whether age-related alterations expressed at the cell surface play a role in either the regulation, determination, or control of cell behavior and aging *in vivo* or *in vitro.*

In contrast, it is well documented that cell cycle kinetics (Macieira-Coelho, 1977; Ryan, 1979; Stein and Yanishevzky, 1979; Kapp and Klevecz, 1976); proliferation of cell populations (Hayflick, 1965; Schneider and Fowlkes, 1976; Mitsui and Schneider, 1976a); intercellular communication (Kelley, 1976); metabolic cooperation (Kelley *et al.*, 1979; Kelley and Perdue, 1980); adhesiveness and motility (Kelley *et al.*, 1980; Crusberg *et al.*, 1979); and the molecular composition and organization of the external surface (Kelley and Skipper, 1977; Kelley *et al.*, 1978; Aizawa and Mitsui, 1979; Aizawa and Kurimoto, 1979; Aizawa *et al.*, 1980a,b; Yamamoto *et al.*, 1977; Schachtschabel and Wever, 1978) are dramatically altered as human diploid fibroblasts age *in vitro.* However, the mechanisms which affect these structural and molecular alterations in the compartments which comprise the cell surface are virtually unknown. Considering these deficits in our understanding of the aging cell surface, the specific aim of this review will be (1) to emphasize the cellular heterogeneity which exists in cultures of early and late passage human diploid fibroblasts; (2) to review what is known about the organization of cell membranes during cellular aging; (3) to examine the nature of alterations in the external cell surface which develop in association with the onset of cellular senescence; and (4) to question whether differentiation of cell membrane organization and surface expression of macromolecules may be correlated with the cytoskeleton. Furthermore, we will discuss whether these variants may affect the alterations in cell behavior observed during the senescent phase of cell viability in culture.

It is our intent in this review to limit our observations to the phase III behavior of human diploid fibroblasts *in vitro.* Furthermore, we wish to bring together information which may provide insights into the functional integration of the cell surface and subjacent cytoplasm as a potential regulatory component in the complex phenomenon termed "cellular aging." It is not our intent to provide a comprehensive documentation of all changes that have been reported in cells as they enter phase III senescence. Rather, we have chosen the more narrow path of using our own investigations of the external surface, cell membrane, and cytoskeleton for the following account, and to refer to recent studies of our colleagues and other investigators where appropriate within that context.

Finally, we wish to note that most of the investigations described in this review

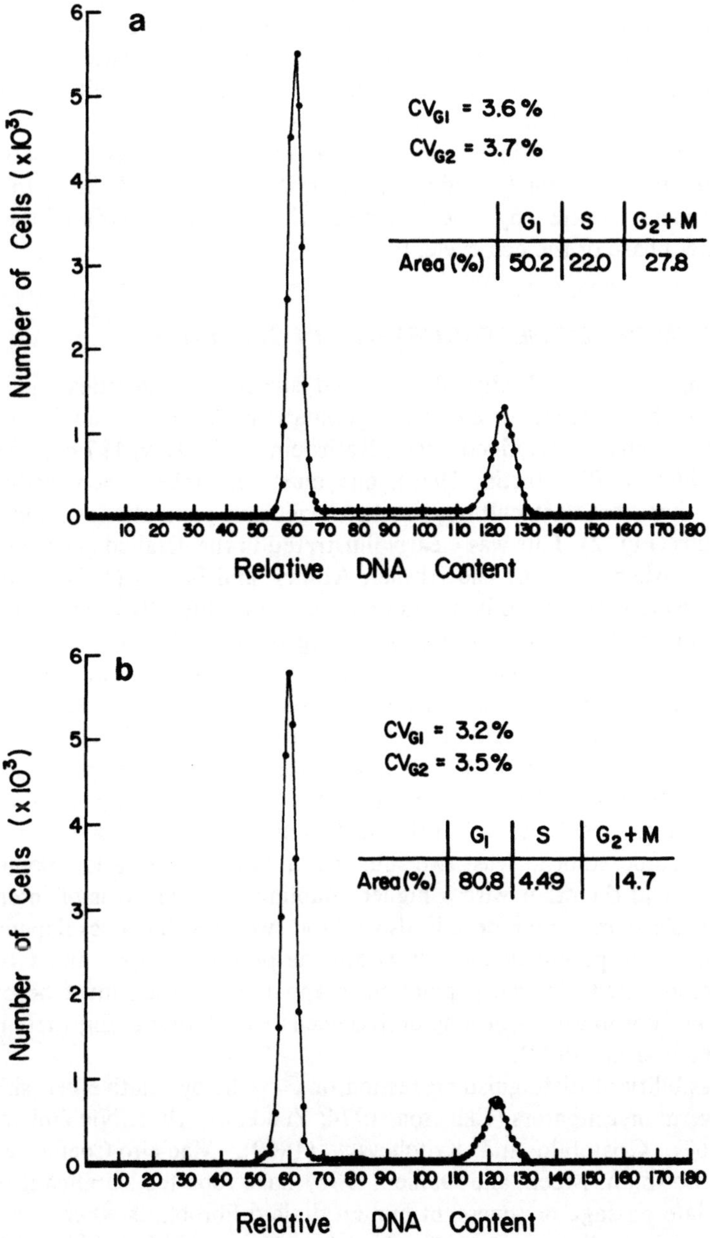

Figure 1. Flow cytometry histograms revealing relative DNA content of cells in cultures of (a) early- (PDL15) and (b) late- (PDL50) passage IMR-90 diploid fibroblasts. The G_1 distribution is set at channel 60 whereas the $G_2 + M$ profile appears approximately at channel 120. S phase distributions are less than 5% in late-passage cultures, indicating a reduction of proliferation by these cells. In addition, cells cycling through $G_2 + M$ are also reduced from approximately 30% to less than 15%.

have been directed to the IMR-90 diploid fibroblast (Nichols *et al.*, 1977). Phase II, a period of exponential growth for multicellular cultures, exists generally through population-doubling levels (PDL) in the mid-40s. However, some variation is noted concerning the time of onset of phase III (senescence). In general, we wait until cultures (a) have reached population-doubling levels in excess of fifty, (b) have failed to exhibit the ability to form monolayers in culture flasks at the same rate as earlier PDL groups seeded at the same cell density, and (c) have exhibited a reduced ability to traverse the cell cycle as demonstrated by flow cytometry (i.e., less than 5.0% of the population entering the S phase of the cell cycle, Fig. 1).

2. DEVELOPMENT OF HETEROGENEITY IN CULTURES

In considering the host of physiological and structural properties that have been reported to be characteristic of aging cell populations (Mellman and Cristofalo, 1972; Rosen, 1978; Grinna, 1977; Good, 1975; Rothstein, 1975; Kay, 1979; Choe and Rose, 1976; Schneider, 1979; Martin, 1979), one must remember when evaluating these observations that fibroblast cultures are heterogeneous populations of individual cells and cell groups (Fig. 2). This was clearly illustrated in the detailed geneological studies of Absher and Absher (1976) and Absher, Absher, and Barnes (1974). These investigators report wide variation in intermitotic time and proliferative capability of individual cells, whether they were taken from young or old cultures. Indeed, cells which exhibit morphological and behavioral characteristics of senescence are found in cultures of all ages. Furthermore, heterogeneity in cell volume (Mitsui and Schneider, 1976b); cell density (Mitsui and Schneider, 1976a); number of nuclei (Mitsui and Schneider, 1976c; Kaji and Matsuo, 1978, 1979); in addition to general morphological variance (Wolosewick and Porter, 1977) has been reported in cultures of diploid fibroblasts progressively subcultivated *in vitro.* Smith and Hayflick (1974) noted that mass cultures appear to be composed of at least two subpopulations: one with a low population doubling potential and the other with a higher doubling potential. It is of interest to note that skin fibroblasts in human dermis also exhibit two growth classes depending on the site of origin of the parent donor cell within the dermis (Harper and Grove, 1979). Although heterogeneity in cell population is apparent, caution must be exercised in comparing skin fibroblasts with lung-derived cells when considering their behavior *in vitro* (Schneider *et al.*, 1977).

The possibility of distinguishing resting or G_0 cells by relative cell size has been raised by several investigators (Clarkson, 1974; Yen *et al.*, 1975; Nicolini *et al.*, 1977). Simons (1967), Cristofalo and Kritchevsky (1969), Macieiro-Coelho and Ponten (1969), and Bowman, Meek, and Daniel (1975) have reported significant increases in cell sizes in late passage or senescent human diploid fibroblasts when compared with similar cultures in earlier passages. Furthermore, Mitsui and Schneider (1976b) demonstrated that separation of senescent populations of diploid fibroblasts on the basis of cell volume revealed that cell fractions with the largest modal volume contained the highest percentage of slow or nonreplicating cells. They concluded that an inverse relationship exists between growth rate and cell volume in cultured human diploid fibroblasts. It is of interest to note that subsequent flow microfluorimetric analyses (Schneider and Fowlkes, 1976) of cell volume and DNA content revealed that enlarged, late-passage cells were not limited to any one phase of the cell cycle but included both

G_1 and G_2 populations. Unfortunately, regardless of the method used for separation of cycling from noncycling cells, separation is not absolute. Inevitably, cells in the noncycling subpopulation (presumably the senescent cells in the culture) exit a latent block in the cycle and resume their normal properties of division, resulting in subsequent growth of the culture.

We have attempted to reduce heterogeneity in cultures by means of flow cytometry and cell sorting, using the basic parameter of light scatter. Building on the observation

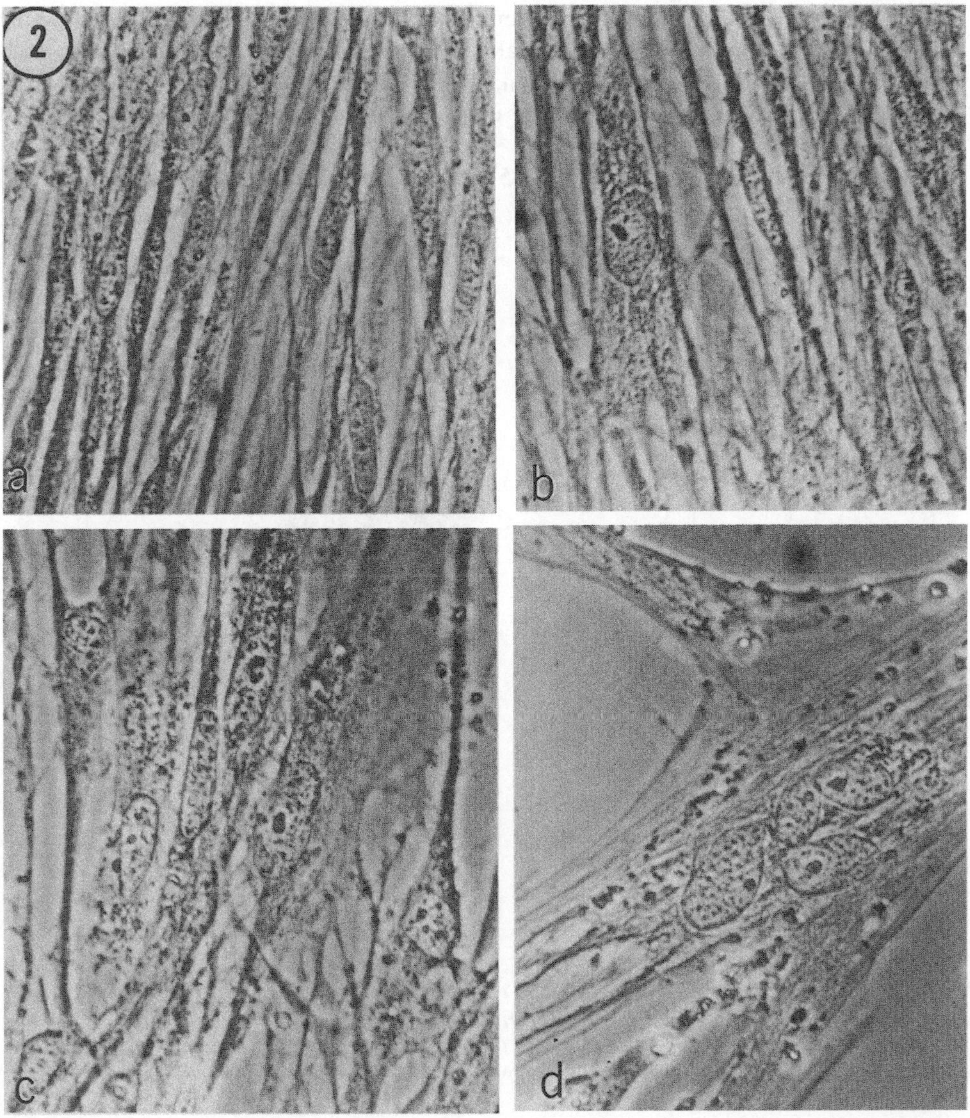

Figure 2. (a–d) Phase contrast micrographs of human embryo fibroblasts (IMR-90) at progressive stages of subcultivation. Cells were cultured on polystyrene in Eagle's Minimal Essential Medium + 10% newborn bovine serum and photographed *in situ* after fixation with 0.2 *M* $HClO_4$. (a) PDL 19, (b) PDL 38, (c) PDL 53, and (d) PDL 57. × 1000.

of Mitsui and Schneider (1976a,b), one would predict that large-volume cells would demonstrate altered properties of forward angle light scatter and could be viably sorted on that basis. The fluorescence-activated cell-sorting (FACS) system uses small forward (1–13°) angle light scatter as an index of cell size (Arndt-Jovin and Jovin, 1978). However, caution must be exercised in the direct application of light scatter to sorting of late-passage, large-volume cells, since scatter is also influenced by such properties as cellular asymmetry, reflectivity, refractive index, granularity, and nuclear size (Herzenberg and Herzenberg, 1978). To this end, increased nuclear sizes (Mitsui and Schneider, 1976c), polyploid cell accumulation (Kaji and Matsuo, 1979), and the altered external and internal structural features associated with late-passage cells (Wolosewick and Porter, 1977; Johnson, 1979) all lend themselves to the successful sorting of a heterogeneous, late-passage culture into subpopulations based on forward angle light scatter.

Cells in early passage (e.g., PDL 15), when removed from culture substrates and

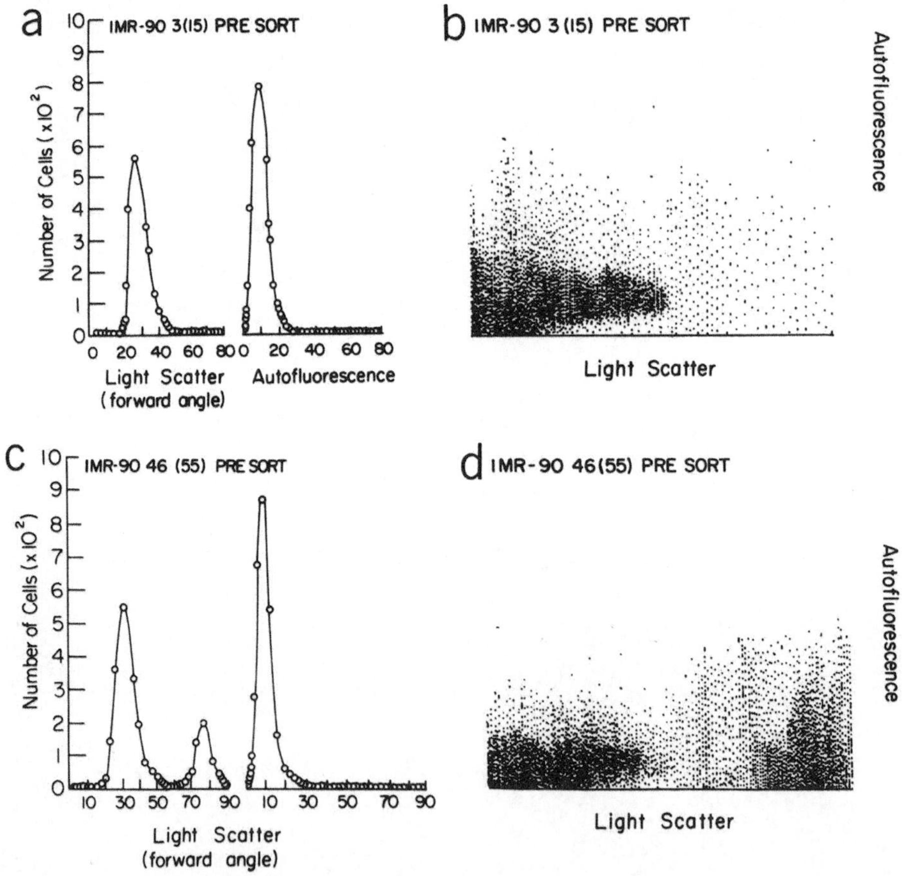

Figure 3. Pulse height histograms of light scatter and autofluorescence exhibiting (a) early- and (c) late-passage cultures prior to flow sorting. (b) and (d) are "dot-plots" of presorted populations which illustrate combined functions of light scatter and autofluorescence. Note the two populations of cells with different properties of light scatter in late-passage cultures (d).

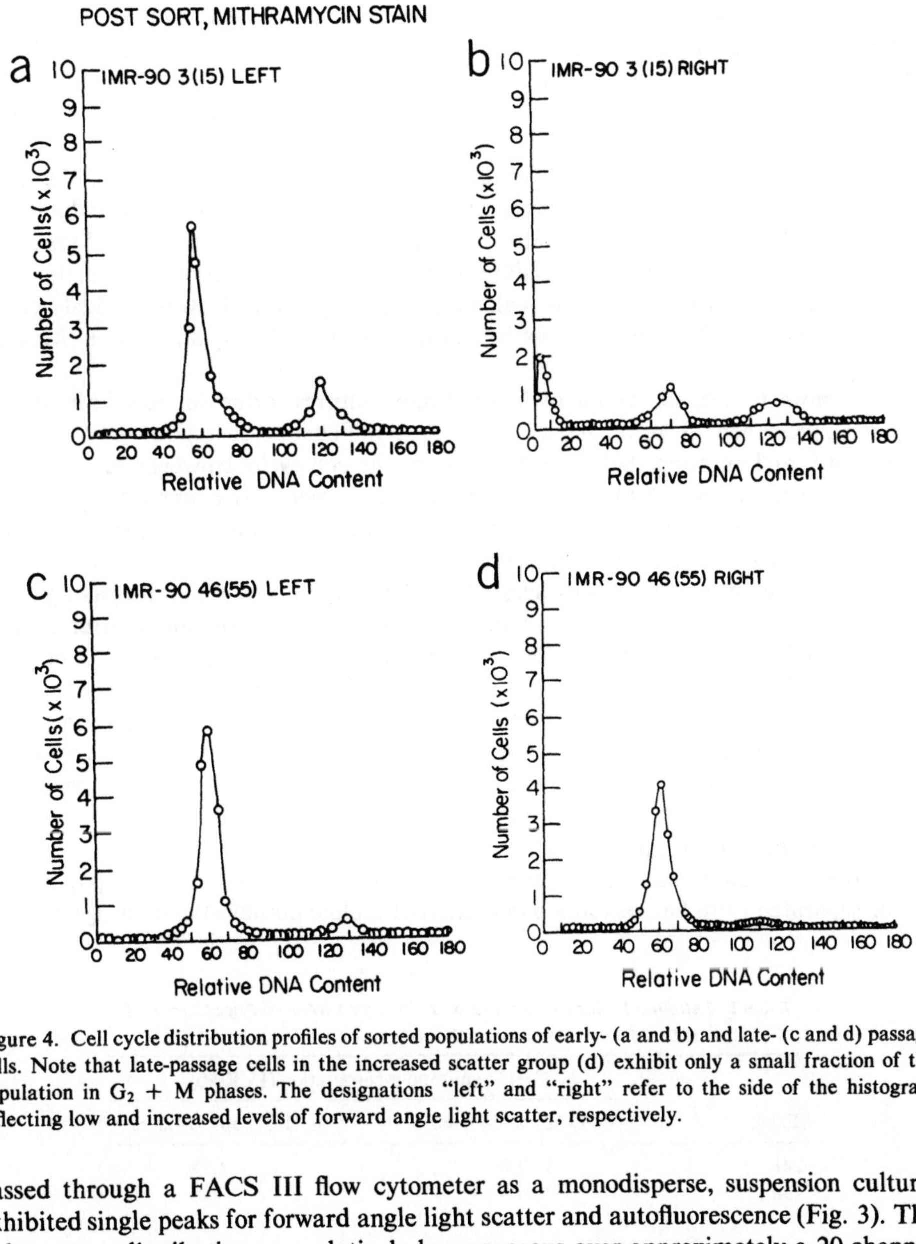

Figure 4. Cell cycle distribution profiles of sorted populations of early- (a and b) and late- (c and d) passage cells. Note that late-passage cells in the increased scatter group (d) exhibit only a small fraction of the population in G_2 + M phases. The designations "left" and "right" refer to the side of the histogram reflecting low and increased levels of forward angle light scatter, respectively.

passed through a FACS III flow cytometer as a monodisperse, suspension culture, exhibited single peaks for forward angle light scatter and autofluorescence (Fig. 3). The light scatter distribution was relatively homogeneous over approximately a 20-channel range. Autofluorescence was similarly homogeneous in distribution for early-passage cells. When both parameters were considered concomitantly, it was clear that a large, relatively homogeneous population of cells shared similar properties of light scatter and autofluorescence.

In contrast, late-passage cultures (PDL 55) exhibited at least two subpopulations of cells with differing light-scattering characteristics. One subpopulation demonstrated lower forward angle light scatter at channels 20–40 when compared to a second population with higher light-scattering properties detected at channels 70–90. Late-passage cultures exhibited patterns of autofluorescence which were not remarkably different from early-passage cells. A "dot-plot" of late passage cultures clearly illustrates that the two subpopulations exist in late passage cultures based upon differences in light scatter and autofluorescence (Fig. 3).

Following DNA staining with mithramycin (Tobey and Crissman, 1975), early passage cells recovered in the prominent, low-scatter-sorted population, exhibited a profile with the prevalent G_1, S, and G_2 + M distribution characteristic of unsorted, early passage cultures (Fig. 4). In addition, the small population of high-scatter cells present in early-passage cultures revealed staining properties which mimicked normal cycle distribution, although the relative number of cells in G_1 and G_2 + M were reduced.

In contrast, subpopulations of late-passage cultures sorted for low light scatter revealed a prominent G_1 peak, and a modest but significant percentage of the population traversing S and entering the G_2 + M phase of the cell cycle. However, the subpopulation with properties of high light scatter exhibited essentially a single G_1 peak with little evidence that a significant percentage of the sorted population was traversing the G_1/S boundary to initiate an additional cycle.

These observations were corroborated by the inability of sorted, late-passage cells exhibiting high-light-scattering properties to incorporate [^{3}H]thymidine after 24 h of incubation (Table I). Only 0.5% of these cultures exhibited labeled nuclei in contrast to 37.8% of the late-passage culture sorted with low-light-scattering properties. It is of interest to note that 59.2% of early-passage cultures sorted for low light scattering incorporate labeled thymidine, a figure remarkably close to the labeling index of an unsorted, early-passage culture after identical periods of incubation. The numbers of early passage cells with high-light-scattering properties were insufficient for subculturing, radiolabeling, and preparation for autoradiography.

From these observations, we have learned that at least two viable subpopulations can be separated from late-passage populations of diploid fibroblasts; the subgroup with

Table I. Labeling Indices of Sorted and Unsorted (Mixed) Populations of IMR-90 Fibroblasts in Both Early and Late Passage[a]

	[^{3}H]Thymidine incorporation (IMR-90)			
Mixed	(PDL 33) % Labeled		(PDL 56) % Labeled	
24h	54.2%		1.5%	
48h	71.6%		5.4%	
	(PDL 20) % Labeled		(PDL 55) % Labeled	
Sorted	Low	High	Low	High
3h	2.7%	—	0.5%	0.0%
24h	59.2%	—	37.8%	0.5%

[a]Sorted populations are further subdivided on the basis of low- or high-forward-angle light scatter. Less than 0.5% of high-light-scatter PDL 55 cells incorporate labeled thymidine after 24 h, indicating the noncycling character of that portion of a mixed cell population.

higher-light-scattering properties (i.e., increased volume and altered structural organization) exhibiting growth characteristics of either resting or noncycling cells (Table I). Light scatter provides a useful basis for the rapid flow sorting of heterogeneous late-passage cultures of diploid fibroblasts and produces viable subpopulations enriched with noncycling cells. In addition, it becomes more clear that considerable caution must be exercised when investigating structural and molecular features of cellular aging using unsorted heterogeneous populations of late-passage cultures.

3. CHANGES IN THE CELL MEMBRANE

Central to modern cell biology is the unsolved problem of how factors interacting with the exterior of the cell membrane and which effect perturbations within its internal structure result in cytoplasmic reorganization and altered cellular metabolism and behavior. Excluding for a moment a consideration of complex carbohydrates, the cell membrane in its simplest form can be considered a two-dimensional mosaic of integral membrane proteins embedded in a fluid lipid bilayer (Singer and Nicolson, 1972). Two important implications for membrane function are compatible with this model: (a) that membrane components may be arranged asymmetrically, permitting certain elements to be associated with either the inner or outer half of the membrane, and (b) that membrane components can move laterally within the plane of the membrane (Edidin, 1974).

Considering the information available on membrane organization and function in eukaryotic cells, relatively little is known of the nature of the cell membrane in aging cells. Nagy (1979) suggests that a time-dependent depreciation of the integrity of the cell membrane is a primary cause of cellular aging. In essence, his hypothesis states that an age-related change in membrane fluidity results in a decrease in the permeability of cell membranes to ions, thus impairing a variety of metabolic functions. Supportive evidence for this hypothesis is varied and often conflicting. Clearly, a great deal more investigation must be performed before the role of the cell membrane in the aging process is understood. However, some factors are being clarified and are noted in the subsequent discussion.

3.1. Lipid Microviscosity

The question of lipid microviscosity is an important one since protein–protein and protein–lipid interactions are affected by the nature of the domains which comprise the cell membrane. The prospholipid and neutral lipid contents of membranes from early- and late-passage WI-38 fibroblasts have been shown to be similar (Polgar, Taylor, and Brown, 1978). In addition, no changes were noted in membrane microviscosity when measured by fluorescence polarization (Polgar, Taylor, and Brown 1978). In contrast, it may be significant to note that lipid microviscosity of lymphocytes from aged mice was reported to be about 20% higher than that of young mice, using the same techniques of fluorescence polarization (Rivnay, Globerson, and Shinitzky, 1979).

Recently, we have initiated an investigation to determine whether cell membranes exhibit differentiation in organization of lipid domains which may affect the various biophysical properties of the plasmalemma. Fluorescence polarization analysis of the lipid probe 1,6-diphenyl-1,3,5-hexatriene (DPH) was applied to IMR-90 diploid fibroblasts *in vitro* in population doubling levels 22 and 55. Fluorescence polarization was quantitated by a program written for the flow cytometer/sorter developed at the Los Alamos National Laboratory by Dr. John Martin. The calculated *P* value in association with fluorescence lifetimes, rotational relaxation rates, and degree of polarization of dye provides an index of membrane "microviscosity" for the microenvironment of the inserted molecules. In both PDL 22 and 55 culture groups, fluorescence intensity increased from time of addition of DPH through a peak intensity at approximately 40 min (room temperature). Late passage cultures exhibited considerable internalization of DPH after 40 min, a factor thought to be complicated by the fragility of these cells in preparation for probe insertion. *P* values for PDL 22 cultures averaged 0.174 whereas comparable values for progressively subcultivated PDL 55 cultures averaged 0.131. Correlation of these data with freeze–fracture images of plasma membranes suggests differentiation of lipid domains within cell membranes which may participate in the senescence of late-passage diploid cultures. However, caution in interpretation is again encouraged, in that a series of factors are known to determine microviscosity: the ratio of cholesterol to phospholipid (Shinitzky and Inbar, 1976); the degree of unsaturation and length of phospholipid acyl chains (Lentz *et al.*, 1976); the ratio of lecithin to sphingomyelin (Schmidt *et al.*, 1977); and the ratio of lipid to protein (Shinitzky and Inbar, 1976). At present, little information is available on lipid composition or the cholesterol to phospholipid ratios in membranes of late-passage cells.

3.2. Intramembrane Organization

Organization of integral protein within cell membranes is an additional index of membrane function and may provide information about the mechanisms of cellular aging. The analysis of protein mobility in these membranes must proceed with the recognition of at least two types of movement; passive diffusion and metabolically driven mobility. The latter is associated with microfilaments and microtubules composing the cytoskeleton and is affected only indirectly by membrane fluidity (Shinitzky and Henkart, 1979). Passive diffusion may be either rotational or translational and the basic question of how changes in lipid fluidity may affect the passive diffusion of membrane proteins is far from simple, since a decrease in fluidity will, on the one hand, oppose mobility but, on the other hand, will decrease the effective volume of the moving proteins (Borochov and Shinitzky, 1976).

We have observed two fundamental variants of intramembrane organization which are associated with the postproliferative phase of diploid cell culture. The first is the reduced ability of cells to exhibit metabolic cooperation as a result of the demise of functional gap junction assembly at points of cell contact. The second is a shift in the partitioning coefficient of intramembrane particles, the distribution ratio of particles which are normally associated with the cytoplasmic half of the lipid bilayer, the P face, versus those affiliated with the exoplasmic half of the membrane, the E face (see Branton *et al.*, 1975, for nomenclature).

3.2.1. Gap Junctions and Cell Contacts

Gap junctions are specialized regions of cell membranes which develop at sites of intercellular contact and which are believed to play a role in the exchange of ions and metabolites between cells (see Staehelin, 1974, for review). The structures consist of intramembrane particles ("connexons," Goodenough, 1974), approx. 9–11 nm in diameter, which form an hexagonal lattice within the plane of each apposed cell membrane (Fig. 5). Center-to-center spacing between particles approximates 9–12 nm. Organization within the lattice varies when physicochemical properties of membranes are altered, a feature which may permit junctions to assemble between contacting membranes but not be permeable to ions and metabolites (Peracchia and Dulhunty, 1976; Peracchia, 1977; Makowski *et al.*, 1977; Caspar *et al.*, 1977). Several investigators have suggested that the permeability of gap junctions to ions and small-molecular-weight metabolites may be important in regulating a variety of cellular activities including proliferation (Bennett, 1973; Cox *et al.*, 1974; Furshpan and Potter, 1968; Gilula *et al.*, 1972; Loewenstein, 1974; Sheridan, 1970).

We have observed that gap junctions develop between IMR-90 fibroblasts in cultures which have been serially subcultivated until reduced proliferative rates are produced. However, distribution is sparse and the junctions exhibit relatively few component subunits (Fig. 6). To investigate whether a concomitant reduction in the ability of senescent cells to maintain metabolic coupling occurs, autoradiography was used to determine the relative exchange of [^{3}H]uridine nucleotides (Pitts and Simms, 1977) between early- and late-passage cells (Fig. 7). In addition, the rate of reappearance of gap junctions between dissociated cells of both sample groups was examined by freeze–fracturing and electron microscopy. Gap junctions formed between young cells approximately 1 h after reassociation in culture. Structurally complete gap junctions were present within 2 h. In contrast, membranes of cells in senescent cultures did not exhibit particle arrays typical of forming junctions until one or more hours of incubation. Small but assembled junctions were observed after 4 or more hours of reassociation. Autoradiography revealed some 70% of young cells to be receiving label from young donors following a 4-h period of metabolic exchange, whereas only 30% of older cells receive label after comparable periods of incubation. Statistical analyses of reciprocal experiments using young-donor–old-recipients and old-donor–young-recipients suggested that recipient cells may have a slightly larger effect on the exchange reaction than donor cells. We concluded that contact-mediated metabolic cooperation is reduced with increased serial subcultivation of cells.

To avoid the possibility that leakage of [^{3}H]uridine was affecting data on metabolic cooperation, similar donor–recipient experiments were conducted between early and late passage cells and mutants which lack a hypoxanthine guanine phosphoribosyl transferase and which are deficient in their ability to incorporate exogenous hypoxanthine (HGPRT$^-$; Kelley and Perdue, 1980). A summary of the relative abilities of early- and late-passage IMR-90 cells to transfer [^{3}H]hypoxanthine into HGPRT+ and HGPRT$^-$ recipients is recorded in Table II. In accord with previous reports on the transfer of [^{3}H]uridine between early- and late-passage cells, the transfer of [^{3}H]hypoxanthine was facilitated rapidly between early-passage donors and recipients, whereas late-passage recipients revealed little detectable label at 1 h of coculture and

Figure 5. Freeze–fracture replica showing gap junction particles in the P face of the cell membrane of an early passage IMR-90 cell. Particles are 9–11 nm in diameter and exhibit center-to-center spacing of approximately 10.5 nm. Note individual aggregates of particles within the entire structure. × 60,000.

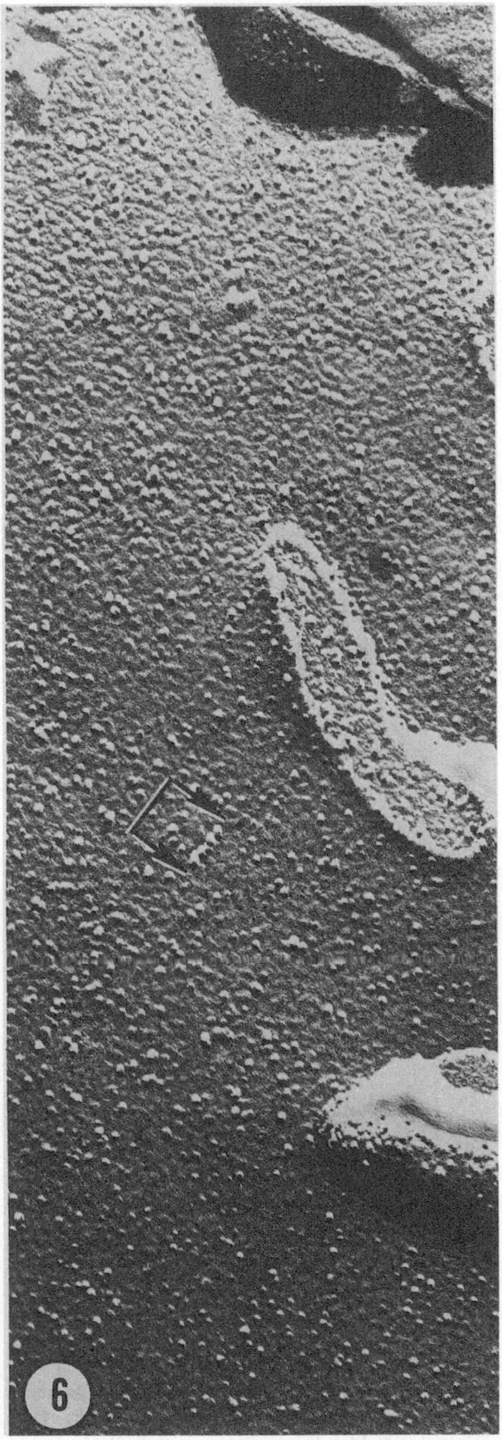

Figure 6. P face of cell membrane of late-passage IMR-90 cell. Gap junctions are markedly reduced when compared to early-passage cells. Typical organization of late passage membranes reveals small aggregates of particles, 9–11 nm diameter (arrows), but few, if any, larger aggregates. × 100,000.

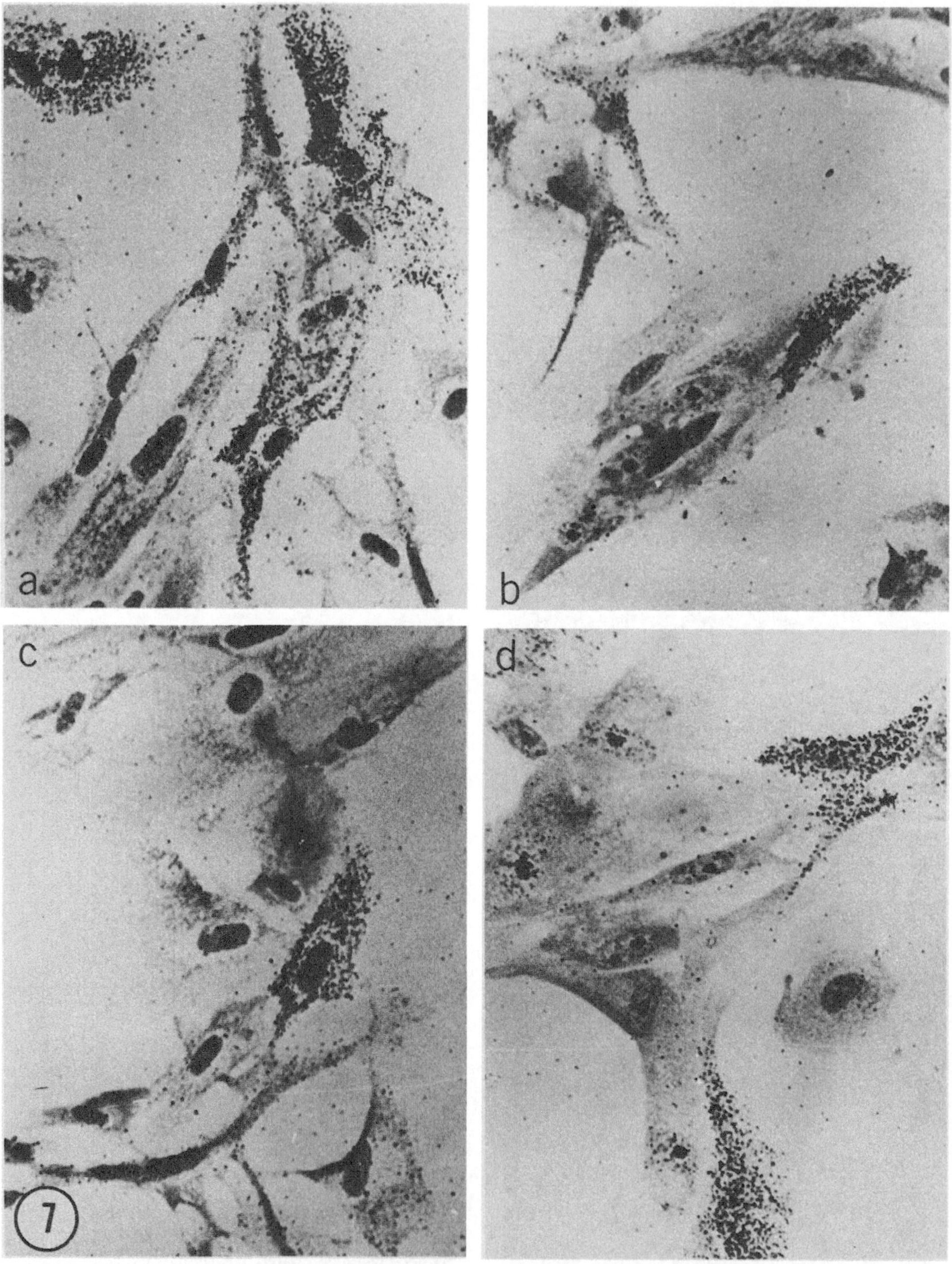

Figure 7. (a–d) Autoradiographs revealing transfer of labeled uridine between contacting cells. Donor cells were prepared by labeling both early- and late-passage IMR-90 cells with [^{3}H]uridine for 3 h and rinsing with balanced salt solution. Unlabeled recipient cells in either early or late passage were added to donor cultures for 4 or 8 additional hours. (a) Young donor–young recipient; (b) young donor–old recipient; (c) old donor–young recipient; (d) old donor–old recipient. × 1000.

Table II. Relative Abilities of Early- and Late-Passage IMR-90 Cells to Transfer [^{3}H]Hypoxanthine into Hypoxanthine Guanine Phosphoribosyltransferase-Postive ($HGPRT^{+}$) or -Deficient ($HGPRT^{-}$) Cells

	Recipient												
	IMR-90 (PDL 14–35)		IMR-90 (PDL 50–55)		BHK21/C13 ($HGPRT^{+}$)			BHK21/TG2 ($HGPRT^{-}$)			L929/A9 ($HGPRT^{-}$)		
Donor	1h	10h	1h	10h	1h	10h	Simultaneous coculture	1h	10h	Simultaneous coculture	1h	10h	Simultaneous coculture
IMR-90 (PDL 14–35)	++	+++	+	+	++	+++	+++	++	+++	+++	–	–	–
IMR-90 (PDL 50–55)	–	+	–	±	–	+	–	–	±	–	–	–	–

sparse, but notable, label after 10 h. Transfer of hypoxanthine from young IMR-90 cultures into both recipient BHK/C13 and BHK/TG2 cells is readily recorded after both 1 h and 10 h. In contrast, little transfer of label from older IMR-90 cells into BHK/C13 cells is observed after 1 h and only modest transfer is prevalent after 10 h of coculture. Similar observations were made on IMR-90/TG2 cocultures. It should be noted that [^{3}H]hypoxanthine is present in HGPRT$^-$ cells only after gap junctions assemble and metabolic coupling is achieved with a wild-type donor cell. Control cultures of IMR-90 (both early and late passage) with L929/A9 mutants exhibited no transfer of label which was detectable in autoradiographs (see figure legend for further description of cell lines).

The variation in metabolic coupling observed between early- and late-passage IMR-90 cells suggests that differences exist in either rates or mechanisms of junction assembly between the two age-related groups. It is important to note that in homotypic and heterotypic cocultures involving older cells, the difference is an apparent retardation in the rate of assembly of formation plaques and larger lattice aggregates which form gap junctions. This reduced rate of assembly has permitted visualization of a step which may immediately follow the formation plaque in this culture system (Kelley and Perdue, 1980). The linear rows of 9- to 11-nm particles are the direct result of lateral aggregation of particles present in the P-face plane of the cell membrane. It is far from clear what conditions determine whether a linear or lattice aggregate forms at this stage. However, the linear aggregate may experience at least two forms of morphogenesis: (1) a bending or curling upon itself to initiate a small aggregation of particles, and (2) a side-to-side pairing with other nearby linear aggregates to achieve the development of a larger lattice aggregate. We have not been able to observe the latter process in older cells and suggest that factors which complete the hexagonal lattice characteristic of gap junctions are either blocked, reduced, or absent in older cells.

3.2.2. Intramembrane Particles

To further explore the question of whether altered membrane organization is expressed with progressive subcultivation, Kelley and Skipper (1977) analyzed the distribution of intramembrane particles in cell membranes of early- and late-passage human embryo fibroblasts. Using techniques of freeze fracture and statistical analyses of particle variance, we have investigated the ultrastructure of membranes from early- and late-passage human embryo fibroblasts with special emphasis on distribution, number, and size of intramembrane particles.

Our observations show that cell membranes of aging human diploid fibroblasts have more intramembrane particles associated with E fracture faces than do cells of earlier passages (Fig. 8). Statistical analysis suggests that total numbers of intramembrane particles may not decline substantially with age (Table III). Therefore, at least one interpretation suggests that alterations in hydrophobic bonding within two membrane halves develop with age in culture permitting more intramembrane structures to fracture with the E face of the membrane. The conclusion of Branton (1966) that techniques of freeze–fracturing cleave sites of weakest molecular bonding would further suggest that forces between lipid domains and intramembrane particles on the P face of the membrane change with age. In addition, a concomitant alteration of similar molecular bonding associated with the E fracture face cannot be overlooked.

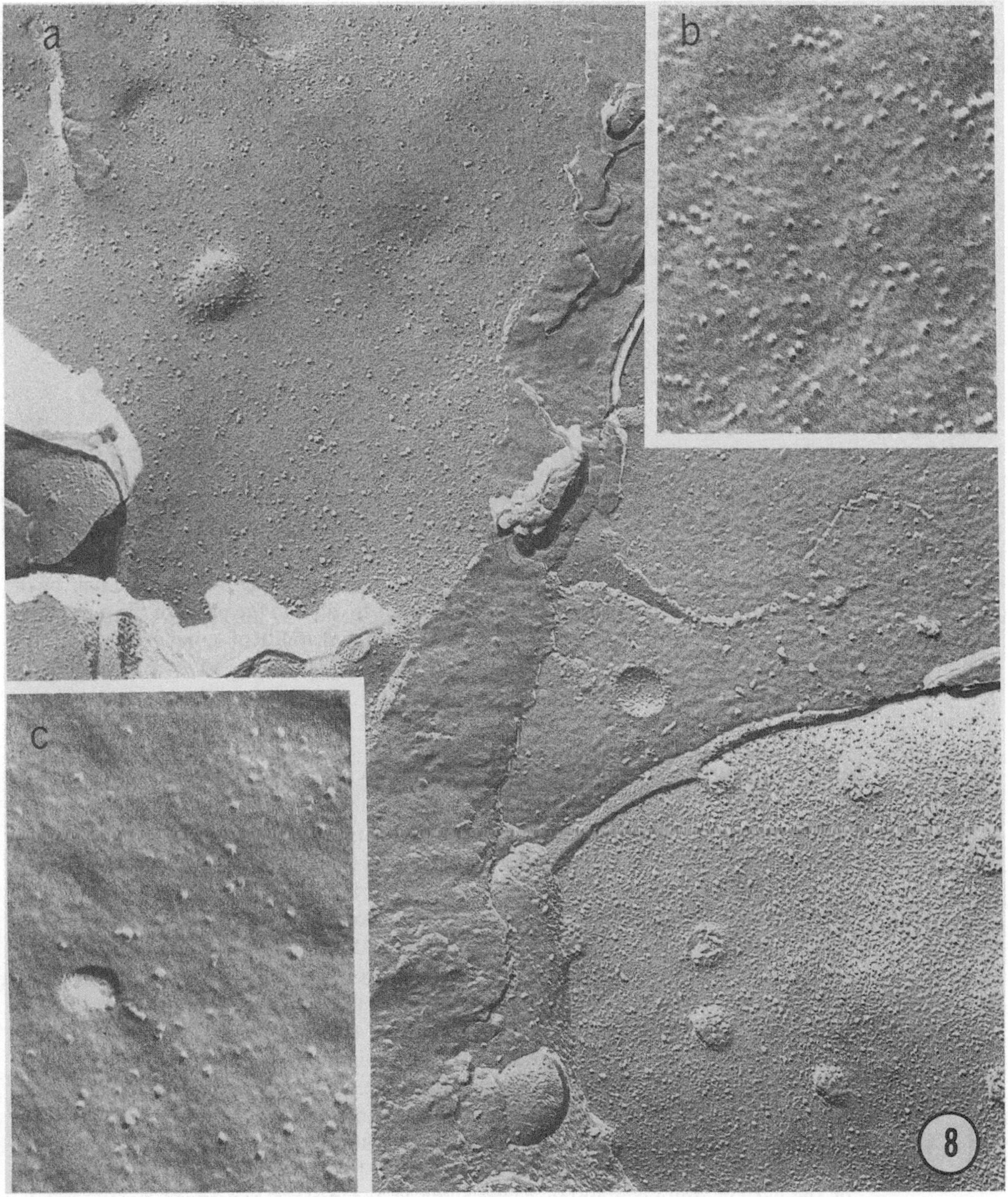

Figure 8. Replicas of freeze–fractured IMR-90 diploid fibroblasts. P faces (arrows) of cell membranes are illustrated from early-passage (a and b) and late-passage (c) cells. Note loss of intramembrane particles associated with the P face of late-passage cell membrane (cf. Table III). (a) × 60,000; (b and c) × 100,000.

Table III. Distribution of Intramembrane Particles Appearing in P and E Fracture Faces[a] Membranes in Early- and Late-Passage Cultures[b]

Fracture face	Early passage cells (mean ± s.e.m.)	Late passage cells (mean ± s.e.m.)	*P* value[c]
P Face	8.24 ± 0.35	5.33 ± 0.29	$P<.001$
E Face	1.43 ± 0.16	3.16 ± 0.22	$P<.001$
Total	9.67 ± 0.38	8.49 ± 0.36	$.025<P<.05$

[a]Particles counted on ten separate P and E fracture faces of cells in either early or late passage. Particles were inclusive in ten randomly selected squares of counting grid (1 square = 1 cm^2) superimposed on micrographs (×100,000) selected only for common angle of tilt to the shadowing gun.

[b]Although total numbers of particles do not diminish significantly, the partitioning coefficients (relative distribution between the P and E faces) shift from approximately 8:1 in early-passage cells to 5:3 in late-passage cells.

[c]*P* values obtained from a two-level nested analysis of variance. The *t* values for total particle variance between early- and late-passage cells were calculated using the difference of the sums of both means for P and E-face particles in both sample groups.

An additional explanation would consider a deterioration in the interaction between submembrane microfilaments and/or microtubules with particles normally resident in the cytoplasmic half of the membrane. On this basis, one would predict that P-face particles would not only lose their anchoring sites to the cytoplasmic half of the membrane but, in addition, would lose the ability to move in the plane of the membrane. Such a loss of lateral mobility may be related to the inability of aging cells to assemble gap junctions, assuming submembrane structures play an active role in junction development.

4. CHANGES IN THE EXTERNAL CELL SURFACE

The precise molecular organization of the cell periphery is not well understood and is undoubtedly variable between different cells. However, its basic composition is known. The surface of a cell is organized into a complex fluid-mosaic of protein, lipid, and carbohydrate (Bretscher, 1973; Singer and Nicholson, 1972; Steck, 1974), the latter contributing to a stainable cell coat associated with the external surface of the plasma membrane (Rambourg and Leblond, 1967). Oligosaccharide chains varying in length from one to twenty sugars are covalently linked to membrane proteins and lipids. In addition, a small percentage of the mass of the cell coat consists of high-molecular-weight, linear carbohydrate molecules, the glycosaminoglycans, some of which may be associated with the cell membrane via core proteins (Kraemer, 1971; Kjellen *et al.*, 1981). Numerous lines of evidence now suggest that the carbohydrate moieties of the cell periphery are involved in cell recognition, adhesion, transformation, and the regulation of cell growth (see Clarkson and Baserga, 1974, for review). In addition, many of the carbohydrate-containing components of the surface have been experimentally coupled to plant lectins which provide useful probes for qualitative and quantitative studies of the cell periphery.

4.1. Surface Charge

The composite nature of the external cell surface results in a net negative surface charge. Bosman *et al.* (1976) have shown that the altered electrophoretic mobility of aging fibroblasts indicated that surface charge becomes less negative with increasing passage number, changing gradually from $-2.82\ \mu/s/V/cm$ at passage 22 to -1.82 at passage 44. Measurement of sialic acid per milligram of protein of purified plasma membranes demonstrated a 60% decrease in sialic acid (Milo and Hart, 1976), suggesting that a decrease in surface negativity could result from reduced sialic acid on surface-associated glycoproteins of late-passage cells.

In an attempt to discover an explanation for reduced sialic acid residues at the surface of senescent cells, the activity of surface-associated sialyltransferase (ectotransferase) was assayed in cultures of WI-38 cells at various passage levels (Spataro *et al.*, 1979). In the absence of exogenous acceptor, the ectotransferase activity was elevated in senescent cells. Additional results indicated that early-passage cells had less endogenous acceptor and received a greater amount of sialic acid after neuraminidase treatment than did the late-passage cells, tending to confirm the presence of asialoglycoproteins at the surface of late-passage cells. Their presence could not be explained by lack of sialyltransferase, however. It is also important to note that no significant shift in the position of glucosamine-labeled membrane glycoproteins on 8% acrylamide gels has been observed (Milo and Hart, 1976).

4.2. Fibronectin

We have observed, as have others (Crusberg *et al.*, 1979), that late-passage fibroblasts spread more slowly than early-passage cells when seeded onto culture substrates. The mechanism of spreading is complex and involves both the cytoskeleton and surface-associated molecules which interact directly with the substratum. Fibronectin is a major surface-associated protein of fibroblasts cultivated *in vitro* (Yamada and Olden, 1978). It has been demonstrated to have a fibrillar organization and may play a significant role in cell adhesion and spreading on culture substrates. Transformed and malignant cells exhibit a reduced accumulation of this protein at the cell surface. In addition, presence of the molecule is reduced during mitosis (Stenman *et al.*, 1977). However, it is important to note that relatively more fibrillar matrix was present on the surfaces of growth-arrested cells than on rapidly growing cells (Hynes and Bye, 1974).

Analysis of fibronectin production by senescent fibroblasts has been initiated in several laboratories. Although one study reported an ^{125}I-labeled glycoprotein with $M_r = 220{,}000$ which was present in a more exposed position, or in greater amounts at the surface of late-passage chick fibroblasts (Courtois and Hughes, 1976), others agree that fibronectin at the surface of human embryo lung fibroblasts decreased during cellular aging (Aizawa *et al.*, 1980c; Vogel *et al.*, 1981a). By indirect immunofluorescence (Fig. 9) the network of fibronectin fibrils which formed over early-passage cells was reduced in midpassage cultures and absent over late-passage cells (Vogel *et al.*, 1981a). Gel electrophoresis of whole-cell preparations also indicated a reduced amount of fibronectin associated with late-passage cultures (Aizawa *et al.*, 1980c). Fibronectin released to

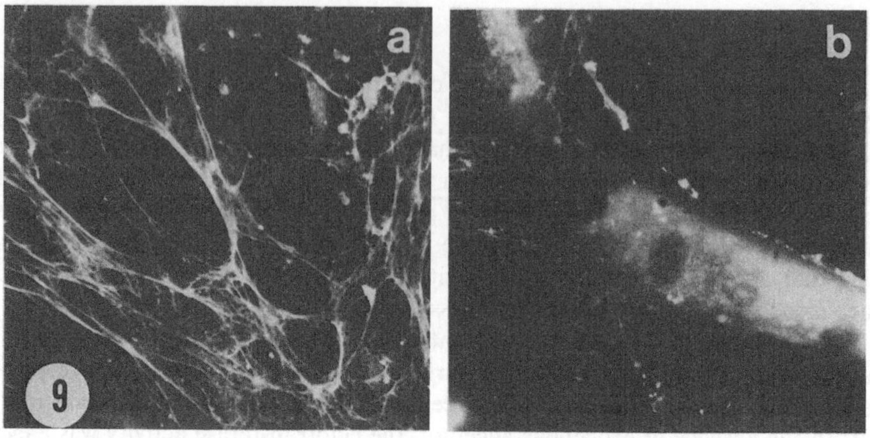

Figure 9. Indirect immunofluorescence microscopy of lung fibroblast cultures at different population doubling levels and stained for fibronectin. (a) PDL 16; (b) PDL 53. Internal fluorescence of late-passage cells was yellow in color and clearly distinct from the FITC-labeled second antibody. Note absence of fibrillar fibronectin over large, late-passage cells. × 1000.

the medium by cell cultures could be purified on a collagen-Sepharose affinity column, and this material demonstrated the usual band position by SDS-polyacrylamide gel electrophoresis (Chandrasekhar and Millis, 1978). Quantitative determination of the amount of fibronectin released by cultures at varying PDLs indicated that although late-passage cultures did not accumulate a surface-associated network of fibronectin fibrils, they did produce and release fibronectin in amounts per milligram of cell protein which exceeded the early-passage cultures (Vogel *et al.*, 1981a). Chandrasekhar and Millis (1980) have reported that fibronectin purified from the medium of late-passage cultures was unable to facilitate adhesion of either early- or late-passage cells to a bacteriological plastic substratum as well as fibronectin from early-passage cultures. These investigators have suggested that the molecule itself may be defective in some manner.

4.3. Glycosaminoglycans and Proteoglycans

Glycosaminoglycans are products of synthesis in cultured fibroblasts and are found in three major compartments: the extracellular medium; the cell surface; and within the cell. Alterations in glycosaminoglycan synthesis and distribution during cellular aging of human embryo lung fibroblasts have been studied in three different cell strains (WI-38, IMR-90, and TIG-1). All investigators have reported a decline in the synthesis of extracellular glycosaminoglycans when related to incorporation of isotope precursors per milligram of cellular protein (Schachtschabel and Wever, 1978; Vogel *et al.*, 1981b; Matsuoka and Mitsui, 1981). However, quantitative alterations in surface and intracellular glycosaminoglycans were much less pronounced. In this regard, it is clear that the distribution of specific glycosaminoglycans in each of the three major culture compartments is not equivalent. Both surface and intracellular compartments are enriched for heparan sulfate (60%–80% of the total) whereas the extracellular medium is enriched for dermatan sulfate (the latter identified as ^{35}S-labeled glycosaminoglycan degraded by chondroitinase ABC but largely insensitive to chondroitinase AC). It is significant that differences in the percentage of each [^{35}S]glycosaminoglycan in the various compartments of IMR-90 fibroblasts were not found in our laboratory, even when cultures reached PDL_{max} (Vogel *et al.*, 1981b). However, others have reported increased relative and absolute amounts of heparan sulfate for senescent WI-38 and TIG-1 cultures (Sluke *et al.*, 1981; Matsuoka and Mitsui, 1981). Turnover kinetics for cell-associated [^{35}S]glycosaminoglycans followed similar patterns in both early- and late-passage cultures, with only a slight delay noted in the late-passage cultures (Vogel *et al.*, 1981b).

It is unclear whether a defect in glycosaminoglycan secretion exists in senescent cells. Early-passage IMR-90 fibroblasts secreted approximately 4.5 μg of glycosaminoglycan to the medium per milligram of cellular protein in 48 h, as determined by uronic acid quantitation of the isolated glycosaminoglycans and corrected for contributions from unconditioned culture medium (Stewart, 1981). Late-passage cultures secreted a reduced amount of glycosaminoglycan when expressed per milligram of cell protein, but the amount was increased if expressed per cell. The sulfation capability of senescent cells is not defective, because addition of exogenous xyloside resulted in a tenfold enhancement of glycosaminoglycan secretion by both early- and late-passage cultures (Vogel *et al.*, 1981b). All workers have noted that even terminally senescent cells continue to secrete glycosaminoglycans, so it is possible that the apparently diminished secretion per mg cell protein is reflecting an accumulation of protein by senescent cells which is unrelated to synthetic processing.

Considering the enormous changes in proliferation as well as other differences between early- and late-passage fibroblasts, the rather minimal nature of alterations in glycosaminoglycan metabolism is somewhat surprising. It seems reasonable to conclude that glycosaminoglycan synthesis and composition are not intimately related to expression of such cellular parameters as proliferation, culture density, fibronectin matrix formation, or development of gap junctions and other cell contacts.

Although major alterations in glycosaminoglyan composition or localization do not develop with progressive subcultivation of fibroblasts, there may be significant alterations in proteoglycan structure. When compared, however, overall differences were relatively modest. In medium from late-passage cultures, the larger ^{35}S-labeled proteoglycans (I and II, Fig. 10) comprised a greater percentage of the total than in early-passage cultures, and the large chondroitin sulfate proteoglycan had a lighter buoyant density in CsCl than did any of the large proteoglycans from early-passage culture medium (Vogel and Pitcher, 1982). Proteoglycan solubilized from the cell layer of late-passage cultures frequently had a major component migrating in the void volume on Sepharose CL-4B, whereas early-passage cells had components only at K_{av} = 0.2 and 0.6 (IV and V, Fig 10). This V_0 component was sensitive to trypsin degradation and contained primarily chondroitin sulfate. Thus, both medium and cell-layer compartments of the senescent cultures demonstrated unique proteoglycan profiles by Sepharose chromatography, and in both cases there was an increased amount of components with larger hydrodynamic volume. Less than 15% of the cell-associated proteoglycan of either early- or late-passage IMR-90 cultures was released from the monolayer by incubation

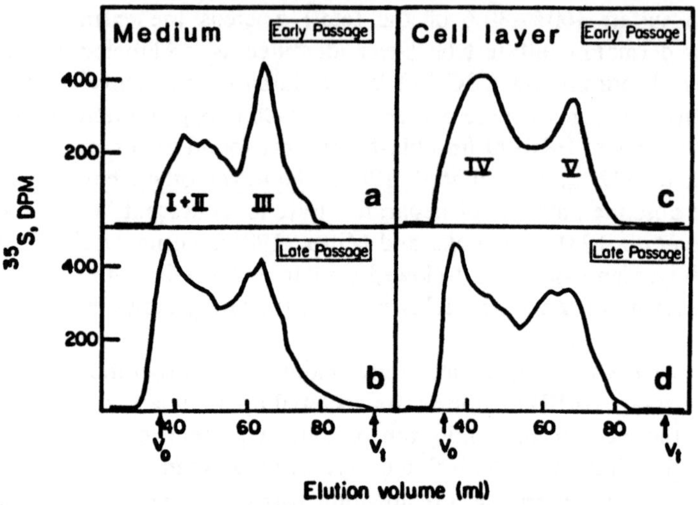

Figure 10. Sepharose 4B profiles of ^{35}S-labeled proteoglycans from medium and cell layer of human embryo lung fibroblast cultures at different population doubling levels. Cultures were labeled with $Na_2{}^{35}SO_4$ for 48 hr and proteoglycans of both the medium and cell layer recovered under dissociating conditions as described in Vogel and Peterson (1981). Macromolecular material obtained after Sephadex G-50 filtration was applied to a Sepharose CL-4B column and eluted with 4 *M* guanidine HCl in 0.1 *M* sodium acetate, pH 5.8. Area I + II of (a) contains a heparan sulfate proteoglycan and a chondroitin sulfate proteoglycan; III is a dermatan sulfate proteoglycan. Area IV of (c) is a surface-associated heparan sulfate proteoglycan plus some chondroitin sulfate; V represents intracellular glycosaminoglycans.

with exogenous heparin, indicating a minimal accumulation of pericellular (presumably matrix-bound) proteoglycans.

The significance of an increased amount of proteoglycan having large hydrodynamic volume in senescent cultures remains to be clarified. The suggestion that senescent fibroblasts are terminally differentiated cells (Bell *et al.*, 1978) would be greatly strengthened if senescent cells could be shown to produce a unique functional product. However, it is not yet possible to conclude that the larger proteoglycans from senescent fibroblasts represent new proteoglycan monomer and thus that senescence is a differentiation process. The proteoglycan profiles from senescent cultures were extremely variable compared with those obtained from early-passage cultures. Replicate cultures assayed only 10 days apart had different profiles, and one PDL_{max} culture completely lacked the V_0 component (but had a profile seemingly identical to early-passage cell layers). For these reasons we remain cautious in interpreting these data. The large components of senescent cultures could indicate aggregation or partial processing of proteoglycans which are normal consitutents of early-passage cells. Whether this might represent self-aggregation, binding within the plasma membrane, or binding to new components or pericellular matrix, is not known. Unfortunately, it is not possible to conclude that the presence of unusually large proteoglycan components correlates with lack of fibrillar fibronectin matrix because all senescent cultures appeared to lack the matrix, whereas some did not demonstrate the large proteoglycan.

4.4. Lectin Binding

Lectins are molecules that can bind specifically to carbohydrate-containing sites on cell surfaces and can be used to probe changes in surface organization and composition associated with altered patterns of growth. A unique probe of alteration in the surfaces of individual human diploid fibroblasts with *in vitro* aging has been described by Aizawa and co-workers (1979; 1980a–c). Termed the concanavalin A (Con A)-mediated RBC adsorption assay, the test is performed either by adsorbing Con A-coated RBCs to fibroblasts (RBC-coated method), or conversely, by adsorbing RBCs to Con A-coated fibroblasts (fibroblast-coating method). By the RBC coating method the number of RBCs adsorbed to fibroblast cell surfaces increased steadily as the PDL of the culture increased (Aizawa *et al.*, 1980a). By the fibroblast-coating method, however, virtually no RBCs were attached until PDL 45 was reached. The amount of [^{3}H]-Con A bound to early- and late-passage fibroblasts did not vary sufficiently to explain this alteration in lectin-mediated RBC binding. Further experiments demonstrated that RBC adsorption was not significantly affected by cell cycle phase, by calendar time in nondividing cultures, or by coculture of different aged cells (Aizawa *et al.*, 1980a), suggesting that it was a genuine probe of surface alteration with progressive cell doublings. In addition, the assay was found not to be correlated with cell size. Smaller cells in senescent cultures demonstrated RBC adsorption similar to larger senescent cells, and significantly greater than small early-passage cells (Aizawa *et al.*, 1980b). The number of RBCs bound by the RBC-coating method was increased 7- to 11-fold by pretreatment of early-passage cultures with trypsin, collagenase, neuraminidase, or hyaluronidase. Similar pretreatment of late-passage cultures increased RBC binding only twofold over the already elevated level (Aizawa *et al.*, 1980c).

Observations of ConA binding to IMR-90 cells reported by Kelley, Azad, and Vogel (1978) support the results from Aizawa's laboratory, although a direct imaging of the lectin was employed rather than an RBC adsorption assay. Our observations suggest that variations in binding patterns were not the result of the changes in numbers of binding sites but may be the result of alterations in the concerted actions of numerous factors which include cell surface topography, surface charge, and molecular composition and the relative distribution of lectin binding sites on the cell periphery.

5. THE CYTOSKELETON

Microfilaments, intermediate (10-nm) filaments and microtubules are among the structural components which form the cytoskeleton of eukaryotic cells. The cytoskeleton is intimately associated with the cytoplasmic surface of the cell membrane. Recently, Crusberg *et al.* (1979) reported that early-passage cells spread 44–52% faster on artifical substrates following reseeding than late-passage cells. Similar observations have been reported by Kelley *et al.* (1980). Since spreading on a substrate is effected by a complex of interactions between components of the cytoskeleton, the plasma membrane and attachment sites on the external cell surface (Pollard and Weihing, 1974; Weihing, 1979), it is reasonable to suggest that a decrease in a cell's ability to spread may be caused either by a diminution of cytoskeletal elements, changes within the cell mem-

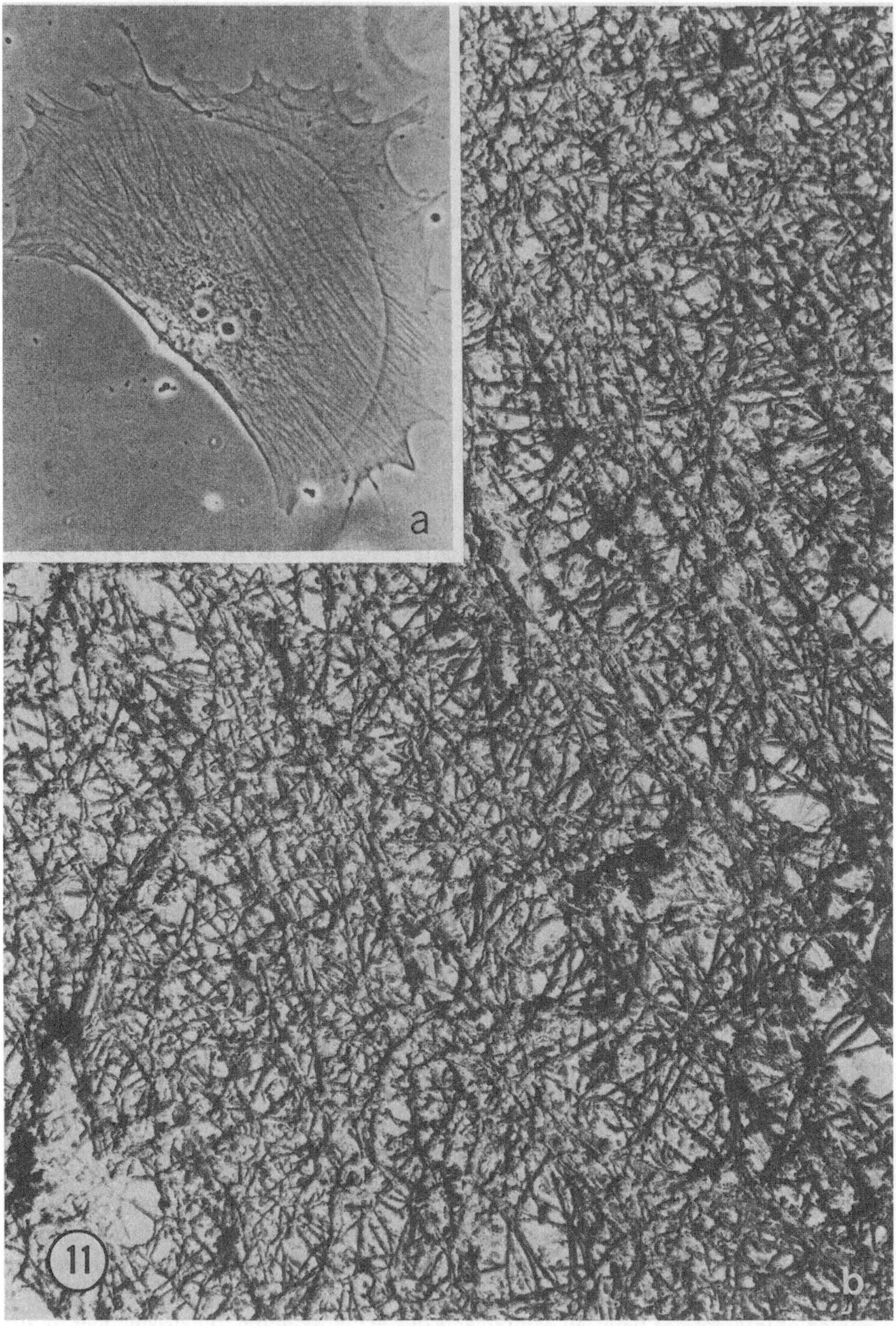

Figure 11. (a,b) Cytoskeleton of late-passage IMR-90 cell is revealed by phase microscopy (a) and detergent extraction (b). Bundles of thin filaments (6 nm in diameter) are evident which are connected by additional filaments (6–10 nm in diameter) arrayed at various angles to individual bundles. (a) × 1000; (b) × 100,000.

brane, alteration in development of attachment sites, or combinations of these and other structural and molecular factors.

We have used direct imaging techniques (Trotter and Kelley, 1979) to examine the organization of cytoskeletons in early- and late-passage cells during spreading following reseeding (Fig. 11). All cells from both young and senescent cultures exhibit a cytoskeletal network of actin microfilaments (Fig. 12a); intermediate (10-nm) filaments; microtubules; and interconnecting thin filaments (6–8 nm in diameter) which do not interact with heavy meromyosin (Fig. 12b). Early-passage cells assemble linear aggregates of actin filaments within 1 h of spreading. By 4 h of incubation, these bundles establish a structural bond with the cell membrane which results in resistance by the plasmalemma to detergent (0.5% Triton X-100) extraction at these sites (Fig. 13). Furthermore, these membrane regions are associated with developing stress fibers of well-spread cells. In contrast, late-passage cells exhibit slower spreading which correlates with a retarded assembly of actin bundles. By 8 h of spreading, cells of older cultures do not exhibit the regions of membrane–actin interaction which impart detergent resistance to the plasmalemma (Fig. 11). We conclude that the ability to reassemble actin–actin and actin–membrane associations during cell spreading is reduced with increased serial subcultivation of cells.

The observation that cells in senescent cultures do not assemble bundles of actin filaments as readily as younger cells following reseeding (Kelley *et al.*, 1980) is an important one, and suggests that older cells may be delayed in developing those factors which contribute either to actin–actin interactions or to the actin–membrane association required for cell spreading. Several proteins have been identified in nonmuscle cells which interact with F-actin and which may in some way alter the organization of actin filaments (Clarke and Spudich, 1977; Korn, 1978). Immunofluorescence microscopy with antimyosin antibodies indicates that myosin is closely associated with actin microfilaments of BHK-21 cells (Goldman *et al.*, 1979). In addition, stress fibers from numerous fibroblastic cell lines contain proteins antigenically similar to alpha-actinin (Lazarides and Burridge, 1975; Lazarides, 1976), tropomyosin (Lazarides, 1975), and filamin, a high-molecular-weight actin-binding protein (Wang *et al.*, 1975; Heggeness *et al.*, 1977). Pure filamin interacts with individual F-actin filaments in solution to form fiber bundles and sheetlike structures, suggesting that, *in vivo,* filamin–actin interactions play an important role in the control of actin filament structure, in cell motility, and in the assembly and stabilization of cell–cell contacts (Heggeness *et al.*, 1977). Clearly, it is important to determine whether these or similar proteins are present in altered amounts (or distribution) in late-passage cells which exhibit reduced rates of assembly of actin bundles. To determine whether potential differences in cytoskeletal proteins develop with progressive subcultivation, early- and late-passage cultures were extracted with 0.5% Triton X-100 for 1 min followed by 1.0% sodium dodecyl sulfate (SDS) prior to separation and characterization of extracted proteins by electrophoresis on 7.5%–15% gradient SDS gels (Fig. 14). Extractions were made of both culture groups 3, 6, and 24 h after reseeding. Although considerable variation in cytoskeletal organization was observed, qualitative differences in gel banding patterns of actin, myosin, and tubulin were not apparent at selected time points. However, late-passage cells at 6 and 24 h did not exhibit either filamin or myosin to be associated with the Triton-insoluble fraction as did early-passage cells (Fig. 15). With the exceptions of filamin and myosin, we conclude that major qualitative variations in the presence of cytoskeletal proteins do not

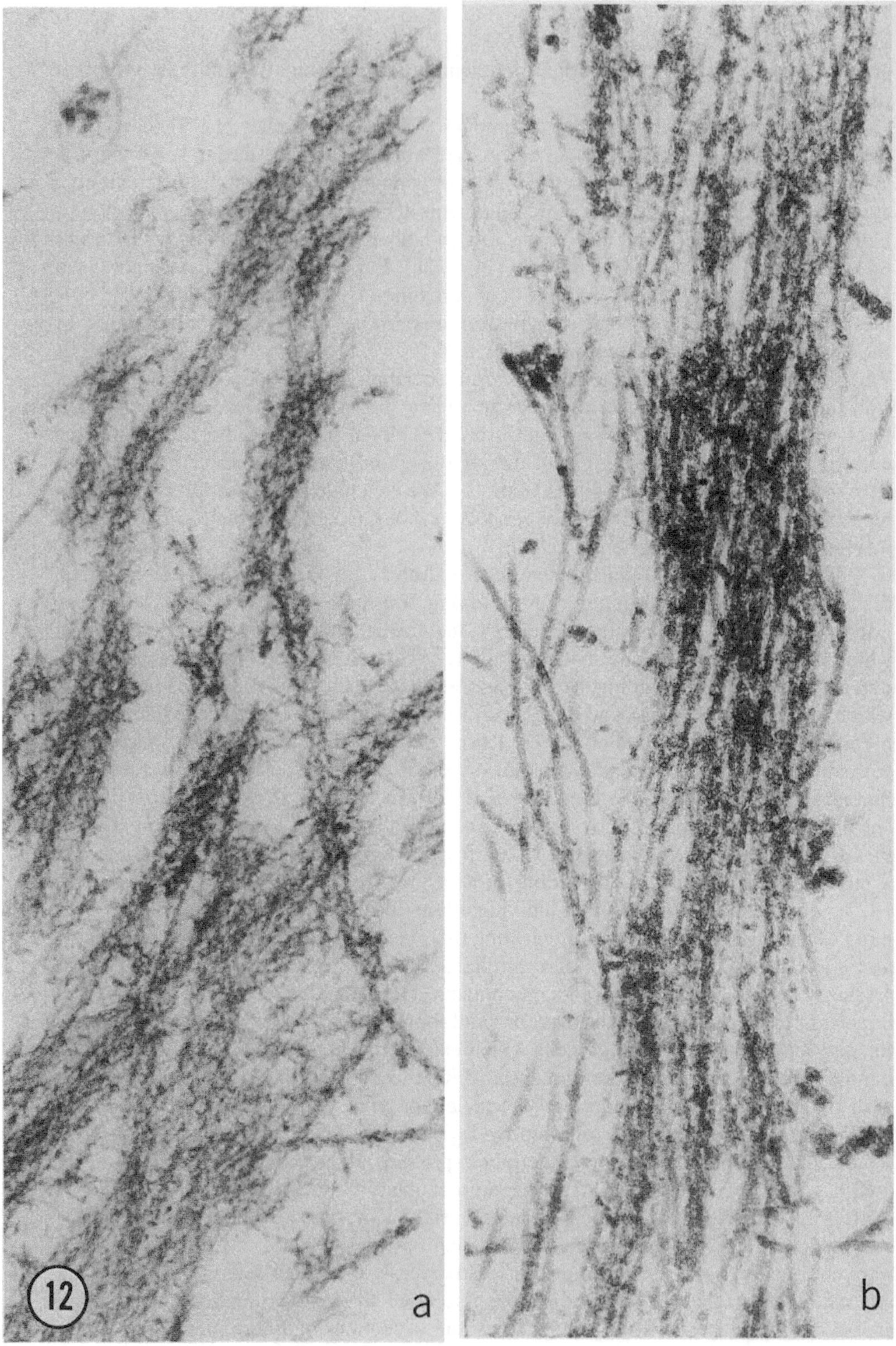
12
a
b

exist between early- and late-passage cells during spreading on a culture substrate. However, it should be noted that cytoskeletal proteins observed by these experiments are residues of preexisting proteins associated with former morphogenetic events. New synthesis of cytoskeletal elements following trypsin treatment may be required to produce the altered structural features observed in late-passage cells.

Finally, many forms of actomyosin-mediated motility in cells involve membranes, and it is widely believed that it is necessary to have some component of the cytoskeleton anchored to the plasmalemma to explain the movements that occur. Kelley *et al.* (1980) have observed that the microfilament–membrane relationship exhibits greater resistance to detergent extraction, once developed, than other regions of the cell surface not in association with bundles of actin filaments. This phenomenon may indicate the presence of structural bonds between these interacting components. Hence, the relative slowness exhibited by late-passage cells in forming detergent-resistant regions of cell membrane adjacent to submembrane bundles of microfilaments suggests that one major variant between early- and late-passage cells may be in the synthesis and assembly of linking elements which form the microfilament–membrane bond. However, it is not yet clear whether associated variation exists in the internal organization of cell membranes or intramembrane particles in detergent-resistant regions.

6. SUMMARY AND CONCLUSIONS

We have presented evidence for alterations in certain characteristics of the surfaces of human diploid fibroblasts as they proceed through their *in vitro* lifespan and reach the senescent state. Within the plasma membrane itself there is a shift in distribution of intramembrane particles revealed by analysis of freeze–fracture micrographs. In addition, a reduction in the number and organization of gap junctions develops as cultures reach senescence. Transfer of intracellular molecules to and from senescent cells is accordingly reduced. The external cell surface loses an organized matrix of fibronectin and increases binding of Con-A-coated red blood cells. Finally, the formation of stable attachment sites between the cytoskeleton and the cytoplasmic face of the membrane is delayed in senescent cells.

In summary, there is considerable evidence that contact interactions between cells of multicellular organisms play vital roles in regulation and control of cellular proliferation (Kaji and Matsuo, 1979). However, a precise relationship of alterations in membrane specialization and reduced junction-mediated metabolic communication to the loss of proliferative capability in late-passage cultures is far from clear. It would be a gross oversimplification to argue that loss of division potential in older cultures results from the inability of cells to contact-feed one another. However, investigators, in attempting to culture mammalian cells, have recognized the perils of seeding cells too

Figure 12. (a,b) The nature of two classes of filaments is revealed by the ability of the 6-nm class to decorate with heavy meromyosin (a), whereas the 10-nm population is void of marker (b). Thin filaments are actin whereas the 10-nm population are intermediate filaments. Both groups coexist in both early- and late-passage cells. (a) × 110,000; (b) × 110,000.

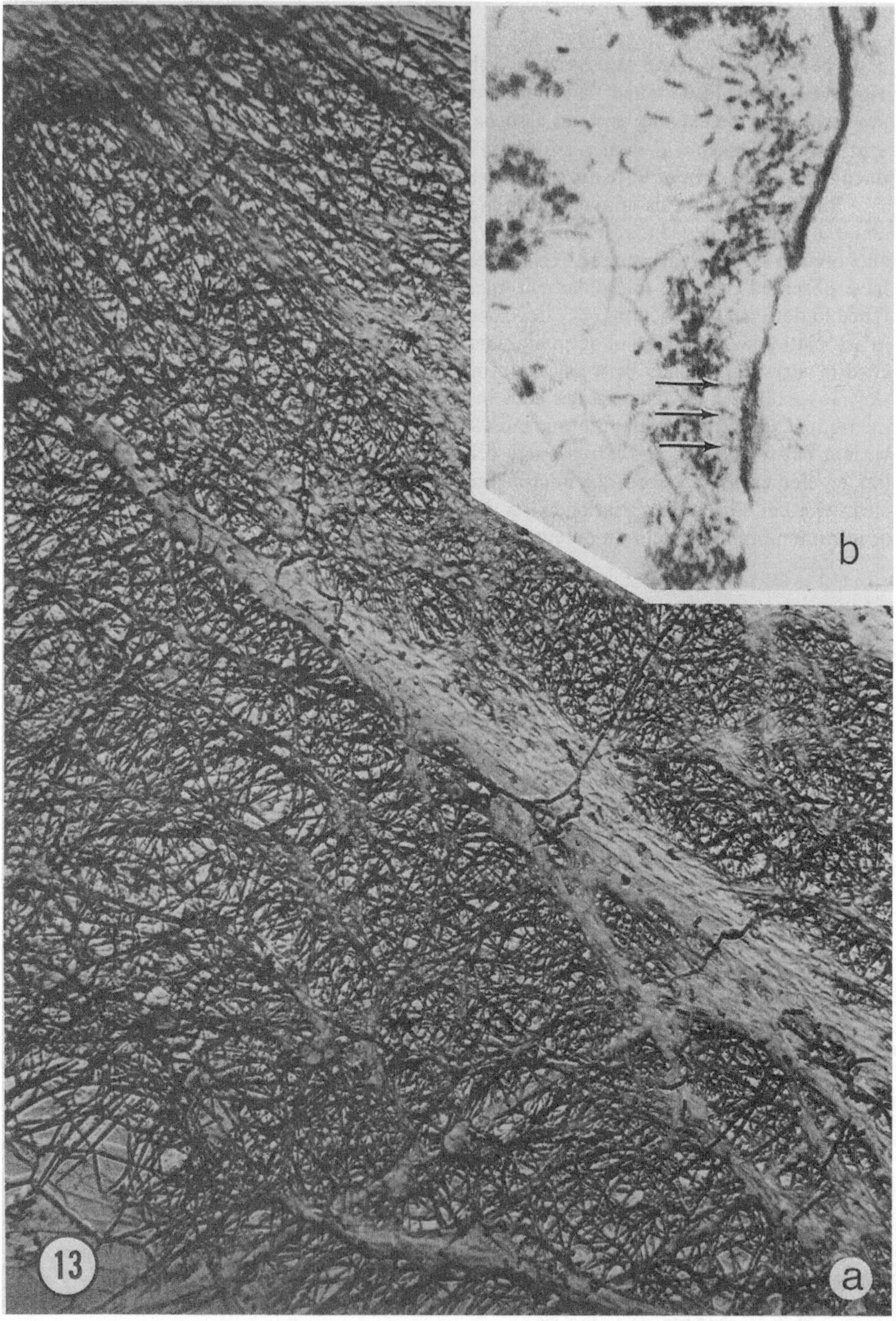
b
13
a

sparsely, thereby depriving them of the apparent growth benefits of direct sharing of metabolites. A somewhat more complicated hypothesis is that diffusion of small molecules through gap junctions provides a mechanism for homeostasis through equalization of ionic and metabolic pools which coordinate cell proliferation. A corollary would be that cells with junctional defects subsequently alter contact-mediated growth control. As an example, studies using the transfer of ions and nucleotides as probes for communication have been conducted on a variety of malignant cells. In general, transformed cells (which have an unlimited *in vitro* lifespan) donate and receive nucleotides less than do their normal counterparts. The dilemma in applying similar information to aging cells which have limited exchange of nucleotides and which have lost, or are in the process of losing, their ability to proliferate is apparent. In addition, one must not overlook the possibility that a complex of membrane–cell surface interactions develops in aging cells which may result in deprivation of both metabolic cooperation and incorporation of critical nutrients from the growth medium. It is not unreasonable to suggest that changes in the fluidity of membrane lipids may alter lipid–protein interactions which further affect assembly of gap junctions and permeability of the cell membrane. To this end, more intensive investigations of the differentiation of membrane fluidity in senescent cells may provide useful information on the development of communicating junctions and the control of growth at the cell surface.

In addition, investigations of the characteristics of molecules at the surface of senescent fibroblasts have demonstrated differences compared with early-passage fibroblasts, but the mechanisms or significance of these differences also remains to be clarified. Fibronectin at the surface of senescent human cells is decreased. However, it is not apparent whether this is due to alterations in the plasma membrane, the membrane–cytoskeletal network, the synthesis or processing of other surface-associated molecules, or to characteristics of the fibronectin molecule itself. Loss of fibronectin by nonproliferative senescent cells and by highly proliferative transformed cells makes it clear that there is no direct correlation between accumulation of a fibronectin matrix and proliferative potential. Furthermore, the interactions which result in formation of a stable pericellular matrix by cells in cultures remain elusive, although collagen, fibronectin, and proteoglycans appear to be involved. Continued investigation of the subtle alterations in these components which accompany cellular aging could result in a generalized understanding of factors which regulate the formation and function of external cell surfaces.

And finally, although both early- and late-passage cells exhibit a prominent cytoskeleton, variation in spreading and contact behavior between age-related culture groups correlates (1) with alterations in organization of cytoskeletal components, and (2) with the establishment of actin bundles and detergent-resistant membrane regions following reseeding of cells on culture substrates. Consequently, those proteins which

←

Figure 13. (a,b) In contrast to late-passage cells, early-passage cultures exhibit membrane regions which are more resistant to detergent extraction. Figure 13a is a replica following extraction which shows extensive regions of residual membrane associated with components of the cytoskeleton. 13b reveals close association of the residual fragment with underlying 6-nm filaments. Thin fibrils (arrows) appear to link the membrane to the adjacent bundle of actin filaments. (a) × 70,000; (b) × 110,000.

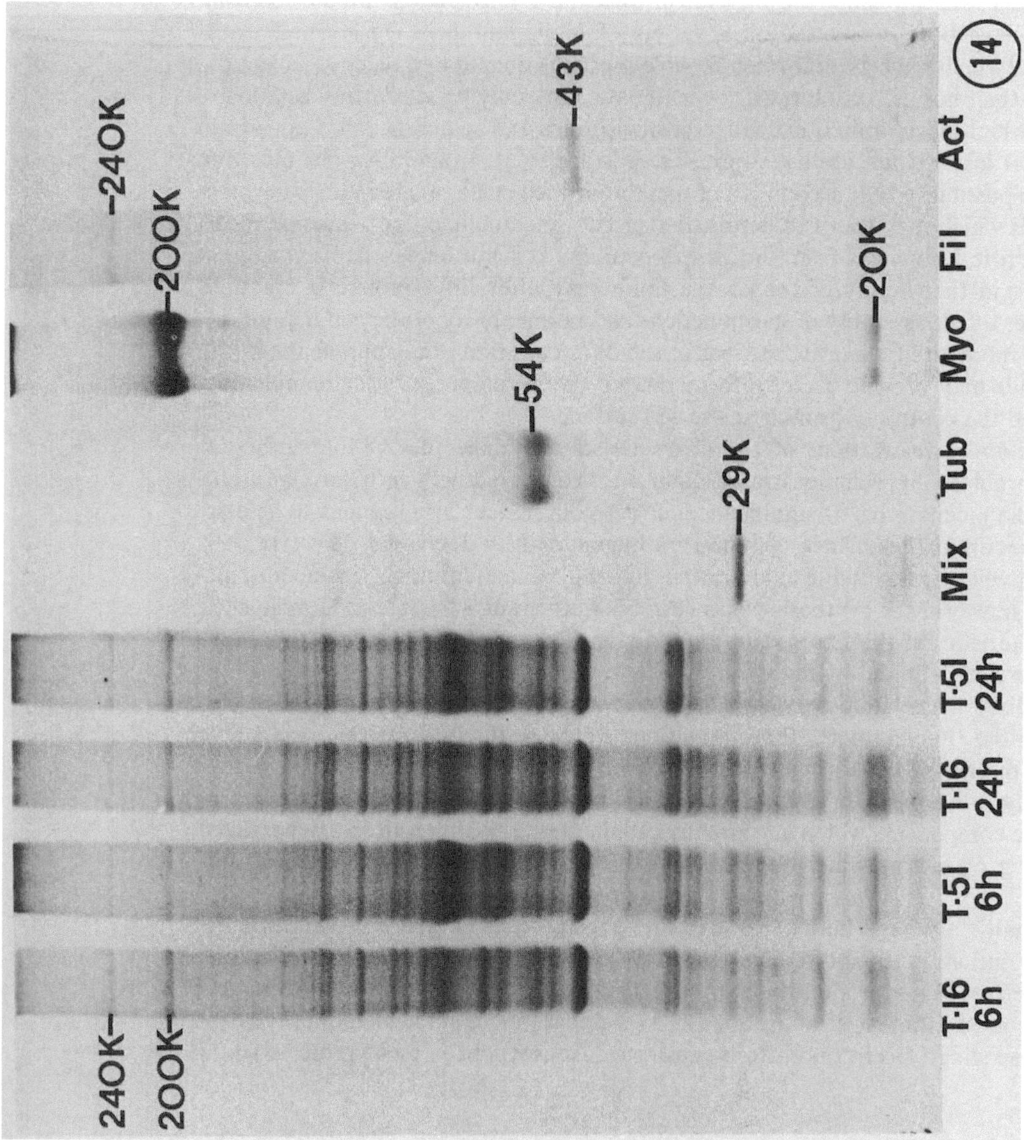

Figure 14. Polyacrylamide gel pattern of Triton X-100 extracted proteins from early- (PDL 16) and late- (PDL 51) passage IMR-90 cells after 6 and 24 h of spreading following seeding on a culture substrate. Marker proteins include tubulin (*tub*); myosin (*myo*); filamin (*fil*); and actin (*act*). Few comparative differences exist in banding profiles with the exception of a band which is present at approximately 20,000 daltons in early-passage cells and which appears to be diminished or absent in late-passage cultures.

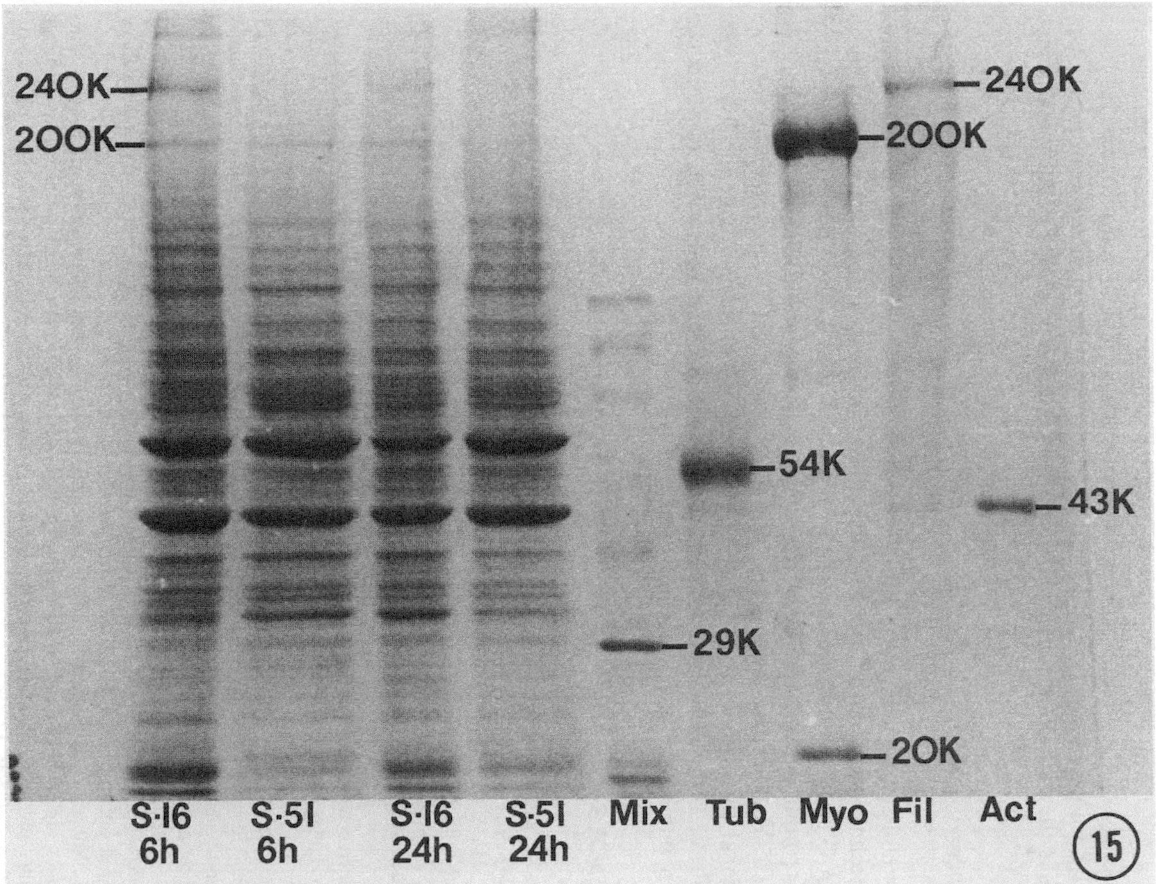

Figure 15. Gel profiles of sodium dodecyl sulfate extraction of the residual material remaining on culture substrates following initial treatment with Triton X-100. Note the reduction and/or absence of bands which comigrate with both filamin (240,000 daltons) and myosin (20,000 daltons). The differences in appearance of these proteins may correlate with the reduced ability of late-passage cells to spread and migrate on culture substrates.

contribute to actin–actin interactions and actin–membrane association may either function more slowly, or are reduced in quantity, following reseeding of late-passage cells.

Collectively, these observations suggest that alterations in the cell membrane, its external surface, and submembrane cytoskeleton are important correlates of cellular aging. However, any cause-and-effect relationships to senescence are far from clear. Clearly, a great deal of critical experimentation will be required to determine the primary etiology of the loss of proliferation potential in diploid cell cultures. However, eventually, we may learn that structural complexes may be implicated as primary participants in the multiple mechanisms which integrate the process of cellular senescence.

ACKNOWLEDGMENTS

Grateful acknowledgement is made to Ms. Judi DeLongo and Mr. Bradley D. Perdue for technical assistance; to Mr. Michael Norviel for assistance with illustrations; to Ms. Jean O'Neill for preparation of the typescript; and to the National Institutes of Health for grant support (AG00191 to Robert O. Kelley and AG01214 to Kathyrn G. Vogel). Dr. Vogel is also the recipient of a Research Career Development Award from the National Institute on Aging. In addition, we wish to thank Academic Press; Pergamon Press, Ltd.; and Elsevier Sequoia, S.A., for granting permission to reproduce previously published illustrations and tables.

REFERENCES

Absher, P. M., and Absher, R. G., 1976, Clonal variation and aging of diploid fibroblasts, *Exp. Cell Res.* **103:**247.

Absher, P. M., Absher, R. G., and Barnes, W. D., 1974, Genealogies of clones of diploid fibroblasts, *Exp. Cell Res.* **88:**95

Adelstein, R. S., Scordilis, S. P., and Trotter, J. A., 1979, The cytoskeleton and cell movement: General considerations, *Meth. Achiev. Exp. Pathol.* **8:**1.

Aizawa, S., and Kurimoto, F., 1979, Effects of glutaraldehyde and other drugs on Concanavalin A-mediated red blood cell adsorption to nonsenescent, senescent and transformed human fibroblasts, *Mech Ageing Devl.* **11:**237.

Aizawa, S., and Mitsui, Y., 1979, A new cell surface marker of aging in human diploid fibroblasts, *J. Cell. Physiol.* **100:**383.

Aizawa, S., Mitsui, Y., and Kurimoto, F., 1980a, Cell surface changes accompanying aging in human diploid fibroblasts. II. Two types of age-related changes revealed by Concanavalin A-mediated red blood cell adsorption, *Exp. Cell Res.* **125:**287.

Aizawa, S., Mitsui, Y., Kurimoto, F., and Matsuoka, K., 1980b, Cell surface changes accompanying aging in human diploid fibroblasts. III. Division age and senescence revealed by Concanavalin A-mediated red blood cell adsorption, *Exp. Cell Res.* **125:**297.

Aizawa, S., Mitsui, Y., Kurimoto, F., and Nomura, K., 1980c, Cell surface changes accompanying aging in human diploid fibroblasts. V. Role of large major cell surface protein and surface negative charge in aging- and transformation-associated changes in Concanavalin A-mediated red blood cell adsorption, *Exp. Cell Res.* **127:**143.

Arndt-Jovin, D. J., and Jovin, T. M., 1978, Automated cell sorting with flow systems, *Ann. Rev. Biophys. Bioeng.* **7:**527.

Baserga, R., 1968, Biochemistry of the cell cycle: A review, *Cell Tissue Kinet.* **1:**167.

Bell, E., Marek, L. F., Levinstone, D. S., Merrill, C., Sher, S., Young, I. T., and Eden, M., 1978, Loss of division potential *in vitro:* Aging or differentiation?, *Science* **202:**1158.

Bennett, M. V. L., 1973, Functions of electrotonic junctions in embryonic and adult tissues, *Fed. Proc.* **32**:65.

Borochov, H., and Shinitzky, M., 1976, Vertical displacement of membrane proteins mediated by changes in microviscosity, *Proc. Natl. Acad. Sci. USA* **73**:4526.

Bosmann, H. B., Gutheil, R. L., and Case, K. R., 1976, Loss of a critical neutral protease in ageing WI-38 cells, *Nature* **261**:499.

Bowman, P. D., Meek, R. L., and Daniel, C. W., 1975, Aging of human fibroblasts *in vitro:* Correlations between DNA synthetic ability and cell size, *Exp. Cell Res.* **93**:84.

Branton, D., 1966, Fracture faces of frozen membranes, *Proc. Natl. Acad. Sci. USA* **55**:1048.

Branton, D., Bullivant, S., Gilula, N. B., Karnovsky, M. J., Moor, H., Muhlethaler, K., Northcote, D. H., Packer, L., Satir, B., Satir, P., Speth V., Staehelin, L. A., Steere, R., and Weinstein, R. S., 1975, Freeze-etching nomenclature, *Science* **190**:54.

Bretscher, M. S., 1973, Membrane structure: Some general principles. *Science* **181**:622.

Caspar, D. L. D., Goodenough, D. A., Makowski, L., and Phillips, W. C., 1977, Gap junction structure I. Correlated electron microscopy and X-ray diffraction, *J. Cell Biol.* **74**:605.

Chandrasekhar, S., and Millis, A. J. T., 1980, Fibronectin from aged fibroblasts is defective in promoting cellular adhesion, *J. Cell. Physiol.* **103**:47.

Choe, B-K, and Rose, N. R., 1976, *In vitro* senescence of mammalian cells, *Gerontology* **22**:89.

Clarke, M., and Spudich, J. A., 1977, Nomuscle contractile proteins: The role of actin and myosin in cell motility and shape determination, *Ann. Rev. Biochem.* **46**:797.

Clarkson, B. B., 1974, The survival value of the dormant state in neoplastic and normal cell populations, in: *Control of Proliferation in Animal Cells,* Vol. 1 (B. D. Clarkson and R. Baserga, eds), pp. 945–972, Cold Spring Harbor Laborabory, Cold Spring Harbor, New York.

Courtois, Y., and Hughes, R. C., 1976, Surface labelling of senescent chick fibroblasts by lactoperoxidase-catalysed iodination, *Gerontology* **22**:371.

Cox, R. E., Krauss, M. R., Balis, M. E., and Dancis, J., 1974, Metabolic cooperation in cell culture. A model for cell-to-cell communication, in: *Cell Communication* (R. P. Cox, ed.), pp. 67–95, John Wiley and Sons, Inc., New York.

Cristofalo, V. J., 1972, Animal cell cultures as a model system for the study of aging, *Adv. Geront. Res.* **4**:45.

Cristofalo, V. J., and Kritchevsky, D., 1969, Cell size and nucleic acid content in the diploid human cell line WI-38 during aging, *Med. Exptl.* **19**:313.

Crusberg, T. C., Haskins, B. B., and Widdus, R., 1979, Spreading behavior and surface characteristics of young and senescent WI-38 fibroblasts revealed by scanning electron microscopy, *Exp. Cell Res.* **118**:39.

Edidin, M., 1974, Rotational and translational diffusion in membranes, *Ann. Rev. Biophys. Bioeng.* **3**:179.

Furshpan, E. J., and Potter, P. D., 1968, Low-resistance between cells in embryo and tissue culture, *Curr. Top Dev. Biol.* **3**:95.

Gilula, N. B., Reeves, O. R., and Steinbach, A., 1972, Metabolic coupling, ionic coupling and cell contacts, *Nature (London)* **235**:262.

Goldman, R. D., Chojnacki, B., and Yerna, M-J, 1979, Ultrastructure of microfilament bundles in baby hamster kidney (BKH-21) cells. The use of tannic acid, *J. Cell Biol.* **80**:759.

Good, P. I., 1975, Aging in Mammalian Cell Populations: A review, *Mech. Ageing Dev.* **4**:339.

Goodenough, D. A., 1974, Bulk isolation of mouse hepatocyte gap junctions, *J. Cell Biol.* **61**:557.

Grinna, L. S., 1977, Changes in cell membranes during aging, *Gerontology* **23**:452.

Harper, R. A., and Groove, G., 1979, Human skin fibroblasts derived from papillary and reticular dermis: Differences in growth potential *in vitro, Science* **204**:526.

Hayflick, L., 1965, The limited *in vitro* lifetime of human diploid cell strains, *Exp. Cell Res.* **37**:614.

Hayflick, L., and Moorhead, P. S., 1961, The serial cultivation of human diploid cell strains, *Exp. Cell Res.* **25**:585.

Heggeness, M. H., Wang, K., and Singer, S. J., 1977, Intracellular distributions of mechanochemical proteins in cultured fibroblasts, *Proc. Natl. Acad. Sci. USA* **74**:3883.

Herzenberg, L. A., and Herzenberg, L. A., 1978, Analysis and separation using the fluorescence activated cell sorter (FACS), in: *Handbook of Experimental Immunology,* Vol. 3 (D. M. Weir, ed.), pp. 22.1–22.21, Blackwell Scientific Publications, London.

Hynes, R. O., and Bye, J. M., 1974, Density and cell cycle dependence of cell surface proteins in hamster fibroblasts, *Cell* **3**:113.

Johnson, J. E., Jr., 1979, Fine structure of IMR-90 cells in culture as examined by scanning and transmission electron microscopy, *Mech. Ageing Dev.* **10**:405.

Kaji, K., and Matsuo, M., 1978, Aging of chick embryo fibroblasts *in vitro*. II. Relationship between cell proliferation and increased multinuclear cells, *Mech. Ageing Dev.* **8**:233.

Kaji, K., and Matsuo, M., 1979, Aging of chick embryo fibroblasts *in vitro*. III. Polyploid cell accumulation, *Exp. Cell Res.* **119**:231.

Kapp, L. N., and Klevecz, R. R., 1976, The cell cycle of low passage and high passage human diploid fibroblasts, *Exp. Cell Res.* **101**:154.

Kay, M. M. B., 1979, An overview of immune ageing, *Mech. Ageing Dev.* **9**:39.

Kelley, R. O., 1976, Development of the aging cell surface: A freeze-fracture analysis of gap junctions between human embryo fibroblasts aging in culture. A brief note, *Mech. Ageing Dev.* **5**:339.

Kelley, R. O., and Perdue, B. D., 1980, Development of the aging cell surface: Structural patterns of gap junction assembly between metabolic mutants and progressively subcultivated human diploid fibroblasts (IMR-90), *Exp. Geront.* **15**:407.

Kelley, R. O., and Skipper, B. E., 1977, Development of the aging cell surface: Variation in the distribution of intramembrane particles with progressive age of human diploid fibroblasts, *Ultrastruct. Res.* **59**:113.

Kelley, R. O., Azad, R., and Vogel, K. G., 1978, Development of the aging cell surface: Concanavalin A-mediated intercellular binding and the distribution of binding sites with progressive subcultivation of human embryo fibroblasts, *Mech. Ageing Dev.* **8**:203.

Kelley, R. O., Vogel, K. G., Crissman, H. A., Lujan, C. J., and Skipper, B. E., 1979, Development of the aging cell surface: Reduction of gap-junction mediated metabolic cooperation with progressive subcultivation of human embryo fibroblasts (IMR-90), *Exp. Cell Res.* **119**:127.

Kelley, R. O., Trotter, J., Marek, L. F., Perdue, B. D., and Taylor, C. B., 1980, Variation in cytoskeletal assembly during spreading of progressively subcultivated human embryo fibroblasts (IMR-90), *Mech. Ageing Dev.* **13**:127.

Kjellen, L., Petterson, I., and Hook, M. 1981, Cell-surface heparan sulfate: An intercalated membrane proteglycan, *Proc. Natl. Acad. Sci. USA* **78**:5371.

Korn, E. D., 1978, Biochemistry of actomyosin-dependent cell motility (A Review), *Proc. Natl. Acad. Sci. USA* **75**:588.

Kraemer, P., 1971, Complex carbohydrates of animal cells: Biochemistry and physiology of the cell periphery, in: *Biomembranes* Vol. 1 (L. Manson, ed.), pp. 67–190, Plenum, New York.

Lazarides, E., 1975, Tropomyosin Antibody: Specific localization of tropomyosin in nonmuscle cells, *J. Cell Biol.* **65**:549.

Lazarides, E., 1976, Actin, α-actinin, and tropomyosin interaction in the structural organization of actin filaments in nonmuscle cells, *J. Cell Biol.* **68**:202.

Lazarides, E., and Burridge, K., 1975, α-Actinin: Immunofluorescent localization of a muscle structural protein in nonmuscle cells, *Cell* **6**:289.

Lentz, R. B., Barenholz, Y., and Thompson, T. E., 1976, Fluorescence depolarization studies of phase transitions and fluidity in phospholipid bilayers. 1. Single component phosphatidylcholine liposomes, *Biochemistry* **15**:4521.

Lowenstein, W. R., 1974, Intercellular communication through membrane junctions and cancer etiology, in: *Membrane Transformation in Neoplasia* (J. Schulz and R. E. Block, eds), pp. 1–18, Academic Press, Inc., New York.

Macieira-Coelho, A., 1973, Aging and cell division, *Frontiers in Matrix Biology* **1**:46.

Macieira-Coelho, A., 1977, Kinetics of the proliferation of human fibroblasts during their lifespan *in vitro, Mech. Ageing Devl.* **6**:341.

Macieira-Coelho, A., and Ponten, J., 1969, Analogy in growth between late passage human embryonic and early passage human adult fibroblasts, *J. Cell Biol.* **43**:374.

Makowski, L., Caspar, D., Phillips, W., and Goodenough, D., 1977, Gap junction structures. II. Analysis of the X-ray diffraction data, *J. Cell Biol.* **74**:629.

Martin, G. M., 1979, Genetic and evolutionary aspects of aging, *Fed. Proc. Am. Soc. Exp. Biol.* **38**:1962.

Martin, G. M., Sprague, C. A., and Epstein, C. J., 1970, Replicative life-span of cultivated human cells, *Lab Invest.* **23**:86.

Matuoka, K., and Mitsui, Y., 1981, Changes in cell-surface glycosaminoglycans in human diploid fibroblasts during *in vitro* aging, *Mech. Ageing Dev.* **15:**153.

Mellman, W. J., and Cristofalo, V. J., 1972, Human diploid cell cultures: Their usefulness in the study of genetic variations in metabolism, in: *Growth, Nutrition and Metabolism of Cells in Culture,* Vol. 1 (G. H. Rothblat and V. J. Cristofalo, eds.), p. 327, Academic Press, New York.

Milo, G. E., and Hart, R. W., 1976, Age-related alterations in plasma membrane glycoprotein content and scheduled or unscheduled DNA synthesis, *Arch. Biochem. Biophys.* **176:**324.

Mitsui, Y., and Schneider, E. L., 1976a, Increased nuclear sizes in senescent human fibroblast cultures, *Exp. Cell Res.* **100:**147.

Mitsui, Y., and Schneider, E. L., 1976b, Relationship between cell replication and volume in senescent human diploid fibroblasts, *Mech. Ageing Devl.* **5:**45.

Mitsui, Y., and Schneider, E. L., 1976c, Characterization of fractionated human diplid fibroblast cell populations, *Exp. Cell Res.* **103:**23.

Nichols, W. W., Murphy, D. G., Cristofalo, V. J., Toji, L. H., Greene, A. E., and Dwight, S. A., 1977, Characterization of a new human diploid cell strain, IMR-90, *Science* **196:**60.

Nicolini, C., Kendall, F., Baserga, R., Dessaive, C., Clarkson, B., and Fried, J., 1977, The G_0–G_1 transition of WI-38 cells. I. Laser flow microfluorimetric studies, *Exp. Cell Res.* **106:**111.

Zs.-Nagy, I., 1979, The role of membrane structure and function in cellular aging: A review, *Mech. Ageing Devl.* **9:**237.

Packer, L., Deamer, D. W., and Heath, R. L., 1967, Regulation and deterioration of structure in membranes, *Adv. Gerontol. Res.* **2:**77.

Peracchia, C., 1977, Gap junctions. Structural changes after uncoupling procedures, *J. Cell Biol.* **72:**628.

Peracchia, C., and Dulhunty, A. F., 1976, Low resistance junctions in crayfish. Structural changes with functional uncoupling, *J. Cell Biol.* **70:**419.

Pitts, J. D., and Simms, J. W., 1977, Permeability of junctions between animal cells: intercellular transfer of nucleotides but not of macromolecules, *Exp. Cell Res.* **104:**153.

Polgar, P., Taylor, L., and Brown, L., 1978, Plasma membrane associated metabolic parameters and the aging of human diploid fibroblasts, *Mech. Ageing Dev.* **7:**151.

Pollard, T. D., and Weihing, R. R., 1974, Actin and myosin and cell movement, *CRC Crit. Rev. Biochem.* **2:**1.

Rambourg, A., and Leblond, C. P., 1967, Electron microscope observations on the carbohydrate rich cell coat present at the surface of cells in the rat, *J. Cell Biol.* **32:**27.

Rivnay, B., Globerson, A., and Shinitzky, M., 1979, Viscosity of lymphocyte plasma membrane in aging mice and its possible relation to serum cholesterol, *Mech. Ageing Dev.* **10:**71.

Rosen R., 1978, Cells and Senescence, *Int. Rev. Cytol.* **54:**161.

Rothstein, M., 1975, Aging and the alteration of enzymes: A review, *Mech. Ageing Dev.* **4:**325.

Ryan, J. M., 1979, The kinetics of chick cell population aging *in vitro, J. Cell Physiol.* **99:**67.

Schachtschabel, D. O., and Wever, J., 1978, Age-related decline in the synthesis of glycosaminoglycans by cultured human fibroblasts (WI-38), *Mech. Ageing. Dev.* **8:**257.

Schmidt, C. F., Barenholz, Y., and Thompson, T. E., 1977, A nuclear magnetic resonance study of sphingomyelin in bilayers, *Biochemistry* **16:**2649.

Schneider, E. L., 1979, Cell replication and aging: *in vitro* and *in vivo* studies, *Fed. Proc. Am. Soc. Exp. Biol.* **38:**1857.

Schneider, E. L., and Fowlkes, B. J., 1976, Measurement of DNA content and cell volume in senescent human fibroblasts utilizing flow multiparameter single cell analysis, *Exp. Cell Res.* **98:**298.

Schneider, E. L., Mitsui, Y., Au, K. S., and Schorr, S. S., 1977, Tissue-specific differences in cultured human diploid fibroblasts, *Exp. Cell Res.* **108:**1.

Sheridan, J. D., 1970, Low-resistance junctions between cancer cells in various solid tumors, *J. Cell Biol.* **45:**91.

Shinitzky, M., and Henkart, P., 1979, Fluidity of cell membranes—Current topics and trends, *Int. Rev. Cytol.* **60:**121.

Shinitzky, M., and Inbar, M., 1976, Microviscosity parameters and protein mobility in biological membranes, *Biochim. Biophys. Acta* **433:**133.

Simons, J. W. I. M., 1967, The use of frequency distributions of cell diameters to characterize cell populations in tissue culture, *Exp. Cell Res.* **45:**336.

Singer, S. L., and Nicolson, G., 1972, The fluid mosaic model of the structure of the cell membrane, *Science* **175**:720.

Sluke, G., Schatchtschabel, D. O., and Wever, J., 1981, Age-related changes in the distribution pattern of glycosaminoglycans synthesized by cultured human diploid fibroblasts (WI-38), *Mech. Ageing Dev.* **16**:19.

Smith, J. R., and Hayflick, L., 1974, Variation in the life-span of clones derived from human diploid cell strains, *J. Cell Biol.* **62**:48.

Spataro, A. C., Bosmann, H. B., and Myers-Robfogel, M. W., 1979, Sialyltransferase activities of aging diploid fibroblasts, *Biochim. Biophys. Acta* **553**:378.

Staehelin, L. A., 1974, Structure and function of intercellular junctions, *Int. Rev. Cytol.* **39**:191.

Steck, T. L., 1974, The organization of proteins in the human red blood cell membrane: A review, *J. Cell Biol.* **62**:1.

Stein, G. H., and Yanishevsky, R. M., 1979, Entry into S phase is inhibited in two immortal cell lines fused to senescent human diploid cells, *Exp. Cell Res.* **120**:155.

Stenman, S., Wartiovaara, J., and Vaheri, A., 1977, Changes in the distribution of a major fibroblast protein fibronectin, during mitosis and interphase, *J. Cell Biol.* **74**:453.

Stewart, C. 1981, Quantitation of glycosaminoglycans and analysis of fibronectin produced by early and late passage IMR-90 fibroblasts. Thesis, University of New Mexico.

Tobey, R. A., and Crissman, H. A., 1975, Unique techniques for cell cycle analysis utilizing mithramycin and flow microfluorometry, *Exp. Cell Res.* **93**:235.

Trotter, J. A., and Kelley, R. O., 1979, A novel technique for high resolution analysis of the cytoskeleton, *Anat. Rec.* **195**:7.

Vogel, K. G., and Peterson, D. W., 1981, Extracellular surface and intracellular proteoglycans produced by human embryo lung fibroblasts in culture (IMR-90), *J. Biol. Chem* **256**:13235.

Vogel, K. G., and Pitcher, D. E., 1982, Sulfated proteoglycans produced by human lung fibroblasts during *in vitro* cellular aging, *Eur. J. Cell Biol.* **29**:61–67.

Vogel, K. G., Kelley, R. O., and Stewart, C., 1981a, Loss of organized fibronectin matrix from the surface of aging diploid fibroblasts (IMR-90), *Mech. Ageing Dev.* **16**:295.

Vogel, K. G., Kendall, V. F., and Sapien, R. E., 1981b, Glycosaminoglycan synthesis and composition in human fibroblasts during *in vitro* cellular aging (IMR-90), *J. Cell Physiol.* **107**:271.

Wang, K., Ash, J. F., and Singer, S. J., 1975, Filamin, a new high-molecular-weight protein found in smooth muscle and non-muscle cells, *Proc. Natl. Acad. Sci USA* **72**:4483.

Weihing, R. R., 1979, The cytoskeleton and plasma membrane, *Meth. Achiev. Exp. Pathol.* **8**:42.

Wolosewick, J. J., and Porter, K. R., 1977, Observations on the morphological heterogeneity of WI-38 cells, *Am. J. Anat.* **149**:197.

Wright, B. E., and Davison, P. F., 1980, Guest Editorial: Mechanisms of development and aging, *Mech. Ageing Dev.* **12**:213.

Yamada, K. M., and Olden, K., 1978, Adhesive glycoproteins of cell surface and blood, *Nature* **275**:179.

Yamamoto, K., Yamamoto, M., and Ooka, H., 1977, Cell surface changes associated with aging of chick embryo fibroblasts in culture, *Exp. Cell Res.* **108**:87.

Yen, A., Fried, J., Kitahara, T., Strife, A., and Clarkson, B. D., 1975, The kinetic significance of cell size. II. Size distribution of resting and proliferating cells during interphase, *Exp. Cell Res.* **95**:303.

2

In Vivo and in Vitro Comparisons of Age-Related Fine Structural Changes in Cell Components

JOHN E. JOHNSON, JR.

1. INTRODUCTION

Although the study of aging cells and tissues dates back many decades, the majority of data gathered are biochemical, pharmacological, and physiological. Histological studies, particularly by electron microscopy, in comparison, have only begun to accumulate. *In vivo* and *in vitro* studies of aging cells have proceeded in parallel over the years with an ever increasingly important issue developing concerning the applicability of data gathered *in vitro* to *in vivo* situations. Very little information exists to support or not support this concept, partially because few have attempted to compare the two disciplines, but also because experimental gerontology is in its early stages of development as a research field, and scientists are still gathering base-line data.

This chapter is a presentation of data on microscopically observed age-related changes in cellular components. Comparisons are made between changes seen in cells taken from animals which have been sacrificed by perfusion fixation (the *in vivo* situation) and cells grown in culture and fixed after having been cultured for particular lengths of time (the *in vitro* situation).

JOHN E. JOHNSON, JR. • Department of Neurology, Johns Hopkins University School of Medicine, Baltimore, Maryland 21205; and National Institute on Aging, NIH, Baltimore City Hospital, Baltimore, Maryland 21224.

2. THE CELL NUCLEUS

2.1. Nuclear Invagination

In most cell types examined from *in vivo* preparations, the nuclei are indented (invaginated) to varying degress. The nervous sytem is very much prone to this (Figs. 1, 2), and evidence suggests that nuclear invagination increases with advancing age.

As early as 1894, Hodge noted that human ganglion cell nuclei became irregular in shape with increasing age. Other laboratories have since reported similar observations (Kuhlenbeck, 1954; Timiras and Vernadakis, 1972; Hasan and Glees, 1973). In a study of aging rats, Johnson and Miquel (1974) found that lateral vestibular neurons had deeper invaginations in older animals.

Not all neurons have the invaginations to the same extent. For example, in the dorsal column nuclei of mice, neurons of nucleus gracilis are much more invaginated than the neurons of nucleus cuneatus (Fig. 3). There appears to be a greater age-related deterioration of axons in nucleus gracilis than in nucleus cuneatus (Johnson, Mehler and Miquel, 1975), and this fact correlates in human neurological studies (Steiness, 1957) showing a larger decrement of sensory information from the legs (axons to nucleus gracilis) than from the shoulders (axons to nucleus cuneatus).

In a recent study of the aging adenohypophysis in macaques, Aschenbrenner (1979) reported that cell nuclei tended to be shrunken and unevenly contoured in older subjects. In our own studies of the aging pituitary (Johnson, 1980a), we found that the alterations in the neurohypophysis are much more extensive than in the adenohypophysis. At the other end of the diencephalon, the pineal gland exhibited some of the most severe age-related changes we have observed, including increased nuclear invagination (Johnson, 1980b). Age-related increases in nuclear invaginations have also been reported in the Purkinje cells and liver cells of old mice (Andrew, 1955, 1962), hippocampus of aging rats (Hasan and Glees, 1973), anterior pituitary cells of old mice (Weiss and Lansing, 1953), chinchilla astrocytes (Cammermeyer, 1963), and dog heart (Munnell and Getty, 1968). Andrew (1964) suggested a relationship of increased nuclear invagination with attenuated blood supply and nutrition. This might imply a decrement in cell metabolism. London *et al.* (1979) examined numerous CNS regions and found that [^{14}C]-2-deoxy-D-glucose utilization was reduced in many areas, particularly the inferior colliculus, medulla, and pons. This would, therefore, support the concept that increased nuclear membrane irregularity in older animals is correlated with a reduction in cell metabolic rates.

In vitro studies also tend to support the idea that nuclear invagination increases with "age." Brock and Hay (1971) found an increased tendency for nuclear invagination in chick fibroblasts at later population doubling levels (PDL). Lipetz and Cristofalo (1972) reported this same phenomenon using WI-38 human fetal lung fibroblasts, and Johnson (1979) confirmed the occurrence in late PDL IMR-90 fibroblasts (Fig. 4–6) which are very similar to the WI-38 strain. Cristofalo and Kritchevsky (1966) found, however, that the rate of respiration was unchanged in late passage WI-38 cells. Goldstein and Trieman (1974), on the other hand, observed *increased* glucose utilization in aging cultured fibroblasts. Thus, although increased nuclear invagination is observed *in vivo* and *in vitro* with age, cell metabolic rate appears to go down on the one hand *(in vivo)* and up, or unchanged, on the other *(in vitro)*. However, it should be noted that

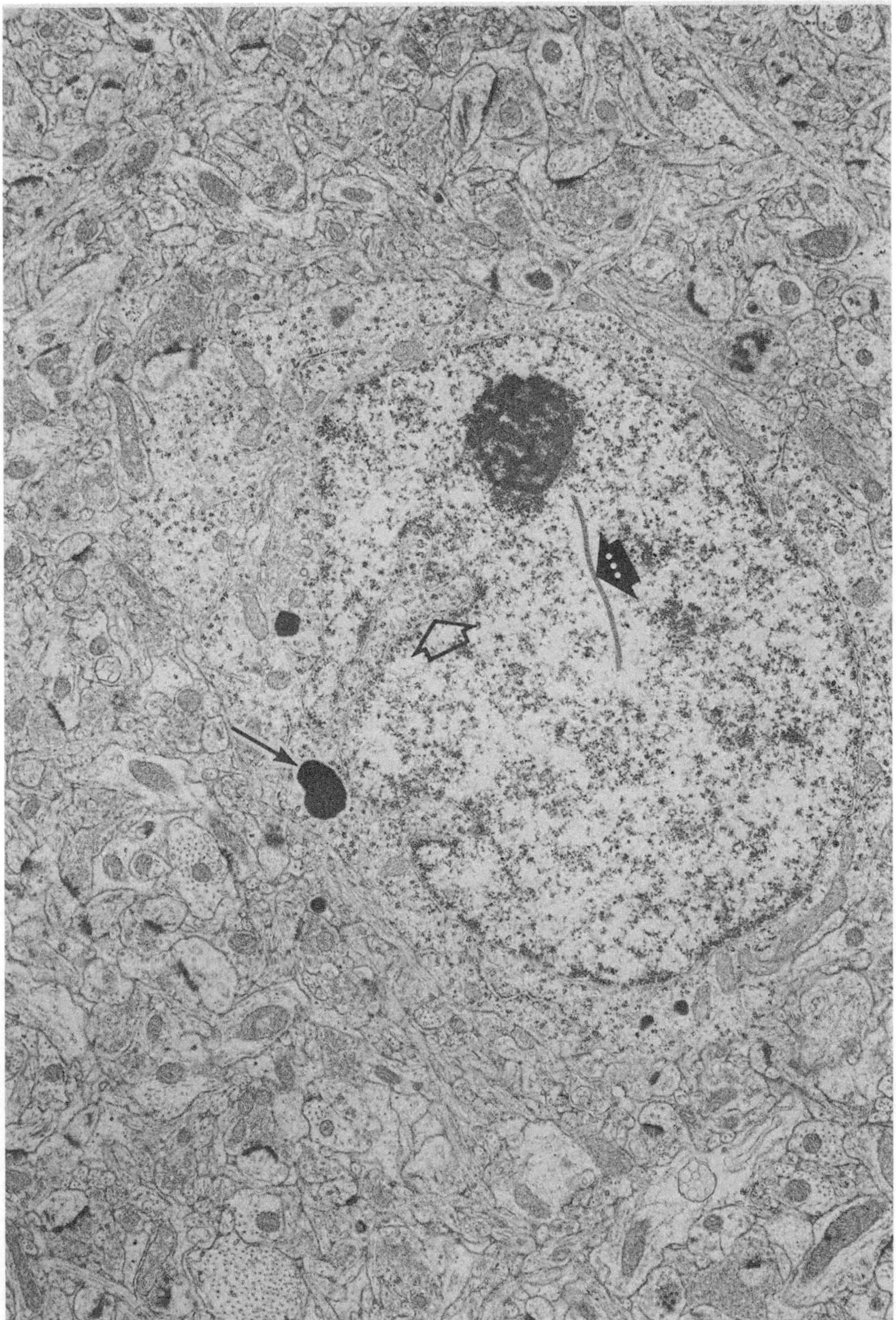

Figure 1. This neuron is from the frontal cortex of the brain of a seven-year-old squirrel monkey. The cytoplasm contains dense bodies (thin arrow). The nucleus is invaginated (hollow arrow) and contains a long narrow inclusion (arrow with three dots). × 13,650, reproduced at 85%.

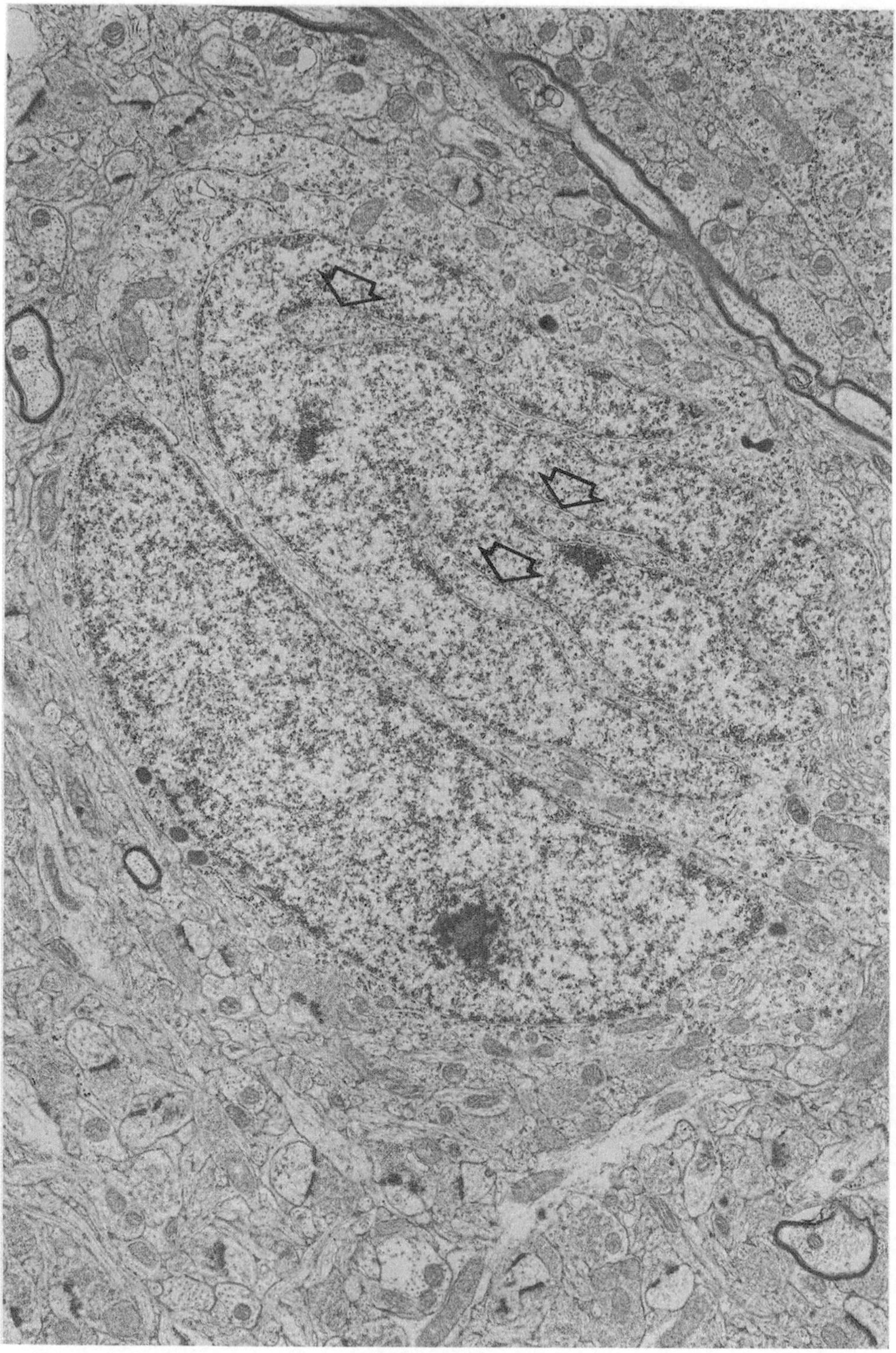

Figure 2. Two astrocytes are illustrated from the frontal cortex of a seven-year-old squirrel monkey. The nucleus of one is smoothly contoured, while the other has deep invaginations (arrows). × 13,650, reproduced at 85%.

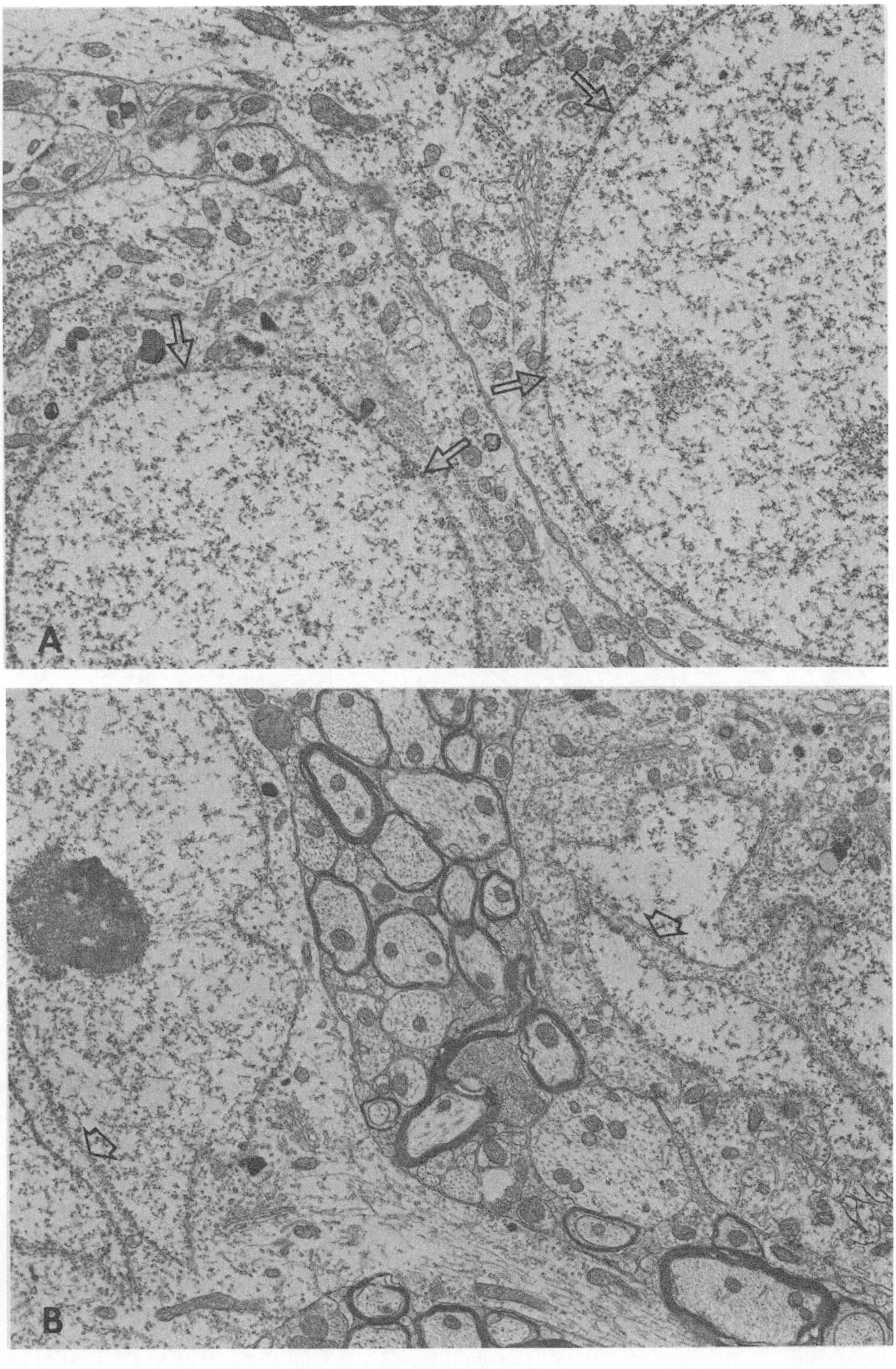

Figure 3. These brain stem neurons from mouse nucleus cuneatus (A) and nucleus gracilis (B) show different nuclear morphology even though they are the same age (three months). Cuneatus neuronal nuclei are smooth (arrows), while gracilis neuronal nuclei have deep invaginations (arrows). × 9600, reproduced at 85%. (From Johnson *et al.*, 1975.)

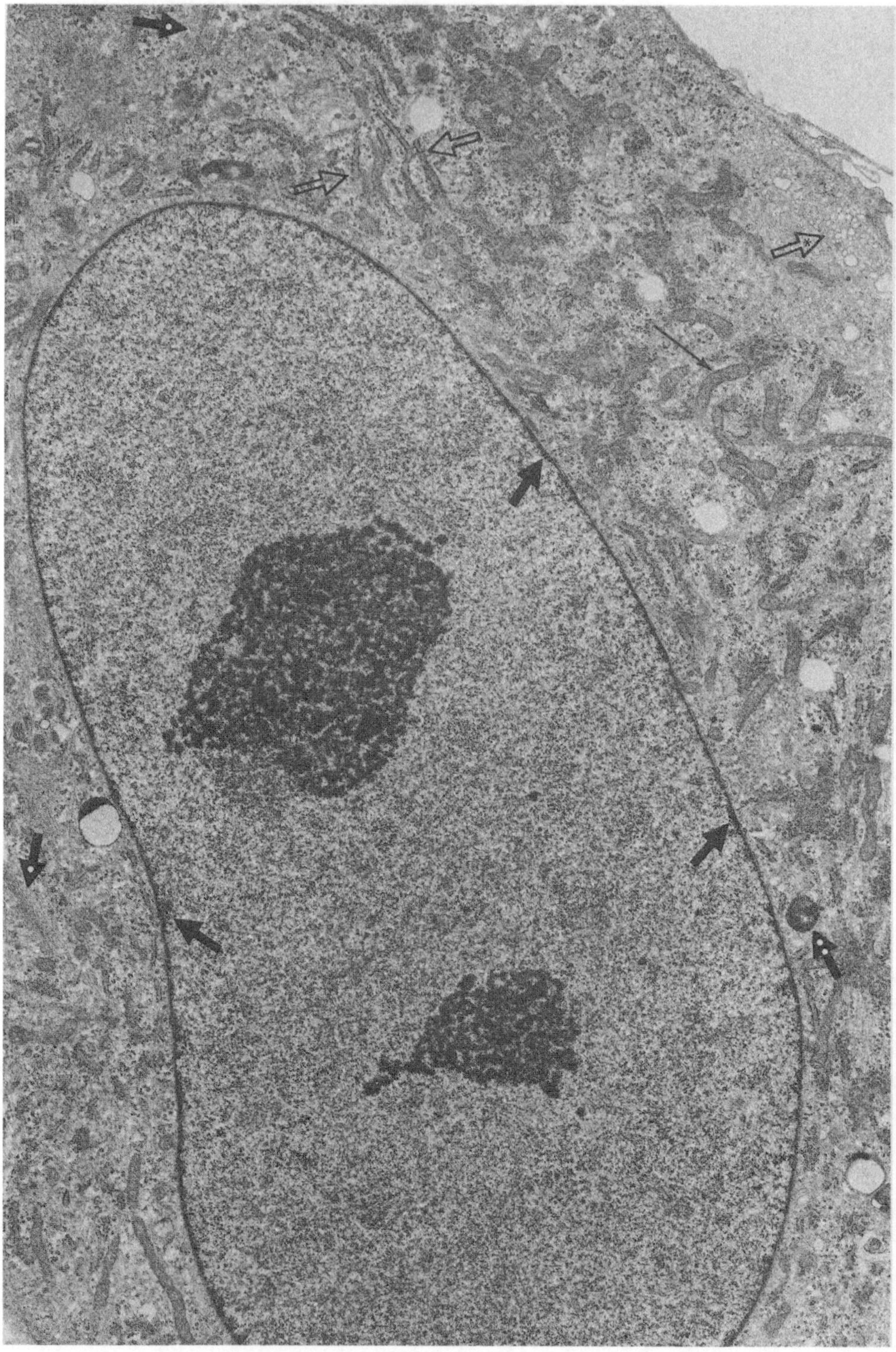

Figure 4. In this IMR-90 human fetal lung fibroblast cultured for 18 population doublings (PDL), the nucleus is smooth, and there is very little chromatin condensation (solid arrows with no dots). The cytoplasm contains well-developed Golgi (arrows with one dot), granular endoplasmic reticulum (hollow arrows with no dots), dense bodies (solid arrow with two dots), numerous mitochondria (long thin arrow), and vesicles (hollow arrow with asterisk). × 9500, reproduced at 85%. (From Johnson, 1979.)

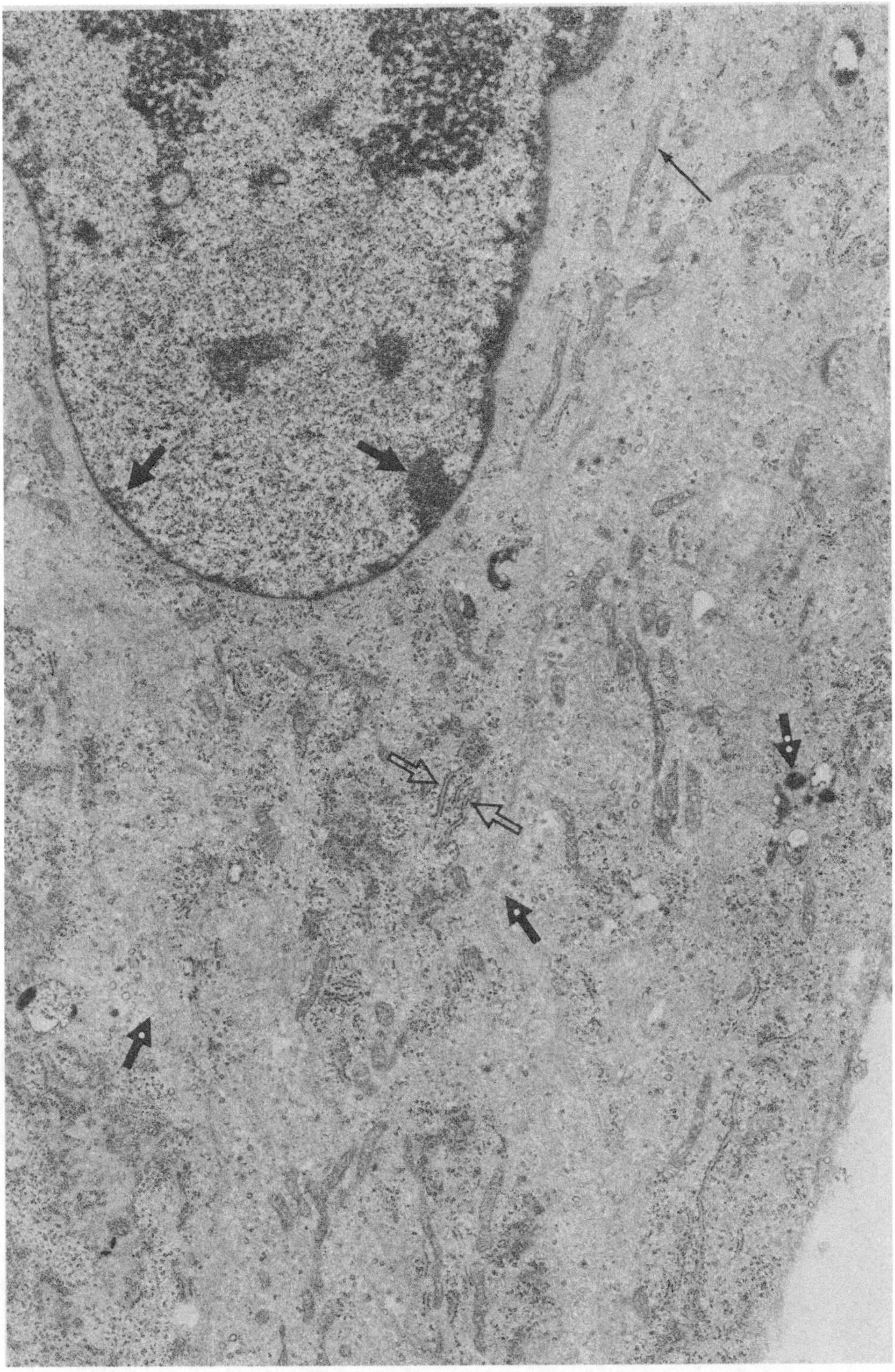

Figure 5. Reduced numbers of mitochondria (long thin arrow) and granular endoplasmic reticulum (hollow arrows with no dots) are seen in this 49 PDL IMR-90 fibroblast. Dense bodies (solid arrow with two dots) and Golgi (solid arrows with one dot) can also be observed. Nuclear chromatin is condensed along the nuclear membrane (solid arrows with no dots). × 9500, reproduced at 85%. (From Johnson, 1979.)

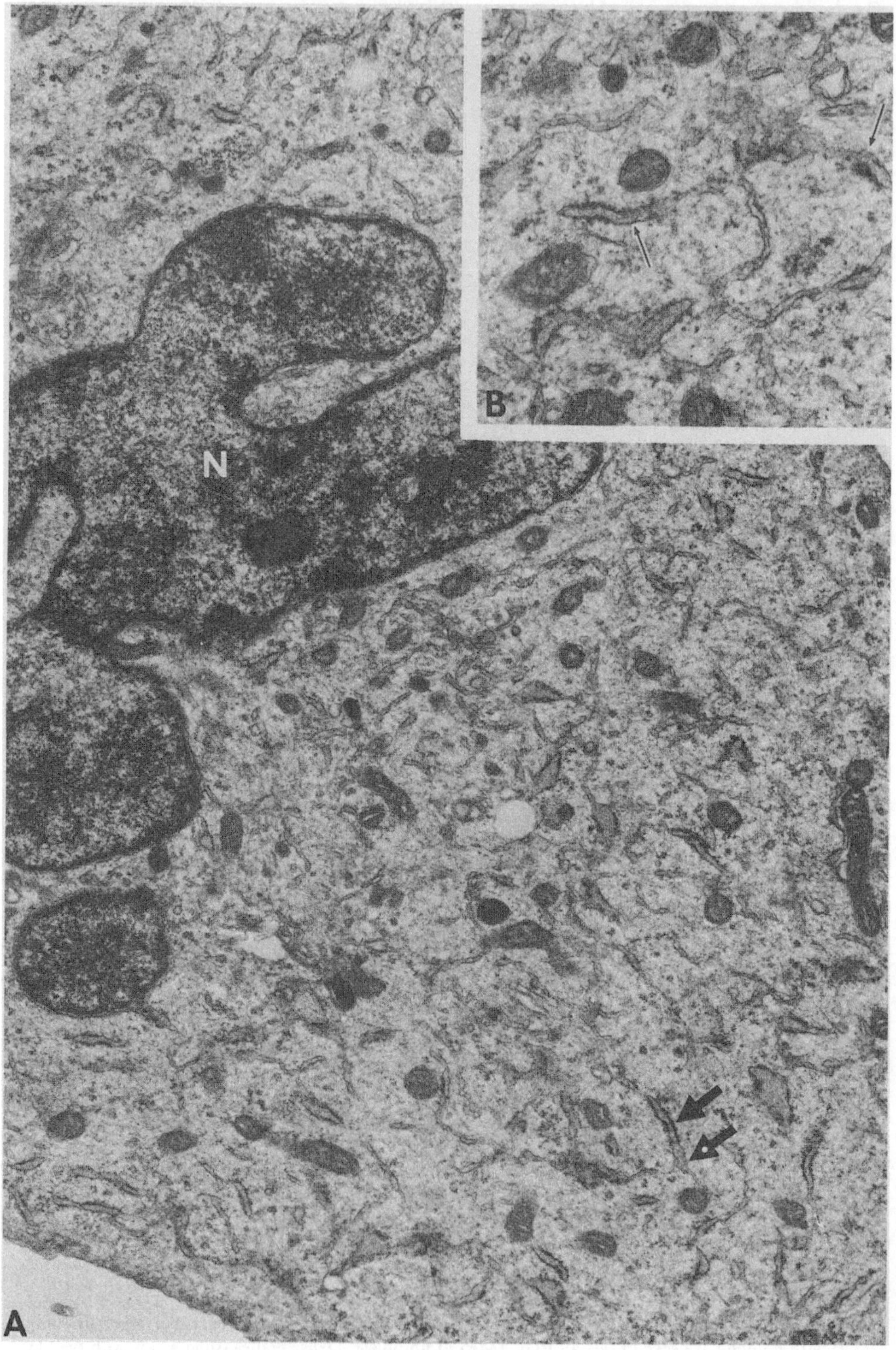
N
B
A

the *in vivo* data represent a consolidation of not only several cell types but also intercellular material, while *in vitro* data represent information gathered on isolated populations of cells of a single type.

2.2. Nuclear Chromatin

Sohal and Allison (1971) and Miquel (1971) have observed chromatin condensation in aging flies. Aschenbrenner (1979) reported that, in the cells of the aging primate pituitary, nuclear chromatin condensation was apparent. Bogart (1970) noted chromatin condensation in the acinar cells of the rat submandibular gland. Studies *in vitro* (Figs. 4–6) indicate that, with increased population doubling levels, nuclear chromatin is also condensed (Brock and Hay, 1971; Johnson, 1979), although Lipetz and Cristofalo (1972) reported negative findings in this regard. Functionally speaking, it has been suggested that cells which contain larger amounts of condensed stainable chromatin (heterochromatin) are less active than cells which contain primarily noncondensed chromatin, i.e., euchromatin (Munnell and Getty, 1968). However, the metabolic studies dealing with this subject, as stated above, are conflicting.

2.3. Nuclear Inclusions

The presence of inclusion bodies within the nuclei of cells (Figs. 1,7) has been a controversial topic since the turn of the century. At that time, of course, electron microscopy was not available. Dimitrova (1901) reported that pinealocytes contained nuclear inclusions and called them "Kernkugeln." Collier (1943) described them in pineals of mice and rats. Quay (1965) did not see them in rodents but concluded that their occurrence, in general, may reflect increased cell activity. Wolfe (1965), using electron microscopy, reported the presence of pinealocyte nuclear inclusions in specimens fixed by immersion. Kevorkian and Wessel (1959), also utilizing electron microscopy, failed to find nuclear inclusions in the human pineal, and concluded that previous light microscopic reports of their presence were based on misconstrued folds in the nucleus. The same conclusion was reached by De Robertis and Pellegrino de Iraldi (1961).

In an electron microscopic study of the aging rat pineal, Johnson (1980b) reported conclusive evidence that pinealocyte nuclear inclusions do exist (Fig. 8). The inclusions, found in pinealocytes but not in the glial cells, consist of filaments 10 nm in diameter which are arranged in bundles. The inclusions were found to be rare in young animals and more frequent with increasing age (Johnson, 1980b). This may explain why the inclusions were not seen in previous reports, since other investigators typically use young subjects in nongerontological experiments. It is difficult to say whether early light

←

Figure 6. A great deal of chromatin condensation can be seen in the nucleus (N) of this 49 PDL IMR-90 fibroblast. The cytoplasm contains shortened segments of granular endoplasmic reticulum. Regions of ribosome-covered (solid arrow with no dot) as well as denuded (solid arrow with one dot) reticulum are present. The inset shows this phenomenon at higher magnification. Main figure × 19,200; inset × 38,000, reproduced at 85%. (From Johnson, 1979.)

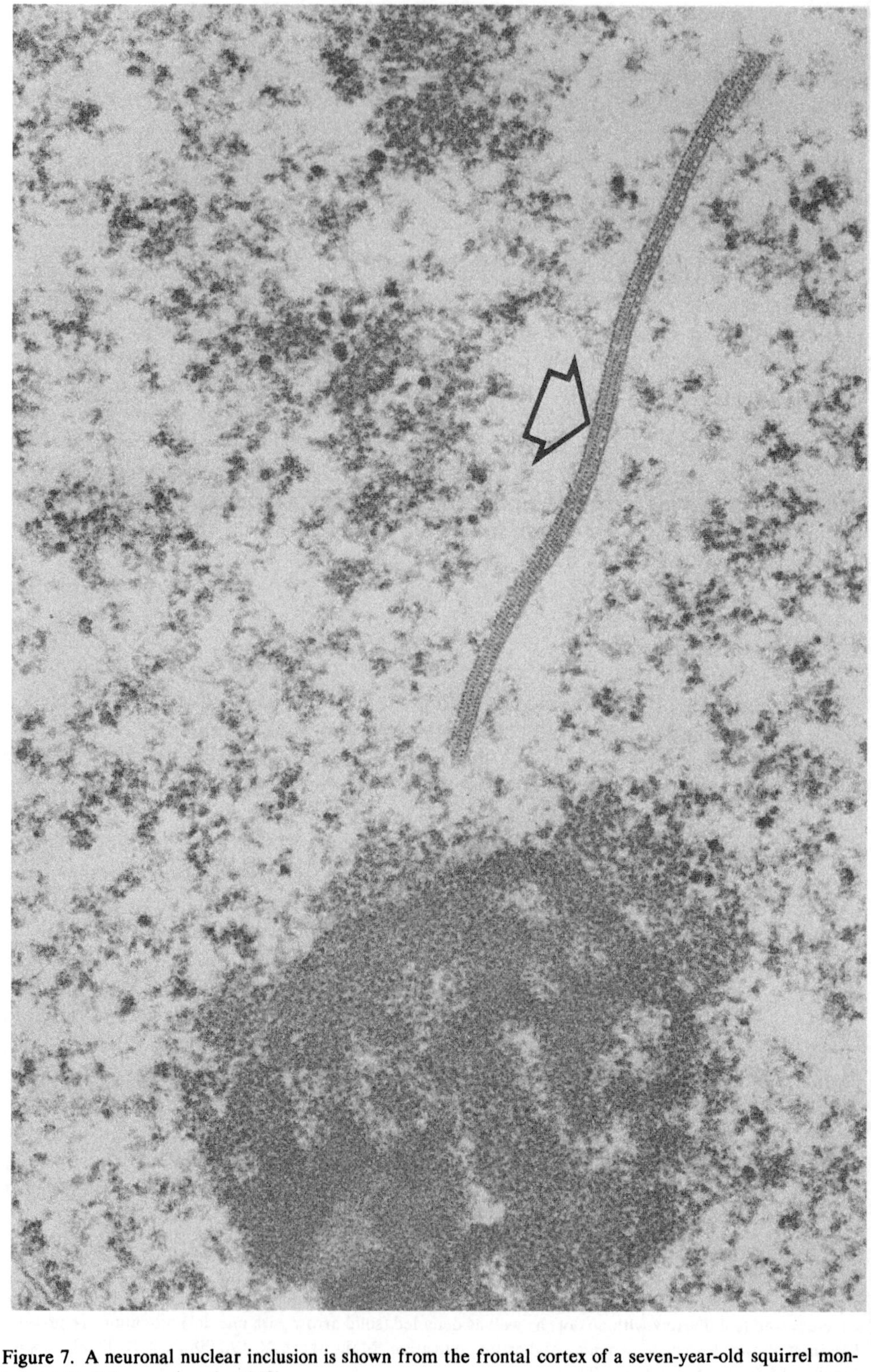

Figure 7. A neuronal nuclear inclusion is shown from the frontal cortex of a seven-year-old squirrel monkey. × 61,500, reproduced at 85%.

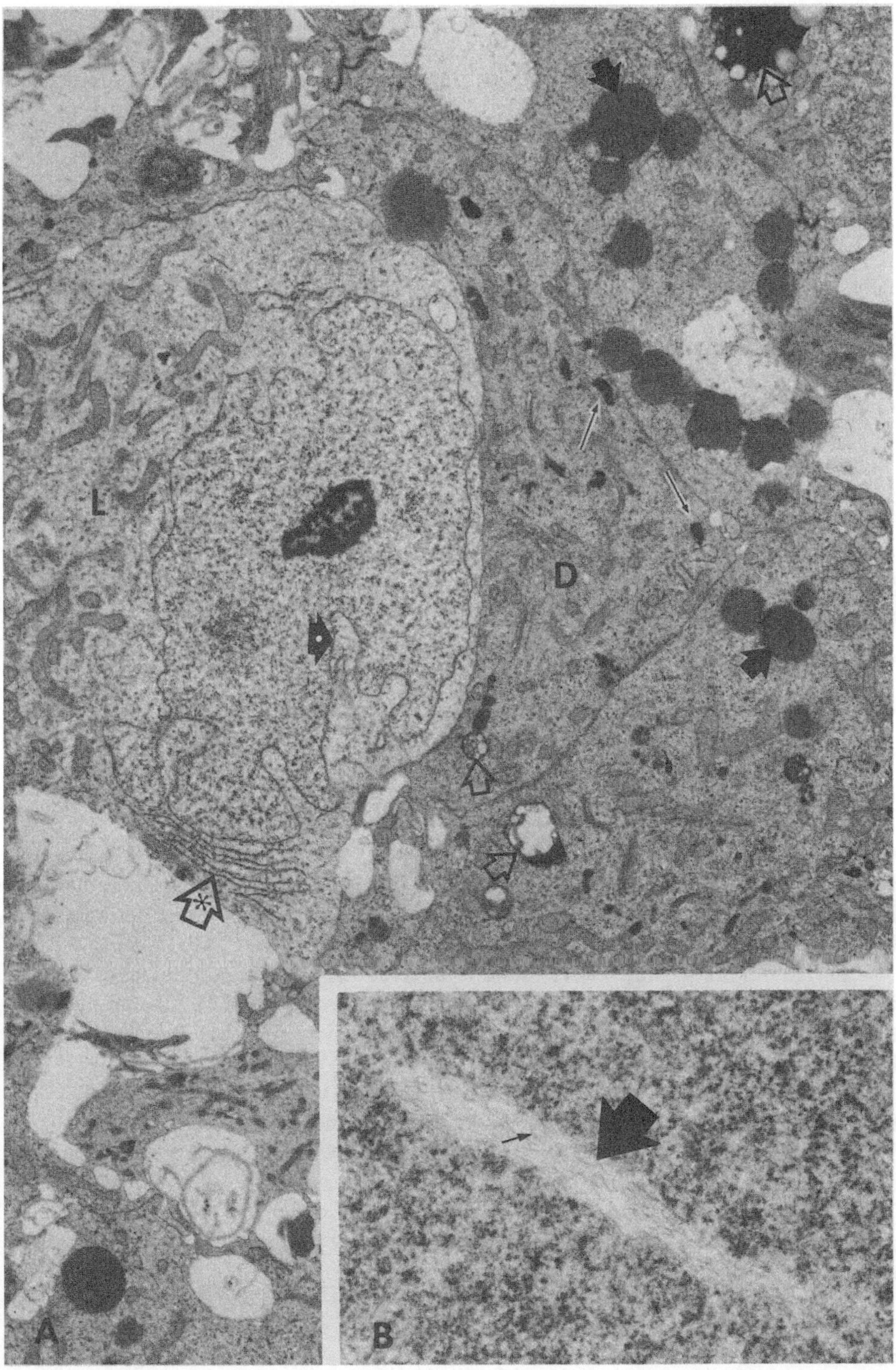

Figure 8. Light (L) and dark (D) pinealocytes are illustrated from a 28-month-old rat. The cytoplasm contains dense bodies (hollow arrows), lipid droplets (solid arrows), dense vesicles (thin arrows), and endoplasmic reticulum (hollow arrow with asterisk). The inset shows a nuclear inclusion from a pinealocyte of a 28-month-old male. Main figure × 7600; inset × 38,000, reproduced at 85%. (From Johnson, 1980.)

microscopic reports of pinealocyte nuclear inclusions represented the same type of structure as seen in Fig. 8.

Nuclear inclusions in other tissues of older animals have been reported previously, primarily in the nervous system (Field and Peat, 1971; Feldman and Peters, 1972; Johnson and Miquel, 1974; Vaughan and Vincent, 1979). Figures 1 and 7 illustrate examples from the aging primate brain. Seite *et al.* (1973) reported that electrical stimulation of the cat stellate ganglion resulted in an increase in the number of nuclear inclusions in only 10 min. They concluded that nuclear inclusions may be formed by intranuclear microtubules and microfilaments, and that their number is related to the activity of the nerve cell. In centrifuged rats, the number of nuclear inclusions in lateral vestibular neurons increased after 4 days of centrifugation (Fig. 9) at 4.15*g* (Johnson, Mehler and Oyama, 1976). Nuclear inclusions were *decreased* in the pineals of hypophysectomized rats (Johnson, 1980c).

In vitro, we have not seen nuclear inclusions such as shown in Figs. 1,7,8 and 9 in any of the cell types we have examined, including IMR-90, BSC-1 kidney cells, chick fibroblasts (Johnson, unpublished observations) or the neuroblastoma lines IMR-32, N-2A, and NB41A3 (Thompson *et al.,* 1982). Some of the neuroblastoma lines do have virus particles, however, which appear to be associated with the nucleus from time to time.

2.4. Other Nuclear Components

In vivo data are published which suggest that the nucleolus changes with age. The nucleolus has been reported to be lost (Hodge, 1894), decreased in pallor (Andrew, 1956), or increased in basophilia (Nandy and Bourne, 1966). However, Bondareff (1959) and Wilcox (1959) did not find the nucleolus to alter. Nosal (1979), using electron microscopy, reported that nucleoli of rat cerebellar Purkinje cells became more irregular and compact. The number of nerve cell nucleoli was observed to increase in human subjects (Bakerman, 1969) and in the mouse (Andrew, 1941, 1959).

Perichromatin granules were found to decrease in the neuronal nuclei of the parietal cortex and cerebellum of aged rats by Zs.-Nagy *et al.* (1977). The granules are thought be an indication of nuclear transcriptional activity and, therefore, such activity may decrease with age.

With respect to *in vitro* alterations in cell nucleoli with increasing population doubling levels, Bemiller and Lee (1976) and Bemiller and Miller (1979) found that, in WI-38 cells, there is an increase in the fraction of cells with one large nucleolus, an increase in the mean nucleolar dry mass and an increase in the mean nucleolar area. The *number* of nucleoli tended to decrease. The increase in nucleolar mass was accompanied by nucleolar aggregation, a process similar to that observed with *Paramecium aurelia* (Sundararaman and Cummings, 1976).

Figure 9. This lateral vestibular neuron is from the brain of a one-month-old rat centrifuged at 4.15*g* for 4 days. The nucleus contains an inclusion. The inset micrograph illustrates a nuclear inclusion from a lateral vestibular neuron in a rat centrifuged for 19.5 months at 2.76*g*. Main figure × 30,780; inset × 27,400, reproduced at 85%. (From Johnson *et al.*, 1976.)

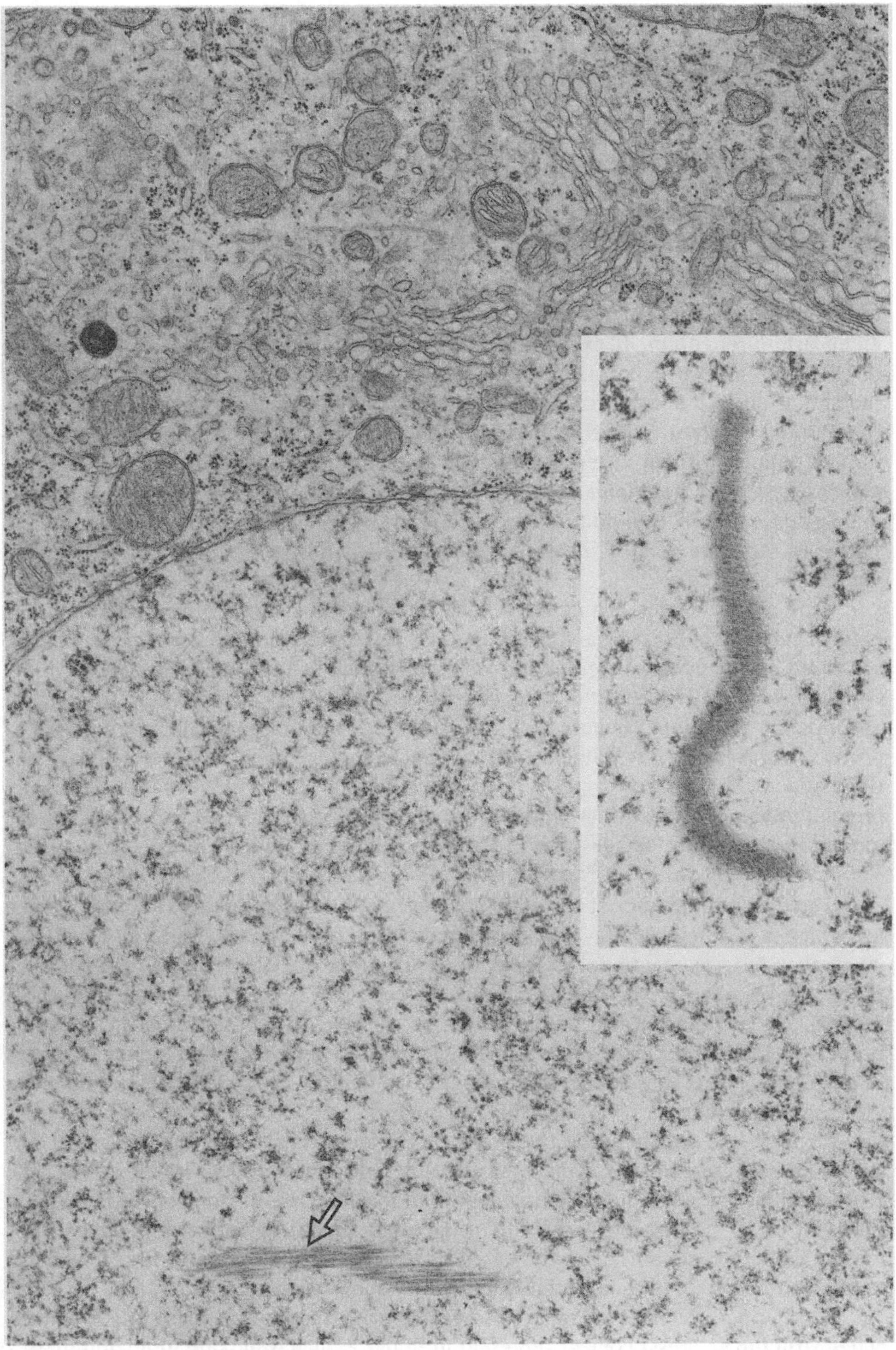

The literature, therefore, suggests that cell aging *in vivo* is accompanied by a tendency for nucleoli to increase in number, while *in vitro*, with successive population doublings, nucleoli decrease in number and increase in size due to aggregation.

3. THE CELL CYTOPLASM

3.1. Endoplasmic Reticulum

Nearly all cells contain a network of membrane bound channels, historically called ergastoplasm by light microscopists, and, with the advent of electron microscopy, the endoplasmic reticulum. It exists in two basic forms, granular endoplasmic reticulum (GER) and smooth endoplasmic reticulum (SER). The GER has ribosomes on the membrane surface and is arranged in the form of tubules and flattened saccular structures called cisternae. The SER has no ribosomes and usually has the form of intermeshed tubules without cisternae. It is now known that the ribosomes of the GER are responsible for the basophilia seen in the light microscope, and that SER was actually undetected until electron microscopy came into being. GER is responsible for protein synthesis, whereas SER is involved with such activities as synthesis of steroids and metabolism of lipids.

Data on age-related changes in cellular endoplasmic reticulum are conflicting. The issue of such changes is certainly an important one since several theories of senescence are based on the fidelity or output of protein synthesis (Orgel, 1963; Medvedev, 1964; Ogrodnik *et al.*, 1975; Strehler, 1966; Strehler *et al.*, 1971). The majority of data on endoplasmic reticulum changes with age deal with the biochemical counterpart, i.e., protein synthesis or RNA related assays. These techniques assay millions of cells at one time, and, in the case of organ studies, sum the data across the different cell types present, unless cell separation methods are used. This fact alone may account for the minimal or no effect of age observed with biochemical studies (Samis *et al.*, 1971; Vaughan and Calvin, 1977; Menzies *et al.*, 1972; McMartin and Schedlbauer, 1975).

The data which do exist are quite controversial. Bowman *et al.* (1976) reported a loss of RNA synthesis in WI-38 cells with increasing population doubling levels (PDL), which they felt was due to decreased synthesis of nucleolar RNA. Schneider and Shorr (1975), on the other hand, found an abrupt *increase* in WI-38 RNA content after about 40 PDL. *In vivo* biochemical studies have shown losses of RNA content in the aging brain (Chaconas and Finch, 1973; Samorajski and Rolsten, 1973) and kidney (Barrows *et al.*, 1962). Heart muscle has shown no change (Shirey and Sobel, 1962) or even an increase (Wayner *et al.*, 1962) in RNA with age.

Bridging the gap between the tissue biochemical assays and the single-cell fine structural analyses are the cytochemical studies. Ringborg (1966) reported that the RNA content per single hippocampal pyramidal cell decreased from 1100 $\pm$ 10 pg in the adult to 53 $\pm$ 1 pg in the old rat. In the aging human spinal cord, Hyden (1967) found that motor neuron RNA increased during the first 40 years of life, leveled off until about age 60, and then declined by 80 and above years of age.

At the microscopic level, many studies have also shown a decrease in RNA associated structures. "Nissl bodies" represent granular endoplasmic reticulum and masses of ribosomes as observed at the light microscopic level using basic dyes (Palay and Palade, 1955). Andrew (1971) reported an age-related decrease in the amount of Nissl material in some neuron types of the cerebrum, cerebellum, and trigeminal ganglion of mice and humans.

At the electron microscopic levels, *in vivo,* numerous studies have documented age related changes in granular endoplasmic reticulum and ribosomes. Andrew (1961) described "close-set" lamellae of the endoplasmic reticulum in cerebellar neurons of aging mice as being partly responsible for the loss of "Nissl" seen at the light microscopic level. In an examination of the hippocampus of rats ranging in age from 4 to 40 months, Hasan and Glees (1973) reported that the GER was well developed and arranged in orderly rows in young animals, while in older subjects it had lost its orderly lamination in the large perikarya and was dispersed. Johnson and Miquel (1974) found a tendency for the GER to be disorganized in the lateral vestibular nucleus of aging rats (Fig. 10). Similar observations have been published in the rat cerebellar Purkinje cells (Nosal, 1979). Most of the above observations on GER were semiquantitative. Morphometric analysis of the GER in the auditory cortex of aging rats (Vaughan and Vincent, 1979) did not show a significant decrease; in fact, an increase in the amount of GER occupying the cytoplasmic volume was noted after 15 months of age. This finding could occur in spite of a tendency for cisternae to no longer be parallel in older animals. In the liver, on the other hand, Schmucker *et al.* (1978) reported that the surface area of smooth endoplasmic reticulum (SER) reached a maximum at 10 months of age and then declined. Granular endoplasmic reticulum (GER) was unaffected by age. However, Garg *et al.* (1979) found the GER to be vesiculated in old rat liver. Smooth ER *increased* in aging mouse neuromuscular presynaptic terminals (Fahim and Robbins, 1982).

At the *in vitro* structural level, several investigators have reported changes in the endoplasmic reticulum. Lipetz and Cristofalo (1972), studying WI-38 human fetal lung fibroblasts, observed that the endoplasmic reticulum cisternae of early (26 and earlier) population doubling levels were dilated, contained amorphous electron dense material and were covered with ribosomes (therefore granular ER). The ER cisternae of later population doubling levels (40 and over) had segments free of ribosomes (smooth), were not as dilated and contained less of the amorphous electron dense material. Johnson (1979) found similar changes in IMR-90 human fetal lung fibroblasts and reported that ER segments tended to be shortened in some cells of later population doubling levels (Fig. 6). It was suggested that there may be a decrease in the total amount of GER in cells at later PDL. In contrast to the above studies, no alteration was observed in the GER of chick fibroblasts with progressive population doublings (Brock and Hay, 1971).

3.2. Mitochondria

Mitochondria, the round to sausage shaped structures present in the cytoplasm of eukaryotic cells, are energy transducing systems that convert energy by phosphorylation into the high-energy phosphate bond of adenosine triphosphate (ATP) (Fig. 11). Thus, the mitochondria are cell power plants which provide energy for such activities as motility, contraction, biosynthesis of cell materials, active transport, transmission of

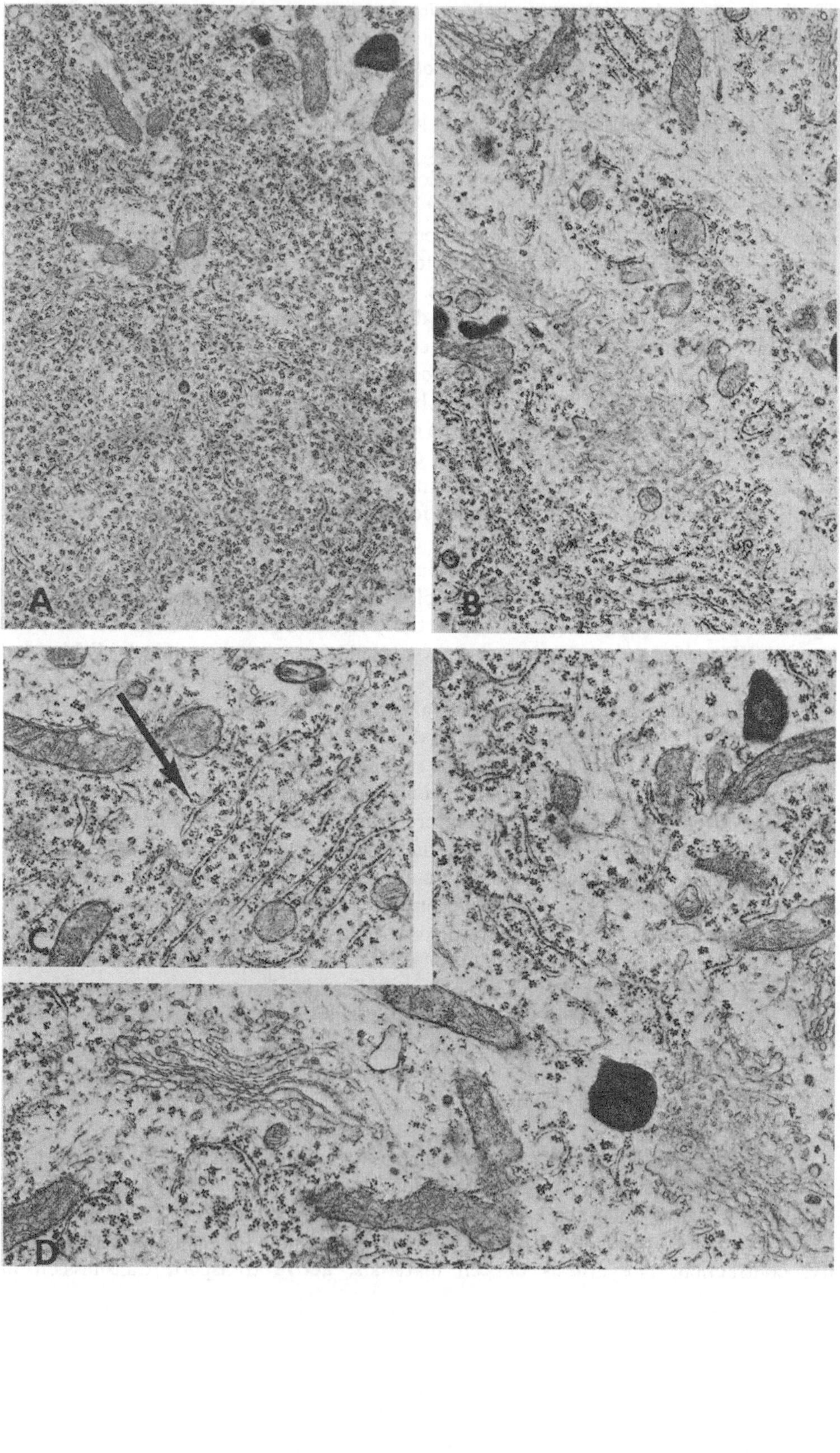

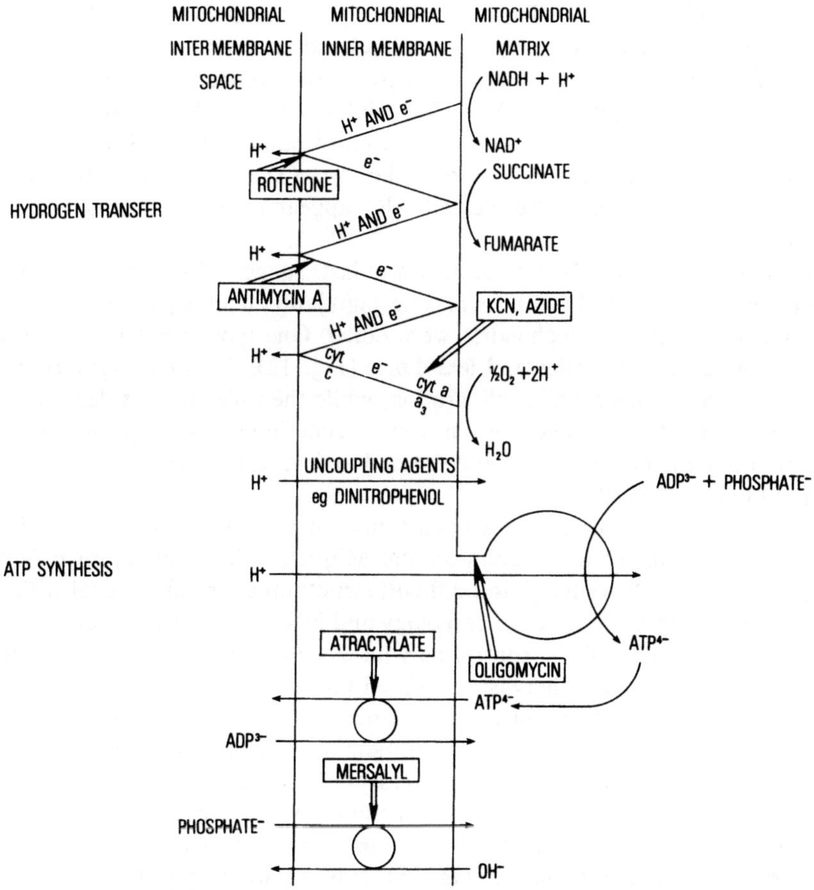

Figure 11. This diagram illustrates the mitochondrial respiratory chain and the points at which a variety of poisons act to disrupt the chain events. Courtesy of Dr. R. Hansford, National Institute on Aging, NIH.

impulses, and bioluminescence (Lehninger, 1962). The number of mitochondria in dividing cells does not decrease by the order of the number of divisions, and the fact that mitochondria are known to contain DNA has led to the suggestion that they divide (Tandler and Hoppel, 1972). Massie, Colacicco, and Williams (1981) reported that, although nuclear DNA remained unchanged in aged *Drosophila,* mitochondrial DNA was dramatically reduced. While mitochondria may divide, they are not totally self-replicating, since some of the mitochondrial soluble proteins, such as cytochrome C, are synthesized in the granular endoplasmic reticulum and then incorporated into the mito-

Figure 10. In one-month-old rat lateral vestibular neurons (10 A), granular endoplasmic reticulum is abundant. At 18 months of age, by comparison (10 B), the GER is reduced. At higher magnification, the one-month-old GER is arranged in parallel segments (10 C, arrow), while the GER in the older animals (10 D, 18 months old) tends not to be arranged this way. A,B × 15,250; C,D × 23,000, reproduced at 85%. (Modified from Johnson and Miquel, 1974.)

chondria (Kadenbach, 1968). Thus, age-related alterations in the GER could theoretically cause age-dependent changes in mitochondrial number. Indeed, Samorajski, Frieda, and Ordy (1971) reported a decrease in the number of mitochondria in pyramidal tract axons of aging mice. Mitochondrial loss has also been noted in presynaptic terminals of neuromuscular junctions (Fahim and Robbins, 1982). Several studies have indicated an age-related loss in liver mitochondria (Shamoto, 1968; Herbener, 1976; Johnson *et al.*, 1978). Heart muscle cells also appear to show a loss of mitochondria (Herbener, 1976).

An *in vitro* study of IMR-90 human fibroblasts suggested a decrease in the number of mitochondria (Figs. 4, 5) with increasing population doublings (Johnson, 1979). In this study, two types of mitochondria were noted. One type was dense and thin (Fig. 12), while the other was wider and less dense (Fig. 12). The dense type is commonly referred to as the "condensed" configuration, while the wider form is designated as the "ortho" configuration. Any one cell was seen to contain only one type of mitochondria. Cells containing the condensed form increased in frequency with successive population doublings (Johnson, 1979).

Mitochondrial *size in vivo* has been found to increase in the liver (Wilson and Franks, 1975) and in the testis (Johnson and Miquel, 1978) of aging mice. Mitochondrial enlargement, or "swelling," is a difficult structural alteration to deal with, since it occurs not only in aging cells but in pathology and inadequately fixed specimens as well. This is particularly true of the central nervous system where mitochondria often have vacuolated regions even in otherwise well-fixed preparations. It is likely that this effect is due to anoxia since the problem is minimized by perfusing a living animal with fixative rather than fixing the tissue by immersion. For autopsy or biopsy specimens, where fixation must be done by immersion (as would be the case for humans), the problem is unavoidable and must be taken into account when making interpretations.

Enlarged mitochondria sometimes have distinct cristae within them, but occasionally the contents appear reticulated (Fig. 13). It is not always possible to ascertain that a particular reticulated body is an altered mitochondrion (Fig. 14), but in having seen them in various stages, the present author feels that many of these bodies do originate from them. Reticulation in association with mitochondria has been observed in normal (Fig. 15) aging animals *in vivo* (Johnson *et al.*, 1975; Johnson, 1980b) and animals *in vivo* under experimental (Fig. 16) conditions (Johnson *et al.*, 1976). With *in vitro* preparations, under anoxic conditions, the mitochondria enlarge (Kim, 1975) but do not quite seem to obtain the reticulated patterns seen *in vivo,* such as in Fig. 13. Even with "normal" treatment of cell cultures, some mitochondria appear vacuolated or swollen, and, thus, it would be very difficult to determine if particular mitochondria swelling or, if found, reticulation, were due to experimental treatment or simply due to the abnormal environment in which cell cultures must live.

→

Figure 12. In populations of IMR-90 fibroblasts, two types of mitochondria are found. Type 1 (arrow 1) is the condensed configuration, while type 2 (arrow 2), the ortho configuration, is larger and less dense. Cells with the type 1 mitochondria appear to be more frequent with increasing population-doubling levels (PDL). The cell in 12 A is at 36 PDL, and 12 B is at 49 PDL. Both micrographs × 19,200, reproduced at 85%. (From Johnson, 1979.)

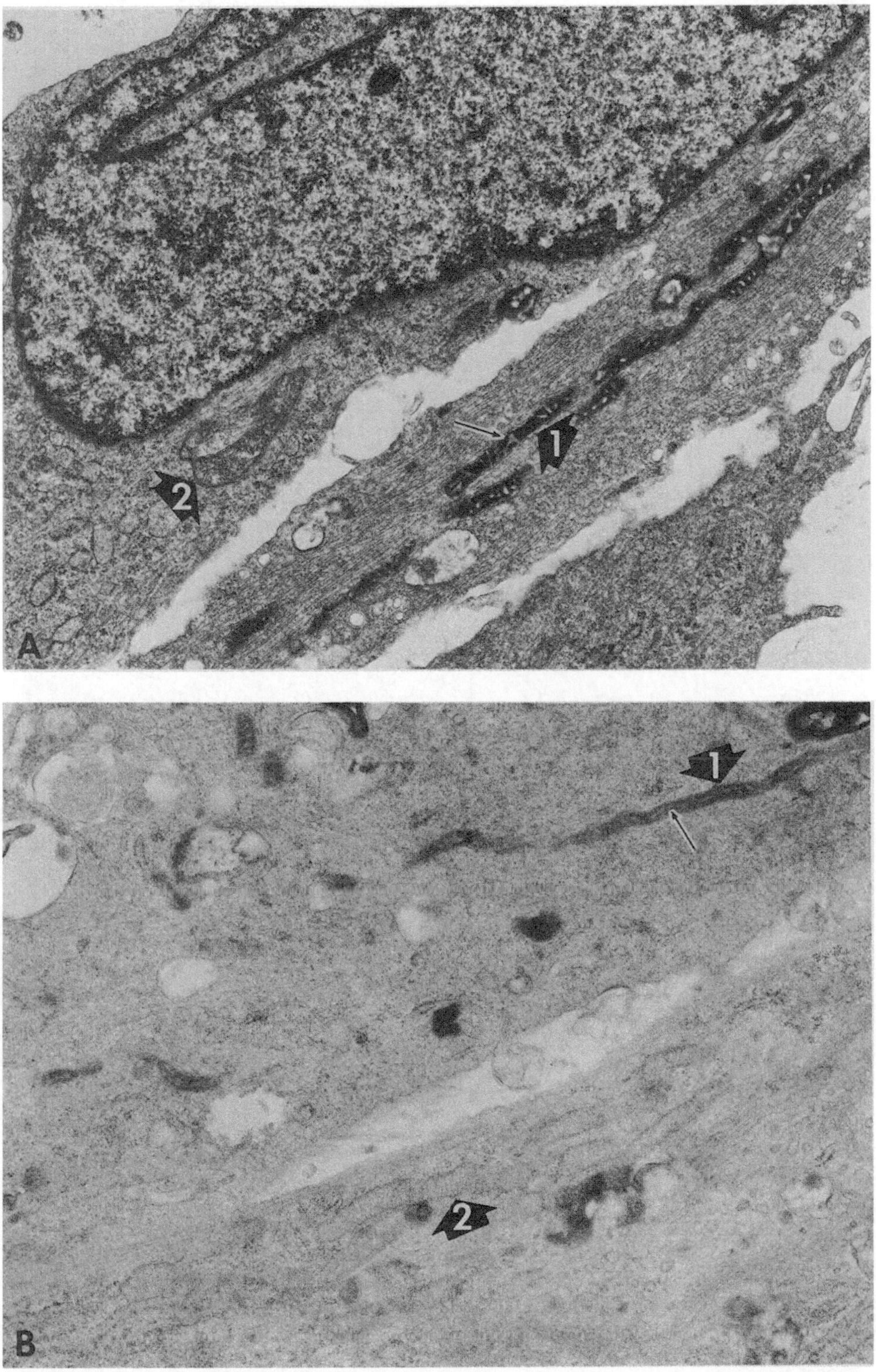
A
1
2
B
1
2

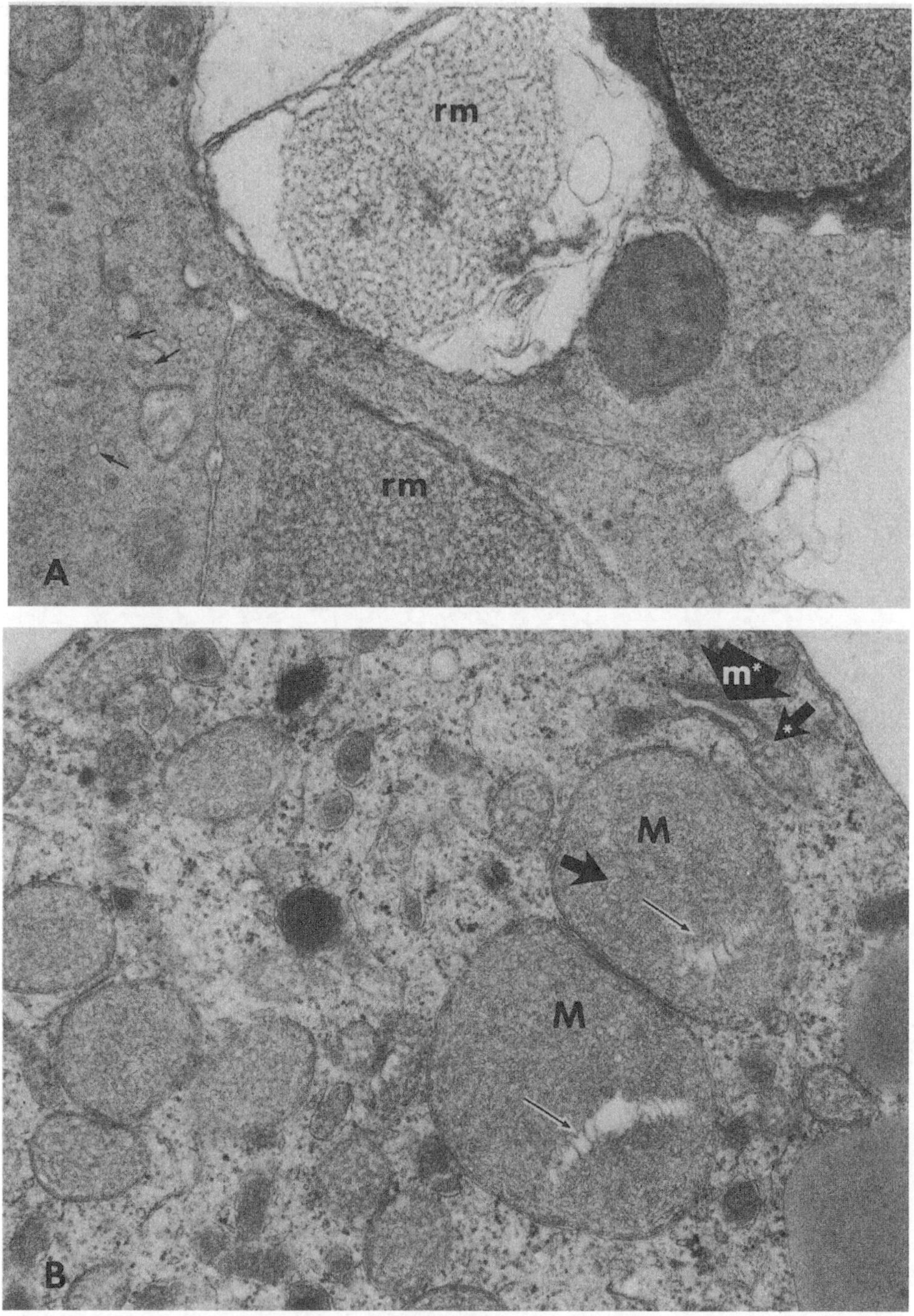

Figure 13. In 13 A, from the pineal of a 23-month-old rat, reticulated membranes (rm) and coated vesicles (arrows) are present. Enlarged mitochondria (M) are shown in 13 B, from the pineal of a 28-month-old rat. The mitochondria have reticulations (solid arrow without dot) and electron lucent areas (long thin arrows). One mitochondrion has longitudinal cristae at one end (arrow with m and asterisk) and reticulations at the other end (arrow with asterisk). Both micrographs × 38,000, reproduced at 85%. (From Johnson, 1980b.)

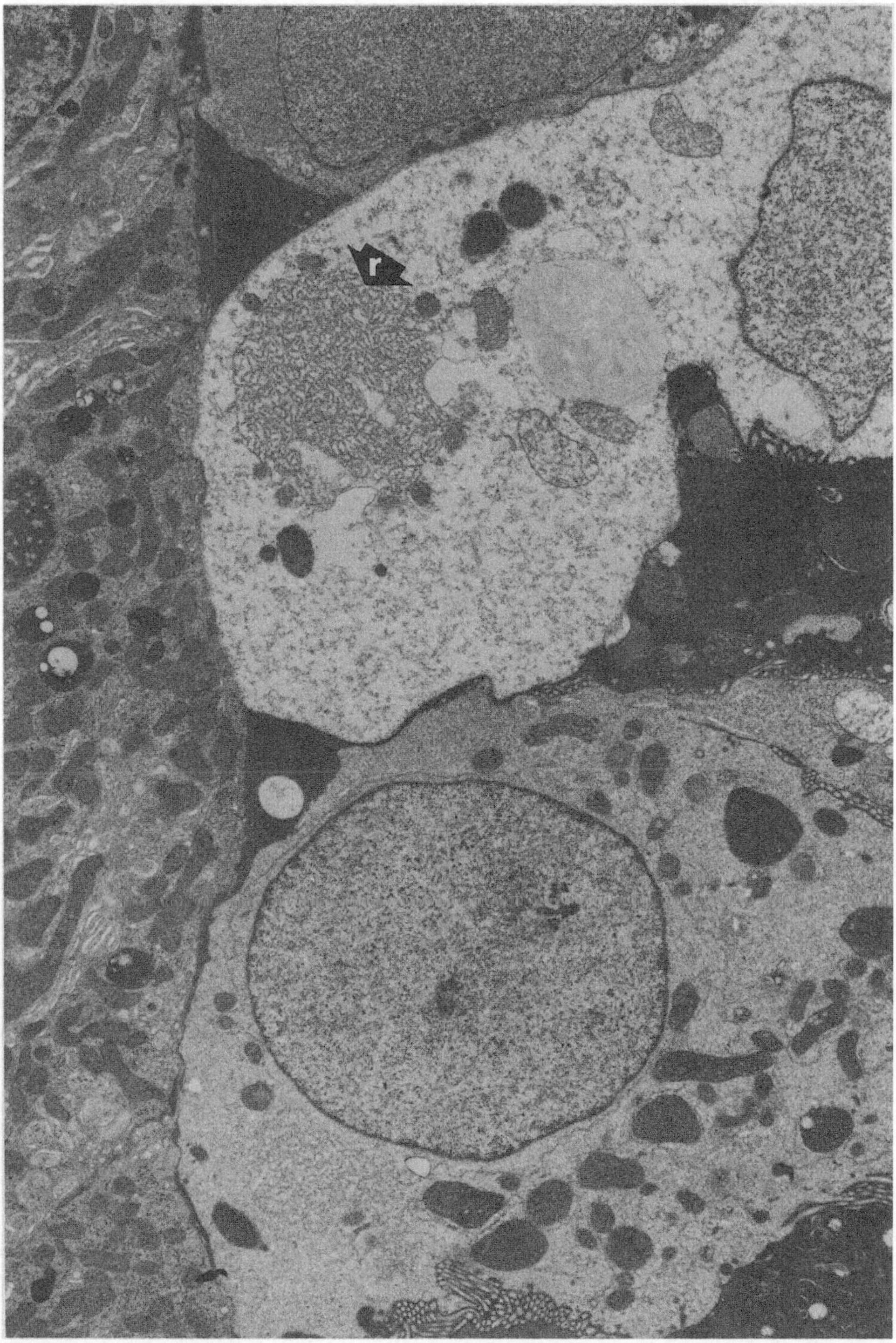

Figure 14. Reticulated membranes (arrow with r) are seen in this example of kidney from a 23-month-old rat. × 7600, reproduced at 85%.

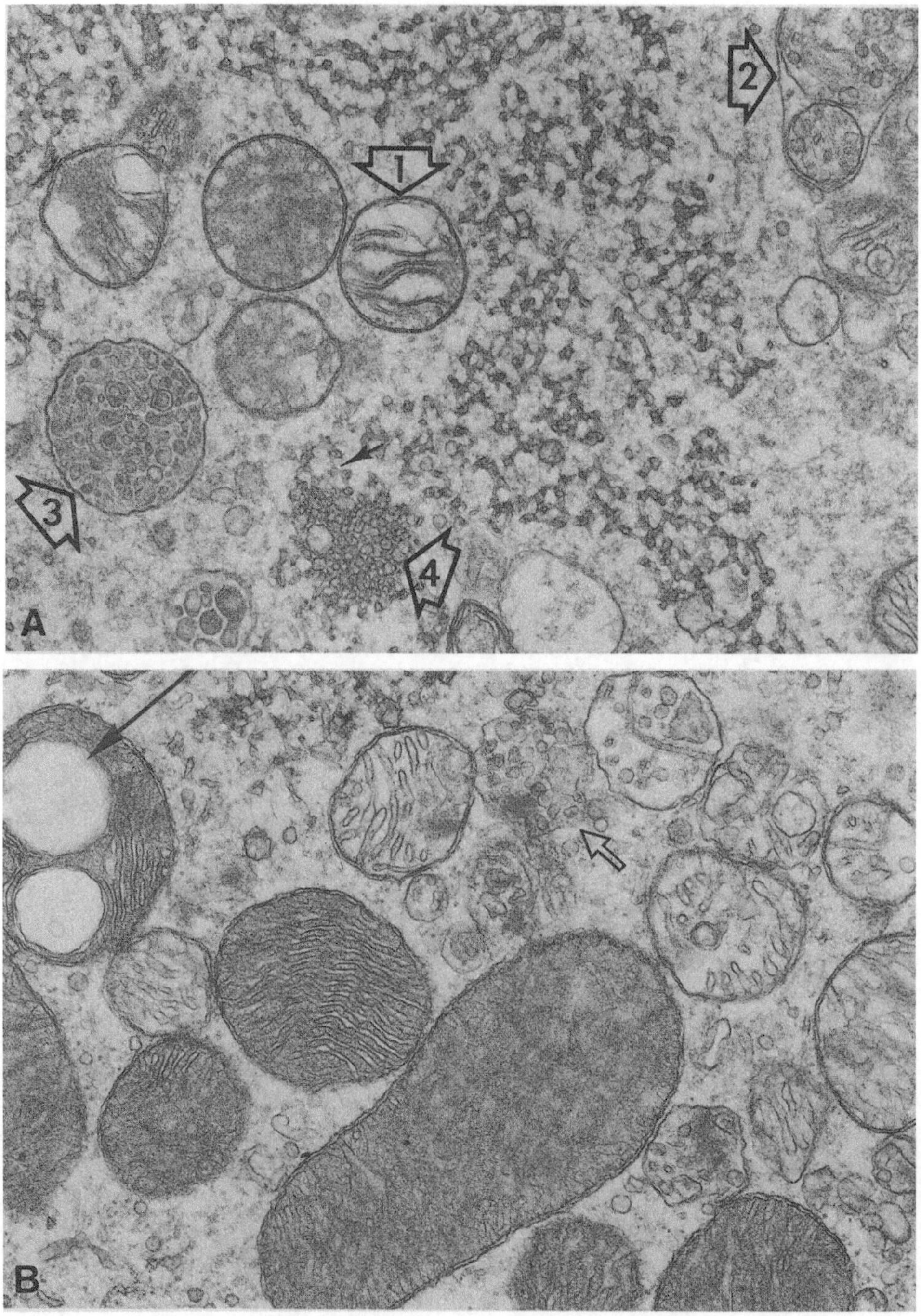

Figure 15. In 15 A, a series of mitochondrial changes is shown from the dorsal column nuclei of the brain of a 23-month-old mouse. The series begins with swelling of the mitochondrion (arrow 1) followed by the formation of multivesicular bodies within (arrow 2). Loss of the outer unit membrane releases the multivesicular body (arrow 3). When the multivesicular body breaks down, a patch of vesicular material results (arrow 4), which may then form a reticular network (small arrow). Vitamin E supplementation to the diet does not appear to prevent the mitochondrial changes from occurring, as illustrated in 15 B, from a 23-month-old mouse. Both micrographs × 48,000, reproduced at 85%. (From Johnson *et al.*, 1975.)

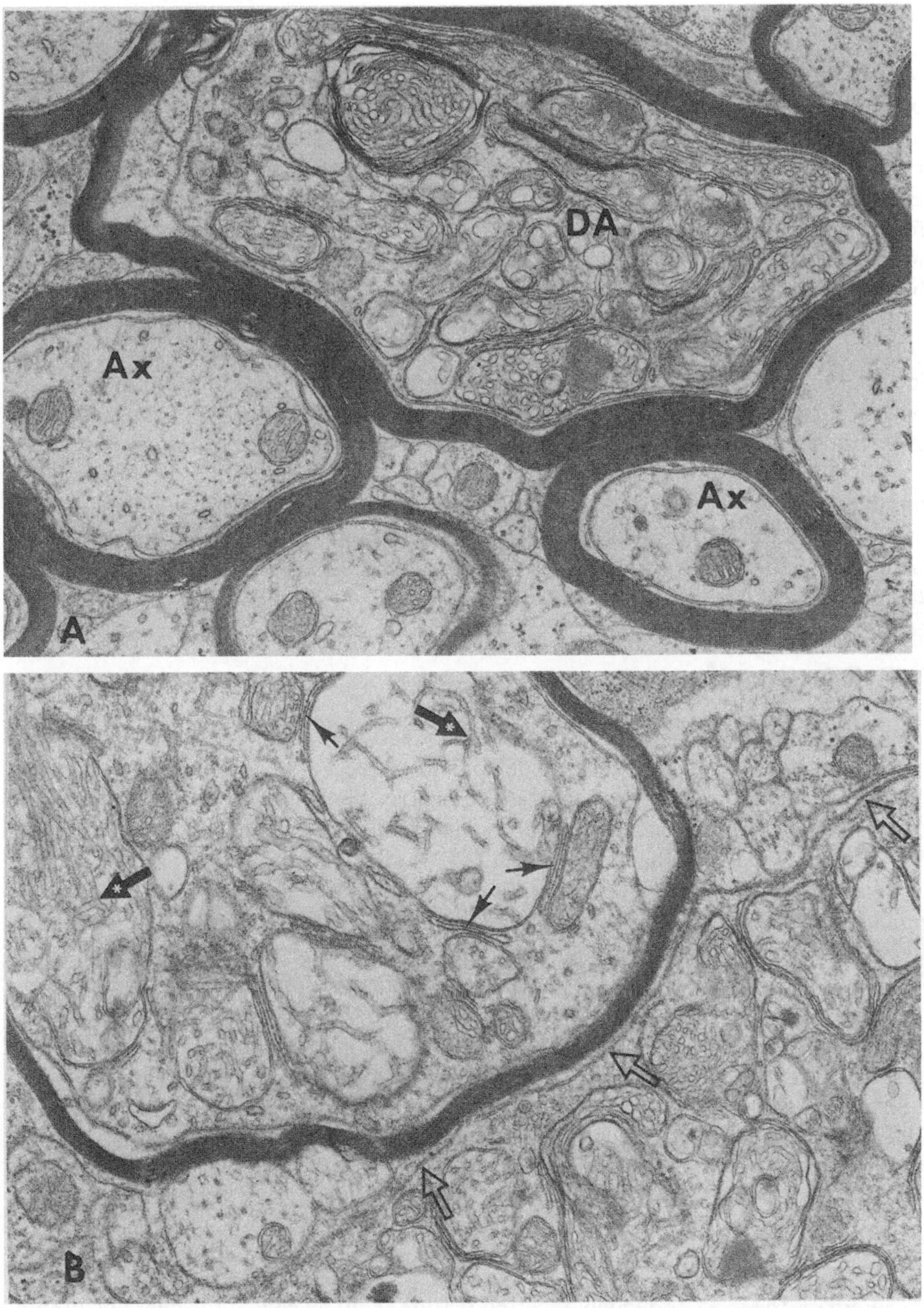

Figure 16. A dystrophic axon (DA) is illustrated in 16 A from the lateral vestibular nucleus of a rat centrifuged for 21 months at 2.76*g*. Numerous reticulated and vesiculated structures lie within. Healthy axons (Ax) are nearby. In 16 B, also from a rat centrifuged for 21 months at 2.76*g*, two dystrophic profiles abut. One is myelinated, and the other is outlined by the hollow arrows. Several of the mitochondria have reticulated contents (arrows with asterisk). Segments of endoplasmic reticulum lie between many of the mitochondria (small thin arrows). Both micrographs × 30,780, reproduced at 85%. (From Johnson *et al.*, 1976.)

From a study of the thoracic spinal ganglia of aged rats, Vanneste and vandenbosch de Aguilar (1981) reported that mitochondria contained tubular or filamentous inclusions, glycogen deposits, and electron dense globules. The authors suggested that, in younger animals, cellular regulation mechanisms eliminated impaired mitochondria. Examining the aging human orbital cortex, Herrlinger, Anzil, and Blinzinger (1974) found mitochondria which contained banded, latticelike and hexagonally patterned inclusions. They could not rule out the possibility that these inclusions were artifacts brought about by the fact that the specimens were obtained by autopsy.

By using 2-4 dinitrophenol, which is an uncoupler of oxidative phosphorylation (Fig. 11), Sahgal *et al.* (1979) were able to produce intramitochondrial paracrystalline inclusions. Antimycin A and oligomycin, which block mitochondrial respiration, caused only swelling and distruption of the mitochondria. They concluded that the presence of such mitochondrial inclusions represents uncoupled, but intact, respiration.

Although changes in the shape (lobulation) of mitochondria have been reported in fibroblasts *in vitro* (Brock and Hay, 1971; Lipetz and Cristofalo, 1972; Johnson, 1979) as well as in the aging mouse pituitary (Munnell and Getty, 1968) and cerebellum (Andrew, 1955), the crystalline type of mitochondrial *inclusion* appears to be a rare event *in vitro* under normal conditions.

3.3. Golgi Complex

The Golgi region of a cell represents a difficult item to study simply because it is not contained within a distinct boundary. One of the newer techniques of analyzing fractional volumes of cell constitutents is to overlay the micrograph with a grid and to count the number of points which overlay a particular cell component (Weibel, 1969). Using this method, Vaughan and Vincent (1979) studied neurons in the auditory cortex of aging rats. Although they found that the relative volume of granular endoplasmic reticulum increased after 15 months of age, the relative volumes of mitochondria, multivesicular bodies and Golgi did not show any clear trends.

Knox, Yates, and Chen (1980) reported an apparent increase in Golgi in neurons of aging rat frontal cortex. Johnson and Miquel (1974) suggested that there was an increase in Golgi in the aging rat lateral vestibular nucleus. In a study of IMR-90 fibroblasts *in vitro,* Johnson (1979) found that the Golgi region of cells at later population doubling levels appeared to have more vesicles associated with it than at earlier PDL (Fig. 17). This substantiates an earlier observation in WI-38 fibroblasts *in vitro* by Lipetz and Cristofalo (1972).

3.4. Dense Bodies

Perhaps the most commonly reported feature of aging cells is the dense body (Fig. 18). In spite of its ubiquitous accumulation with increasing age in most types of animal cells (Johnson and Miquel, 1974; Samorajski and Ordy, 1967; Miquel *et al.,* 1978;

Figure 17. At 18 PDL (17 A), the Golgi regions of IMR-90 fibroblasts have a few associated coated vesicles (arrows), while at 49 PDL (17 B) the coated vesicles are more frequent (arrows). Both micrographs × 38,000, reproduced at 85%. (From Johnson, 1979.)

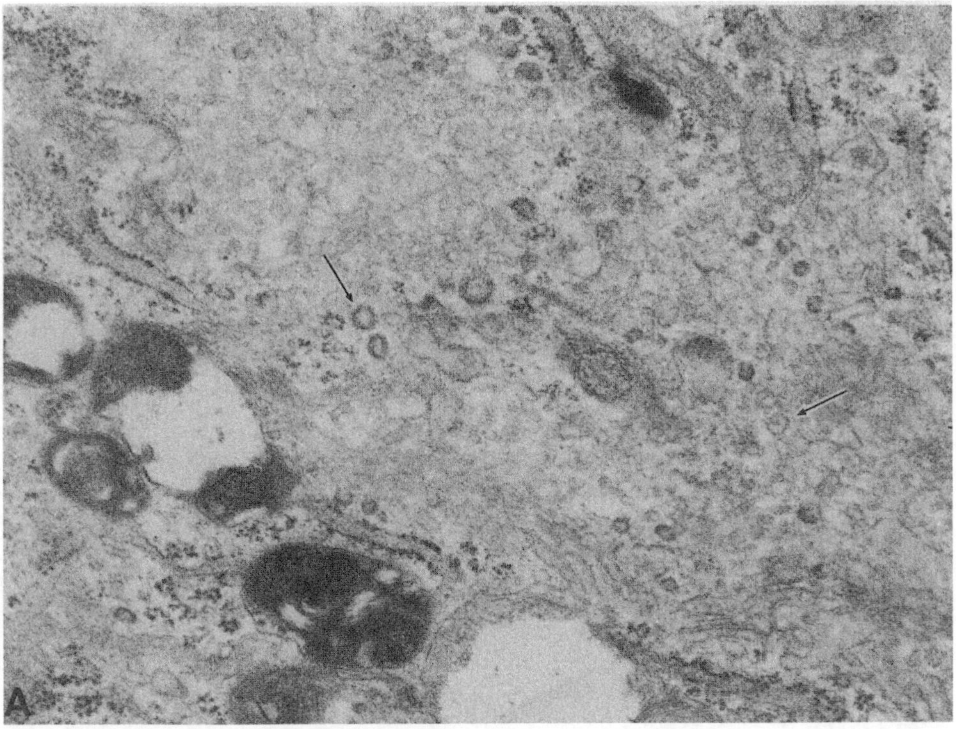
A

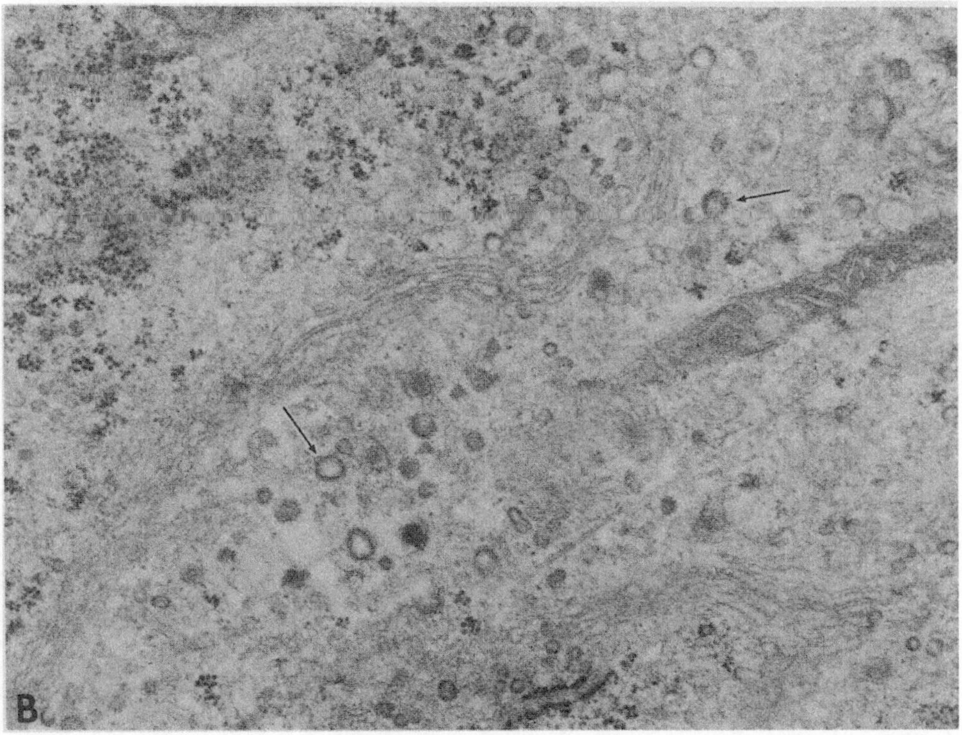
B

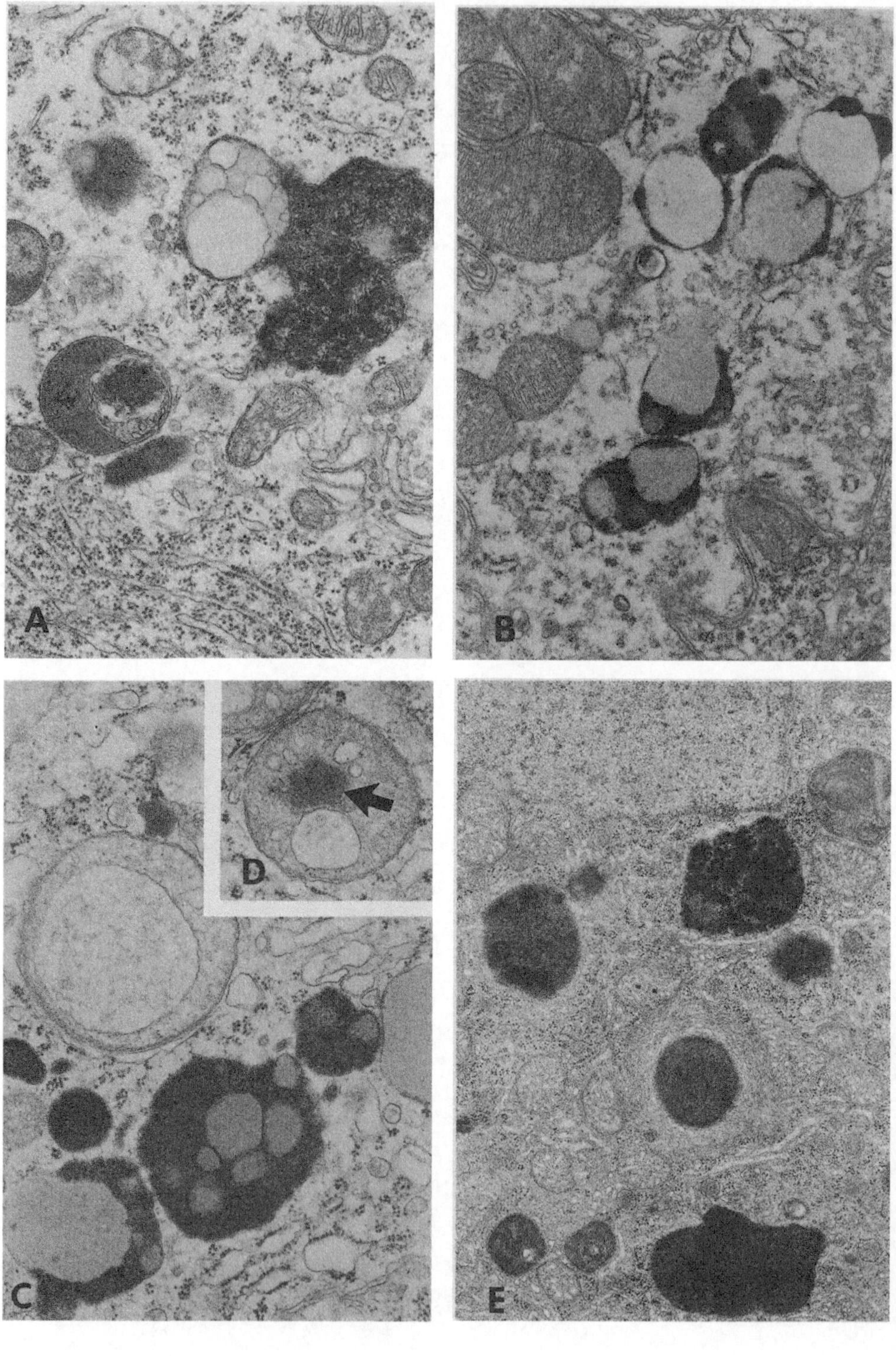
A
B
C
D
E

Spoerri and Glees, 1973; Gresik *et al.*, 1982; Miquel *et al.*, 1983), its etiology is misunderstood. The term "dense body" is rather all-inclusive. Lipofuscin and ceroid are terms very often used by histologists, but other names have been applied such as hemofuscin, lipochrome, cytolipochrome, chromolipoid, yellow pigment, age pigment, senile pigment, interceroid, preceroid, hemoceroid, hyaloceroid, residual bodies, lamellar bodies, and others (Miquel *et al.*, 1977).

Although there have been scores of research publications on this subject, beginning in the 19th century, the origin of dense bodies is still controversial. Dolley (1917) believed the nucleus to be involved in age pigment formation. In spite of the fact that age pigment tends to accumulate at one pole of the nucleus (Brizzee and Johnson, 1970; Johnson and Miquel, 1974), there is no clear evidence of nuclear origin. Neurofibrils (Sosa, 1952) and endoplasmic reticulum (Strehler, 1964; Novikoff, 1967) also have been suggested as sources of age pigment, but these theories have not gained support. The evidence for Golgi apparatus as an origin of age pigment has been based on similarities in staining properties between the Golgi and age pigment as well as spatial proximity of these structures (Gatenby and Moussa, 1950; Bondareff, 1957).

A great deal of evidence suggests that lysosomes and mitochondria are sources of age pigment. Both lysosomes and age pigment have a single limiting membrane, and both have a positive acid phosphatase reaction (Brizzee and Johnson, 1970; Brizzee and Ordy, 1981). Occasional age pigment granules have a double limiting membrane as do mitochondria. Also, mitochondria have been observed to become dense under aging (Fig. 18) (Miquel *et al.*, 1978) and experimental (Fig. 19) (Johnson *et al.*, 1976) conditions. Mitochondrial enlargement, followed by what appear to be dense bodies with cristae inside (Figs. 20,21) were observed in IMR-90 fibroblasts *in vitro* (Johnson, 1979).

The fine structure of age pigment varies between different tissues. Some pigment bodies have light staining droplets (Figs. 18,22–25), and some are very dense throughout (Figs. 18E,24B). Although the neurohypophysis of aging rats contains a great deal of age pigment within some of its cells (Fig. 25), the adenohypophyseal cells (Fig. 26) do not contain nearly so much (Johnson, 1980a). One of the hypophyseal target tissues, the adrenal gland, was found to contain age pigment and other unusual inclusions (Fig. 27).

Pigment seen *in vitro* is not always identical morphologically to that seen *in vivo*, although there is a basic similarity. Collins and Brunk (1976) reported residual bodylike structures in growth-inhibited human glial cell cultures plated at high density. Their study, as well as experiments by Barr-Nea and Wolman (1977), using cultured chick hepatocytes, indicated autofluorescent properties similar to age pigment observed *in vivo*. Lipofuscin was reduced *in vivo* (Nandy and Bourne, 1966; Spoerri *et al.*, 1974; Meier, 1974) and *in vitro* (Spoerri and Glees, 1974; Nandy *et al.*, 1978) following administration of centrophenoxine. This drug is thought to act by shunting of glucose metabolism towards the pentose cycle (Nandy, 1968).

←

Figure 18. "Age pigment" is illustrated from nucleus hypoglossus in the brain of a 23-month-old mouse (18 A), kidney in a 31-month-old mouse (18 B), Sertoli cell in the testis of a 29-month-old mouse (18 C,D), and Leydig cell in the testis of a 30-month-old mouse (18 E). In 18 D, mitochondrial densification has occurred. A,C,D, × 30,780; B,E, × 20,550, reproduced at 85%.

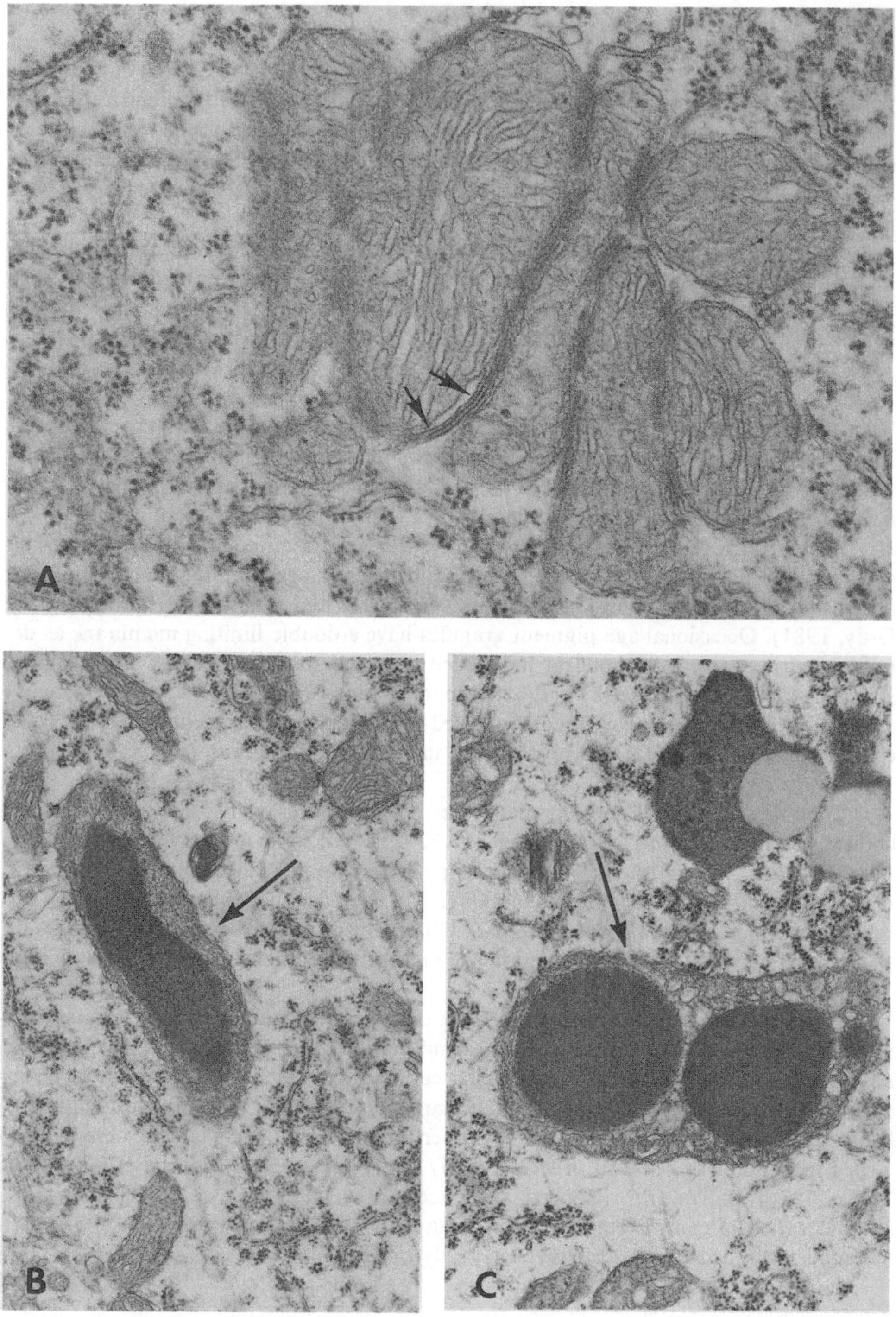

Figure 19. In 19 A, from the lateral vestibular nucleus of a rat centrifuged for 16 months at 4.15*g*, more than the usual two unit membranes per mitochondrion occur at mitochondrial boundaries (arrows). Structures (indicated by arrows) thought to have originated from mitochondria are shown in 19 B (centrifuged for 16 months at 4.15*g*) and 19 C (centrifuged for 19.5 months at 2.76 *g*). Dense material is seen within. A, × 61,500; B,C, × 30,780, reproduced at 85%. (From Johnson, Mehler and Oyama, 1976.)

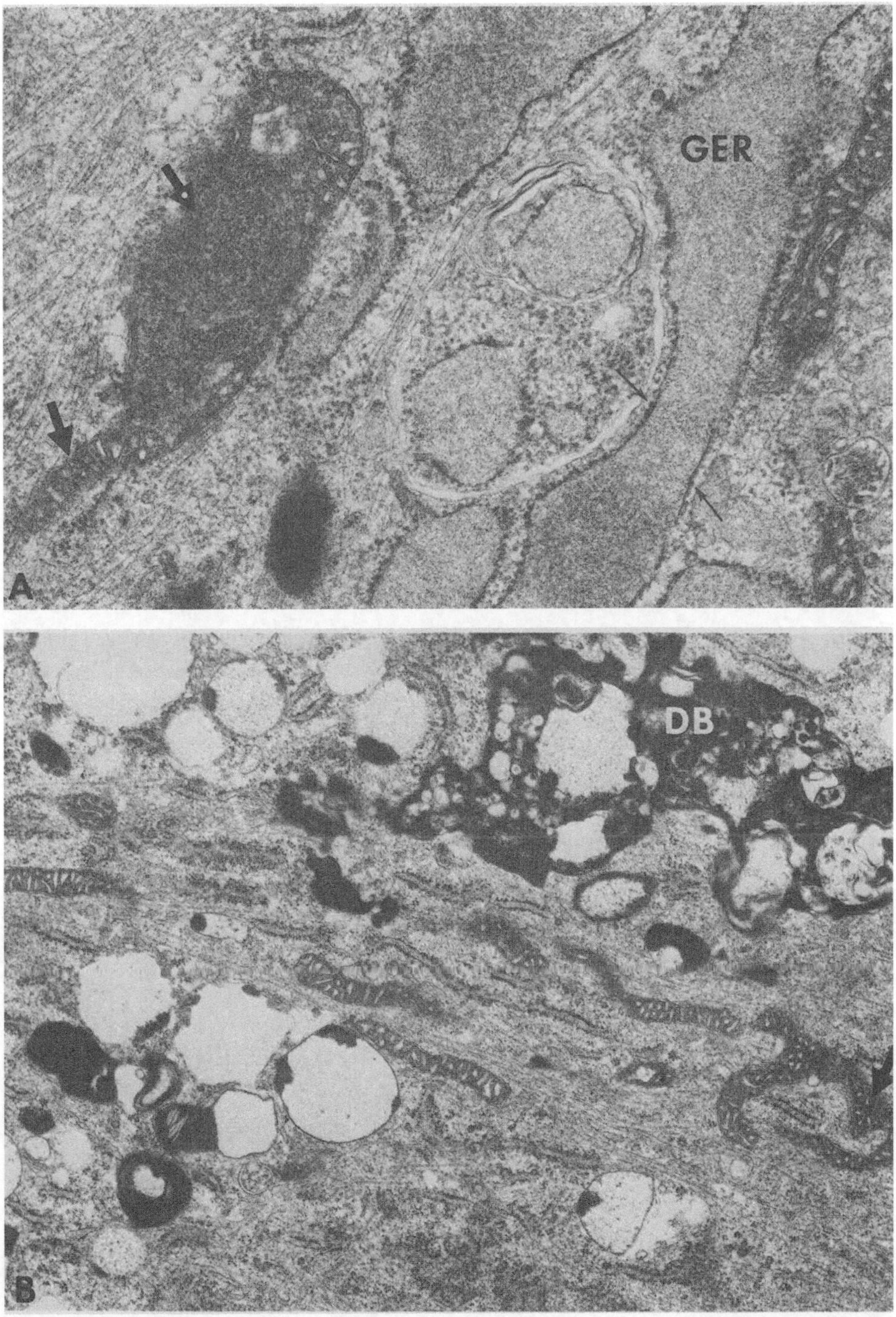

Figure 20. The 36 PDL IMR-90 fibroblast in 20 A has dilated granular (small thin arrows) endoplasmic reticulum (GER) filled with fine granulated material. A mitochondrion (arrow without dot) is enlarged at one end (arrow with dot). In 20 B (36 PDL IMR-90 fibroblast), a number of mitochondria are seen, one of which has an enlarged region (arrow). A very large dense body (DB) is also present. A, × 38,000; B, × 19,200, reproduced at 85%. (From Johnson, 1979.)

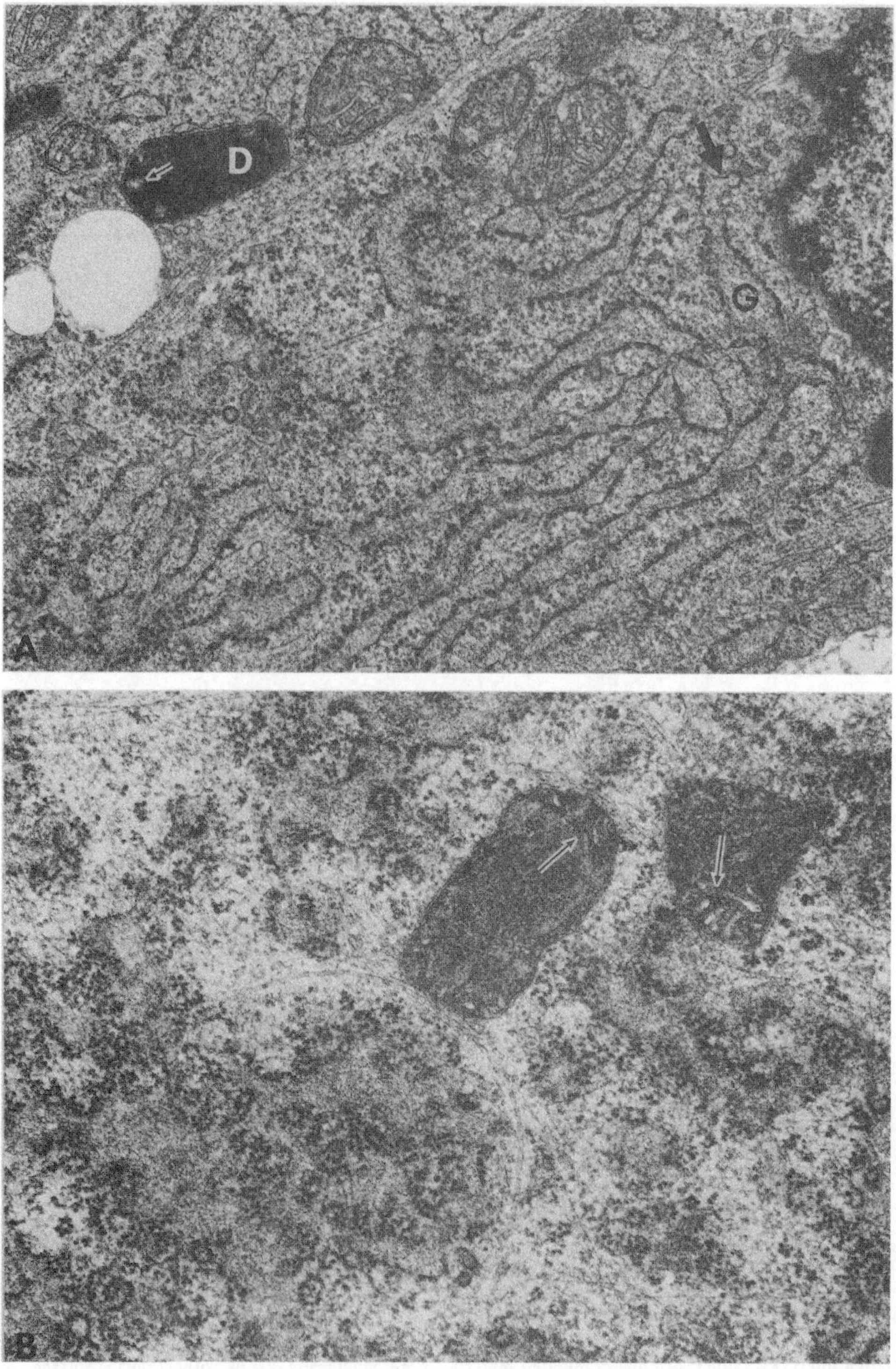

Figure 21. The cytoplasm of a 36 PDL IMR-90 fibroblast in 21 A contains dilated granular endoplasmic reticulum (G). At one end of a cisternal segment, a vesicle appears to be pinching off (large arrow). Nearby, a dense body (D) appears to have remnants of mitochondrial cristae at one end (small arrow). Dense bodies in 21 B (36 PDL IMR-90 fibroblast) contain remnants of mitochondrial cristae (arrows). Both micrographs × 38,000, reproduced at 85%. (From Johnson, 1979.)

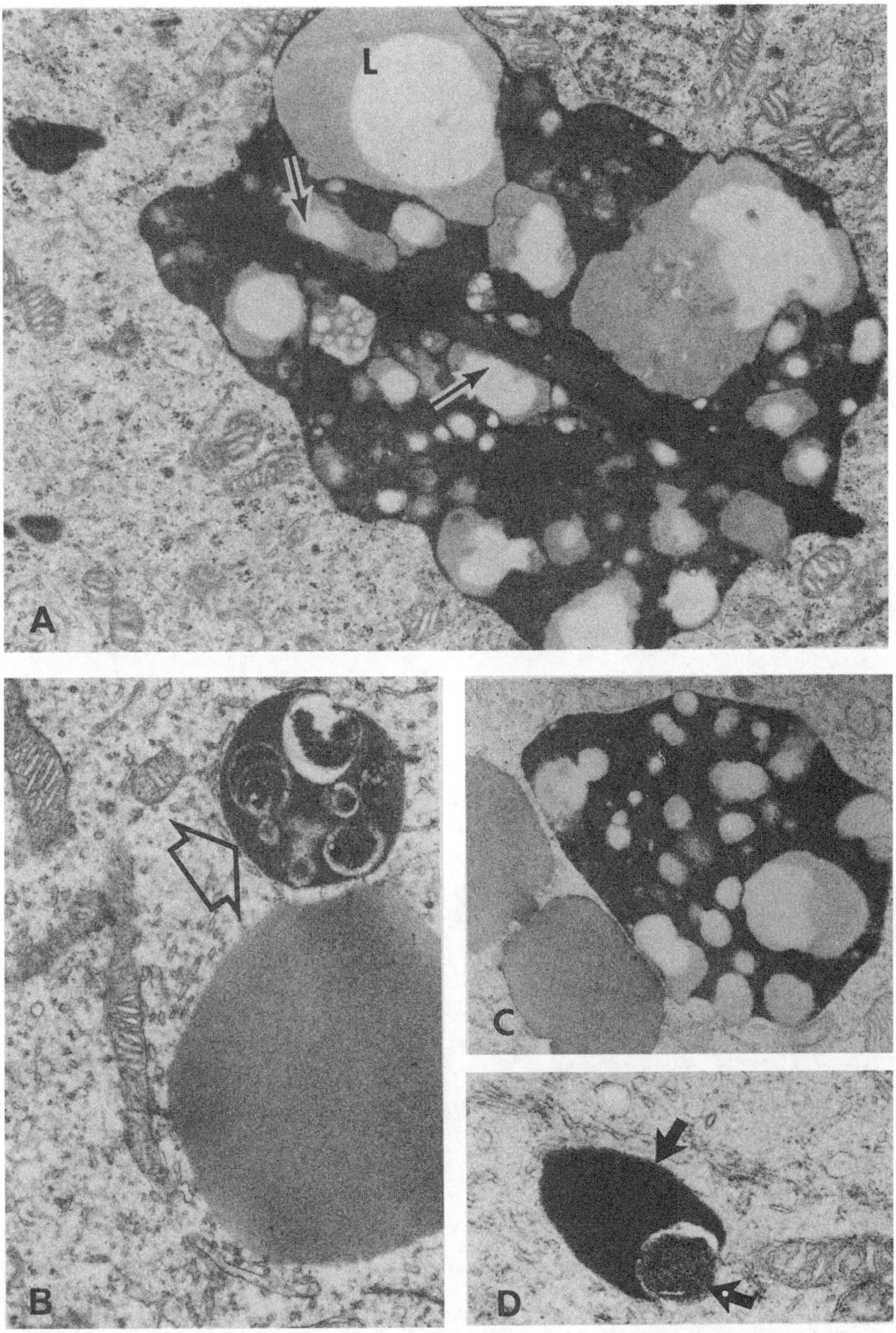

Figure 22. These micrographs illustrate dense bodies from aging rat pinealocytes. The large one in 22 A (from a 28-month-old male) has a long straight inclusion (arrows) and lipid droplets (L). In 22 B (23-month-old female), a dense body (arrow) has the appearance of an autophagic vacuole. The dense body in 22 C (23-month-old female) looks like the one in 22 A (28-month-old male). The autophagic vacuole in 22 D (23-month-old female) has a very dense portion at one end (arrow without dot) and a fine granular portion at the other end (arrow with dot). A,B,C, × 19,200; D, × 38,000, reproduced at 85%. (From Johnson, 1980b.)

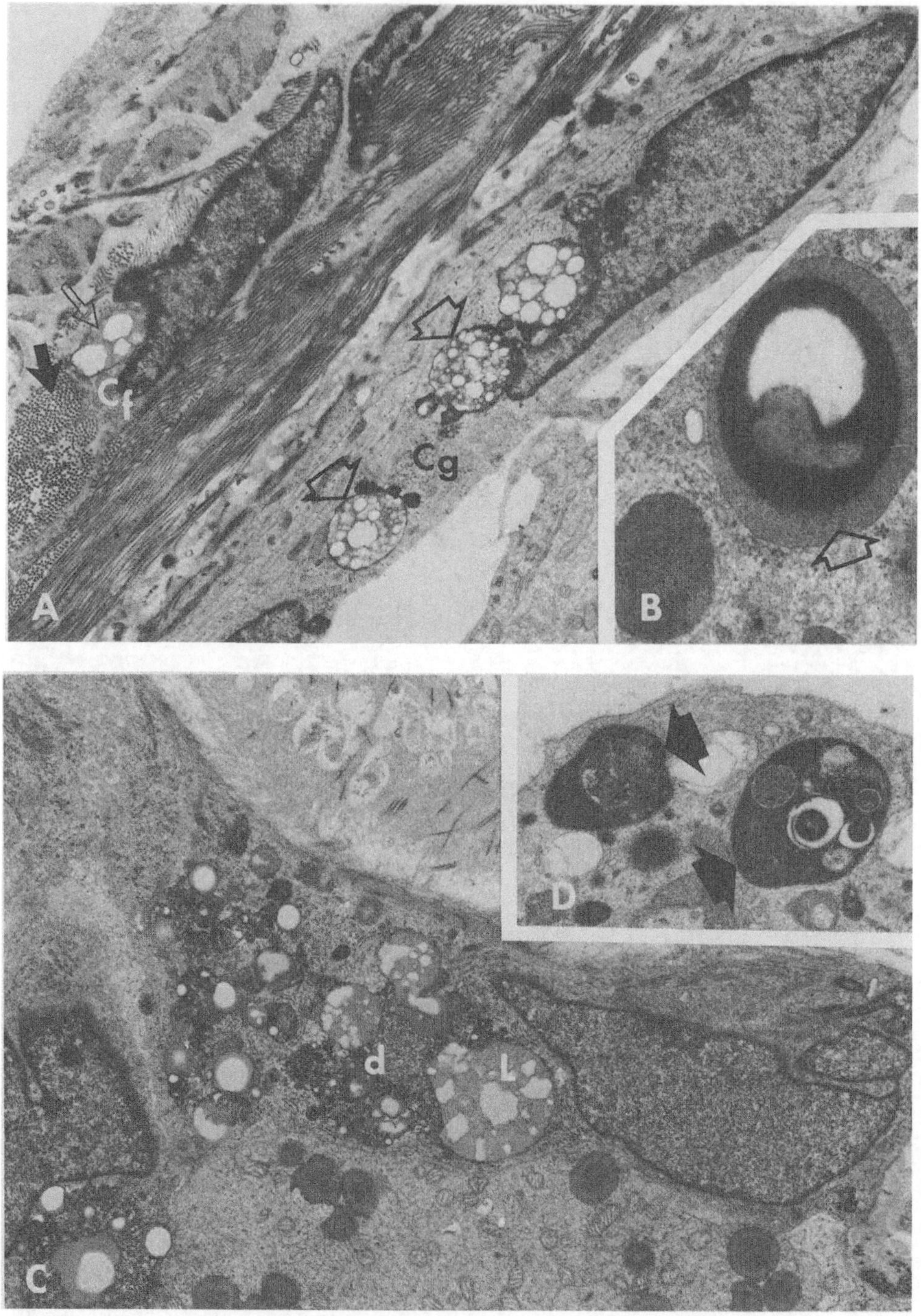

Figure 23. Glial cell dense bodies are shown from the aging rat pineal. In 23 A (28-month-old male), one cell (Cf) is thought to be a fibroblast because of nearby collagen (solid arrow). A single dense body lies within the cell (small hollow arrow). Adjacent to the parenchyma lies another cell (Cg) which is probably glial and contains several dense bodies (large hollow arrows). The glial cell dense body in 23 B (23-month-old female) seems to be made up of concentric layers, each with a different density. Two glial cells are shown in 23 C (28-month-old male). Dense body aggregates (d) contain lipid droplets (L). Two autophagic vacuoles (arrows) are shown in a glial cell in 23 D, from a 23-month-old female. A, × 7600; B, × 38,000; C,D, × 19,200, reproduced at 85%. (From Johnson, 1980b.)

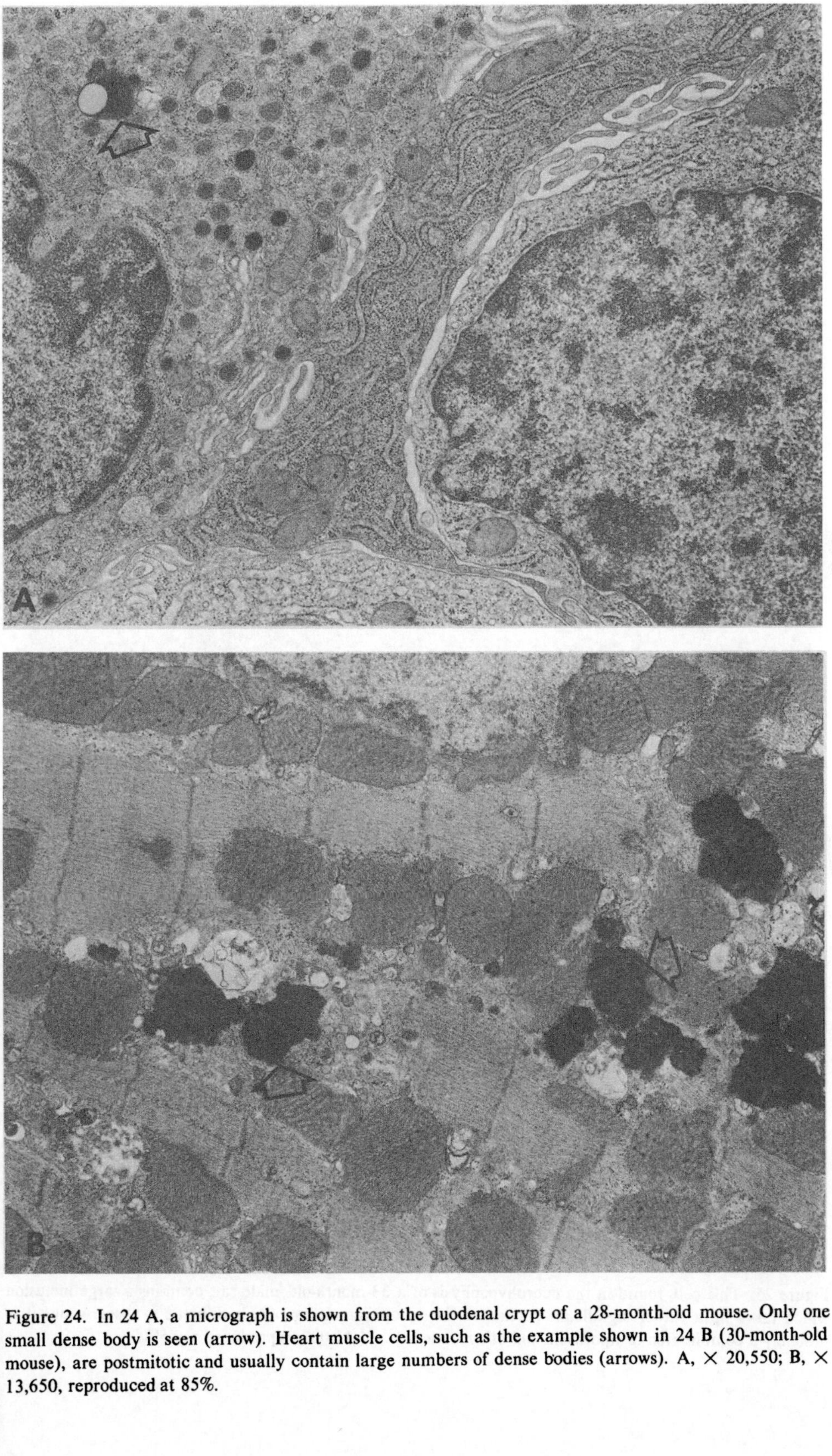

Figure 24. In 24 A, a micrograph is shown from the duodenal crypt of a 28-month-old mouse. Only one small dense body is seen (arrow). Heart muscle cells, such as the example shown in 24 B (30-month-old mouse), are postmitotic and usually contain large numbers of dense bodies (arrows). A, × 20,550; B, × 13,650, reproduced at 85%.

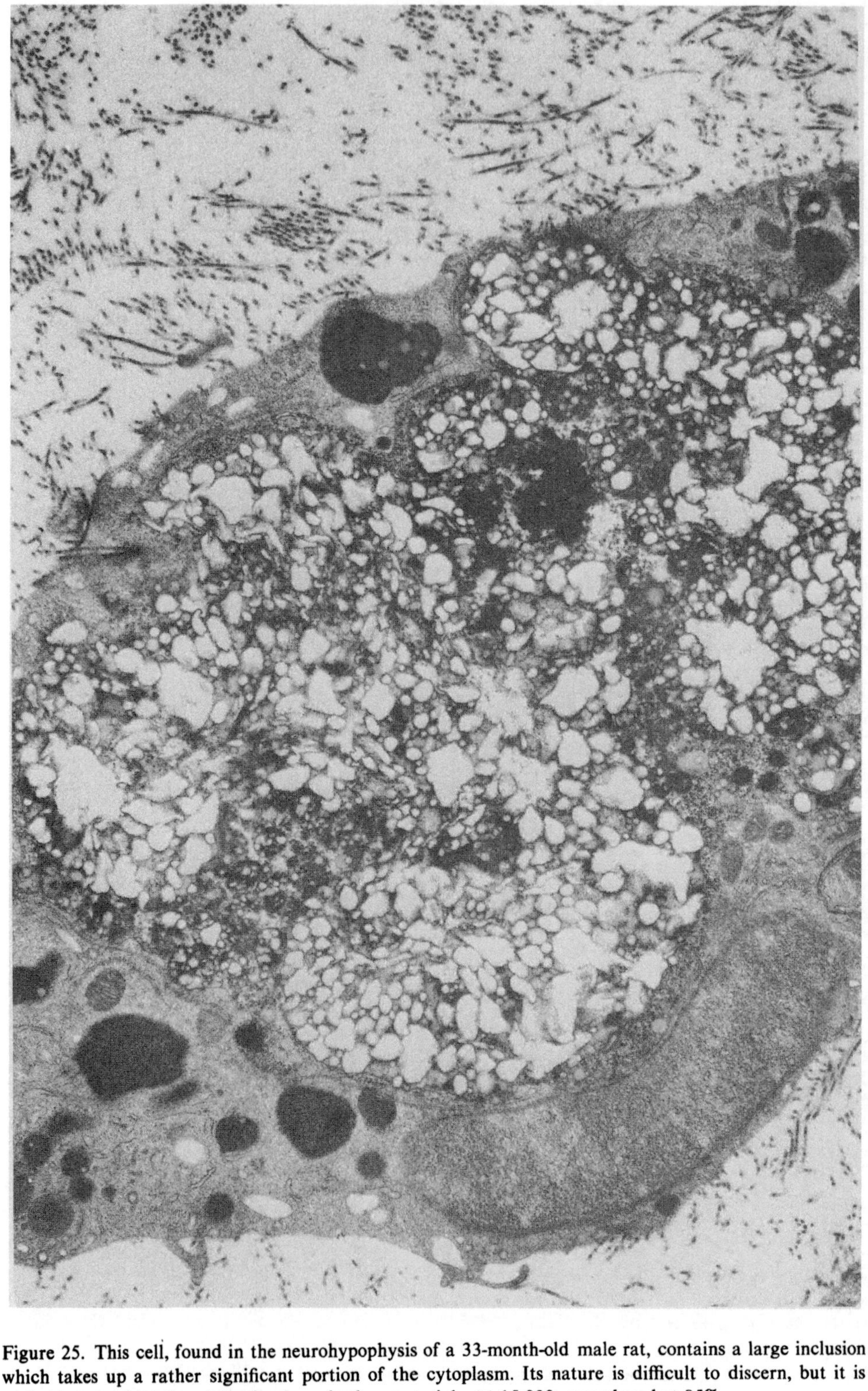

Figure 25. This cell, found in the neurohypophysis of a 33-month-old male rat, contains a large inclusion which takes up a rather significant portion of the cytoplasm. Its nature is difficult to discern, but it is probably a combination of lipofuscin and other materials. × 15,000, reproduced at 85%.

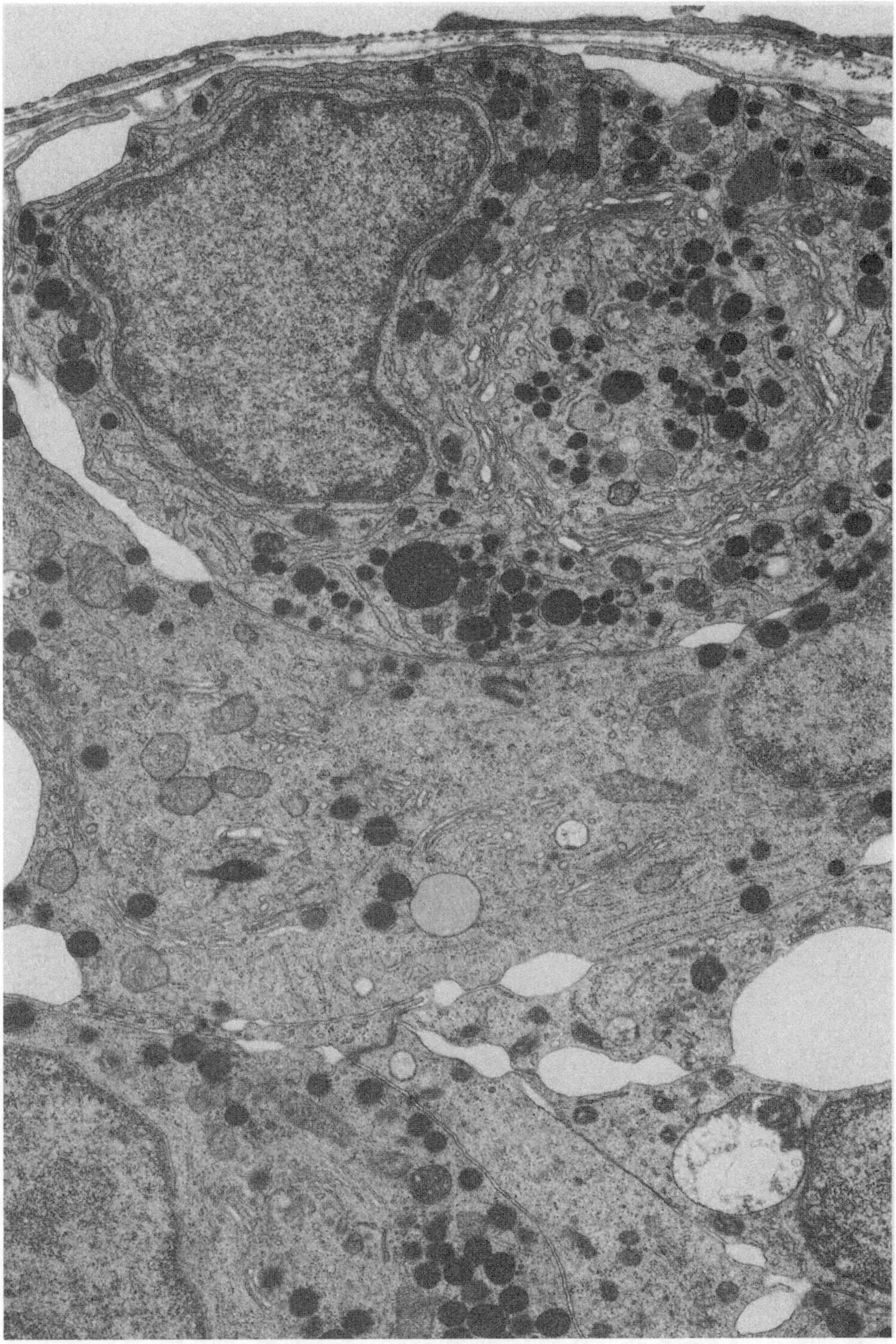

Figure 26. Anterior pituitary cells in this illustration (33-month-old male rat) contain little, if any, evidence to suggest they are from an old animal. × 15,000, reproduced at 85%.

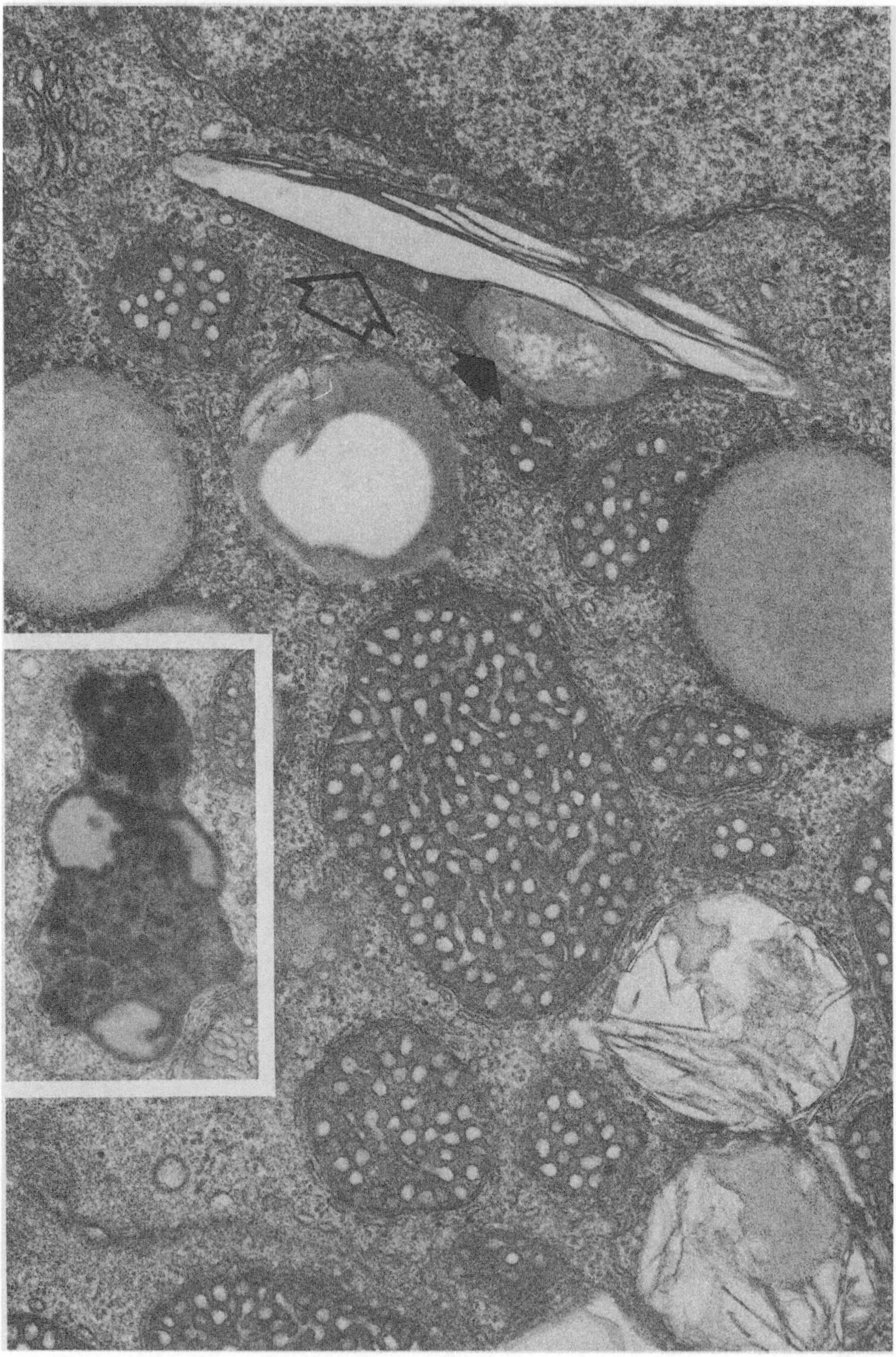

Spoerri and Glees (1973) reported pigment accumulation in rat spinal ganglia which were cultured up to four weeks. The ganglia were taken from 1-month-old animals. We have also observed significant numbers of pigment granules in cultured nervous system of young rats (Fig. 28). An important issue arises from these observations; since so much dense body pigment is seen in cultures from young animals (total age would be about two months, including culture time), is this accumulation an aging phenomenon at all? Probably not. This is supported by the fact that many of the pigment granules are similar to those seen in pathological conditions (Fig. 28) and even in apparently healthy young animals (Figs. 29,30). This type of dense body has also been induced *in vivo* using silica particle injection in an attempt to elucidate the causes of Alzheimer's disease (Rees and Cragg, 1983). Cell lines such as neuroblastoma N2-A, which have "infinite" life spans *in vitro,* may be found only to contain a few "age" pigment granules (Figs. 31,32). In any case, the lipofuscin pigment present in both cultured cells and *in vivo* may very well be a result of the same process, i.e., lipid peroxidation by free radicals. The reaction *in vitro* may proceed at a faster rate because the cells lack protective mechanisms which would be present if the cells were in a normal environment. Therefore, although *in vitro* models of pigment accumulation may be useful, improvements in culture conditions will be necessary before such studies can be taken full advantage of.

4. THE CELL SURFACE

Internal cellular fine structure is examined primarily by transmission electron microscopy (TEM). The scanning electron microscope (SEM) represents a powerful tool for studying the external morphology of cells. In an initial study of IMR-90 fibroblasts *in vitro,* it was found that the number of cells having extensive amounts of microvilli ("microprojections") on their surface (Figs. 33,34) increased at later population doubling levels (Johnson, 1979). Further studies, using the kidney glomerular podocytes (Fig. 35) as a model, indicated that the *in vivo* "microprojection index" increases in animals with advancing age (Johnson and Barrows, 1980). Dietary restriction, which increases laboratory animal lifespan (McCay *et al.,* 1939; Leto *et al.,* 1976) resulted in reduced microprojection indices (Fig. 35) (Johnson and Barrows, 1980).

It has been suggested that the beneficial effects of dietary restriction on lifespan act through a process of "nutritional hypophysectomy" (Everitt and Burgess, 1976). Hypophysectomy, followed by thyroxine administration in old rats has resulted in biochemical and physiological measurements returning to levels found in younger animals (Denckla, 1974; Bilder and Denckla, 1977) as well as lengthening life span (Everitt *et al.,* 1980). The microprojection index in rat kidney was reduced following hypophysectomy (Johnson and Cutler, 1980), correlating with the dietary restriction data (Johnson and Barrows, 1980).

Figure 27. The main figure shows an unusual inclusion (hollow arrow) found in a cell from the adrenal zona reticularis of a 36-month-old male rat. An attached light staining droplet (solid arrow) has the same appearance as the contents of other secretory droplets in the cytoplasm. The inset illustrates a dense body in a zona reticularis cell of a 33-month-old male rat. Main figure × 45,000; inset × 30,000, reproduced at 85%.

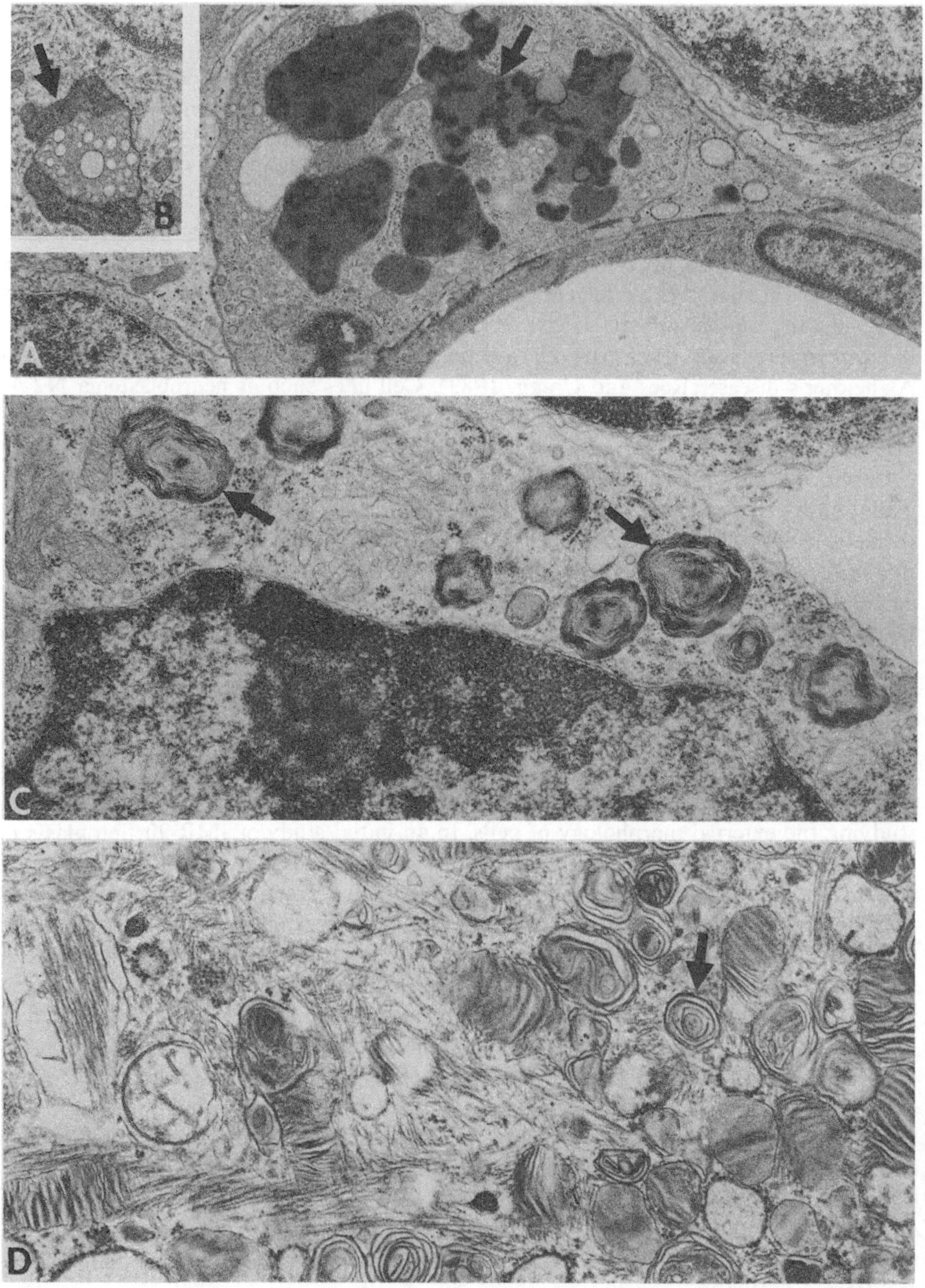

Figure 28. Part of a pericapillary cell from a 1-month-old rat cerebellum (perfusion fixed) is illustrated in 28 A. The dense bodies (arrow) are characteristic of brain and some other tissues. In 28 B, a brain astrocyte dense body is shown from a 7-year-old squirrel monkey superior colliculus (perfusion fixed). A rat brain cerebellar explant, taken at birth and grown in culture for one month, is shown in 28 C. The dense bodies are ringed (arrows). In 28 D, a micrograph is shown from a patient with Tay–Sachs disease (courtesy of Dr. Mary Herman and Dr. A. Dekaban of Stanford University). Note that the dense bodies are similar to those in 28 C. A,B, × 13,650; C, × 30,780; D, × 32,500, reproduced at 85%. (From Johnson, 1979.)

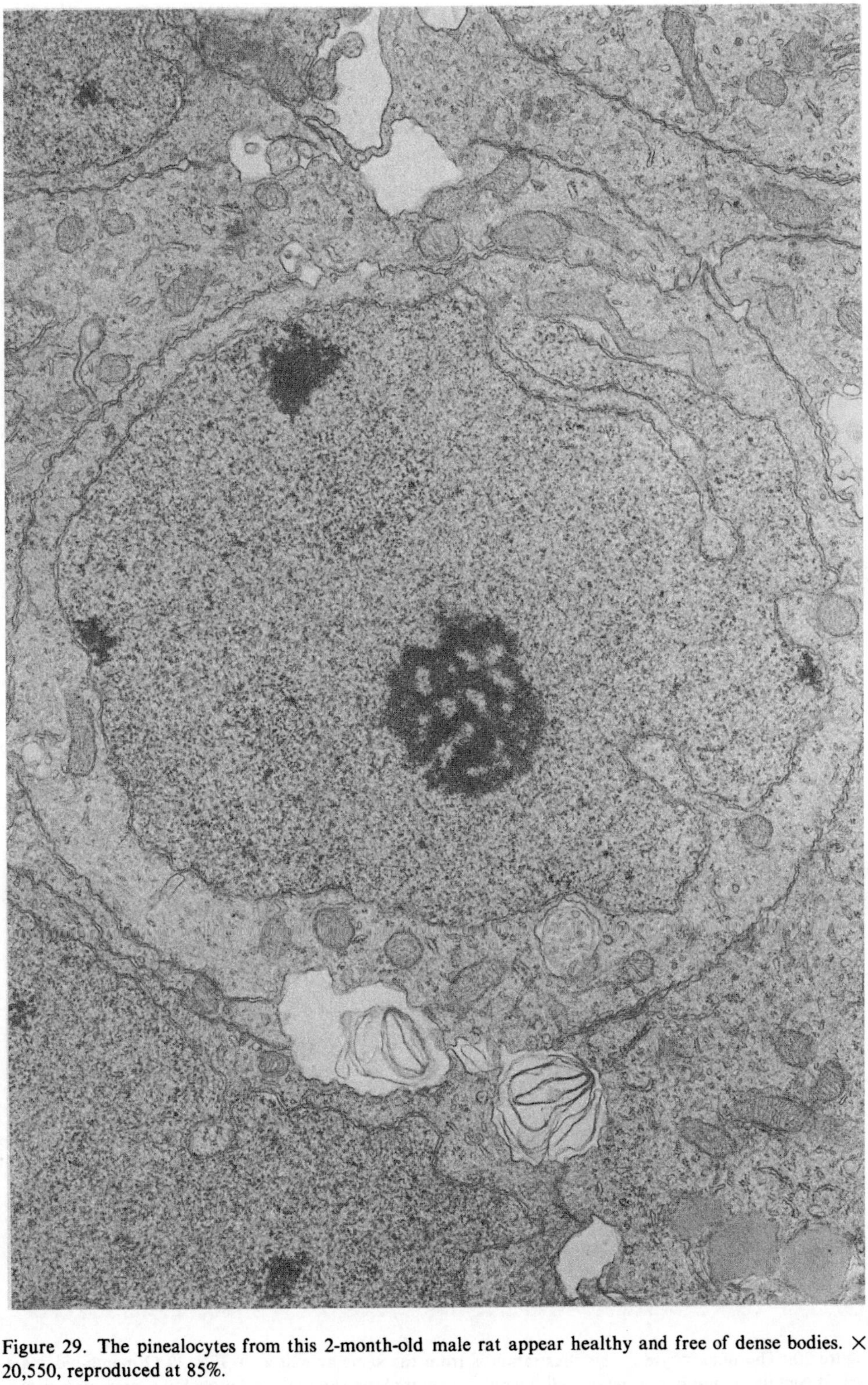

Figure 29. The pinealocytes from this 2-month-old male rat appear healthy and free of dense bodies. × 20,550, reproduced at 85%.

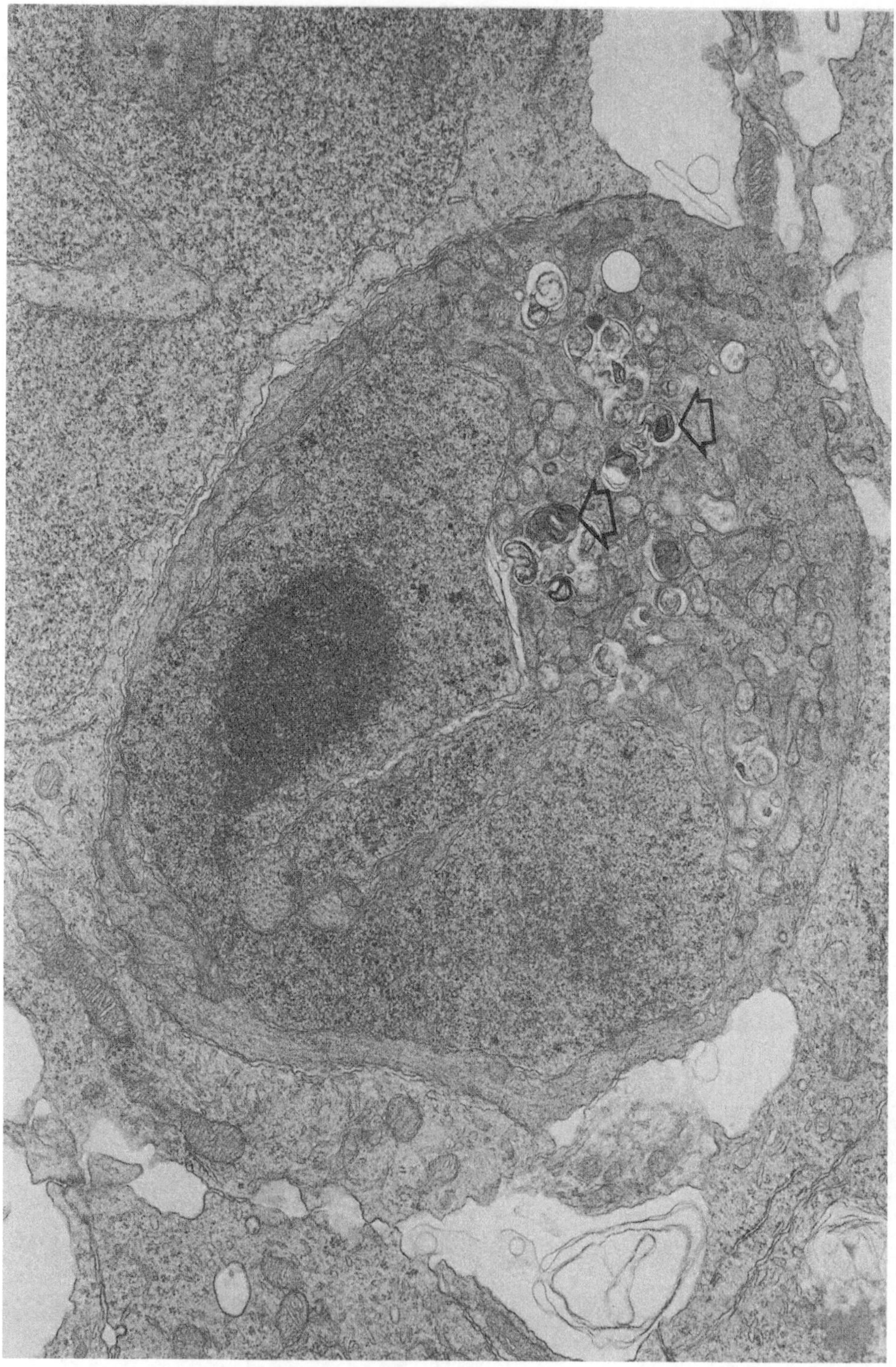

Figure 30. The pinealocyte in this illustration is from the same animal as in Fig. 29 (2-month-old male rat). It contains a number of dense bodies within the cytoplasm (arrows) and probably represents a developmental phenomenon rather than aging. × 20,500, reproduced at 85%.

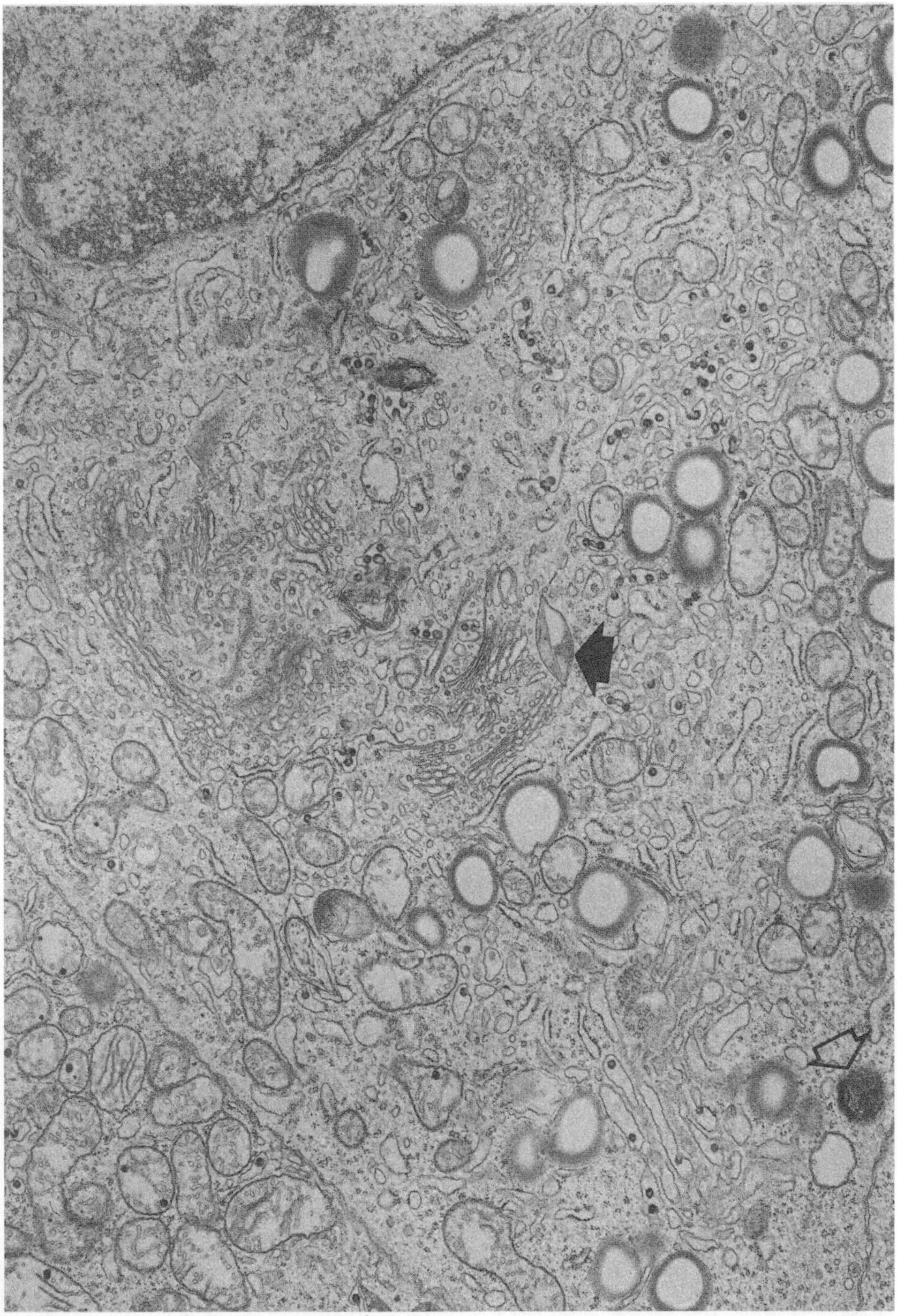

Figure 31. This N2-A neuroblastoma cell (a rapidly dividing "immortal" cell line) contains only one dense body (hollow arrow). The rather curious looking inclusion pointed out by the solid arrow was not seen in every population of N2-A cells which was examined. I was given samples grown at U.C. Berkeley and others from Harvard and NIH. The possibility must be considered that long-lived lines obtained from a common source probably change slightly as they are subcultivated many times over the years in various laboratories. × 18,200, reproduced at 85%.

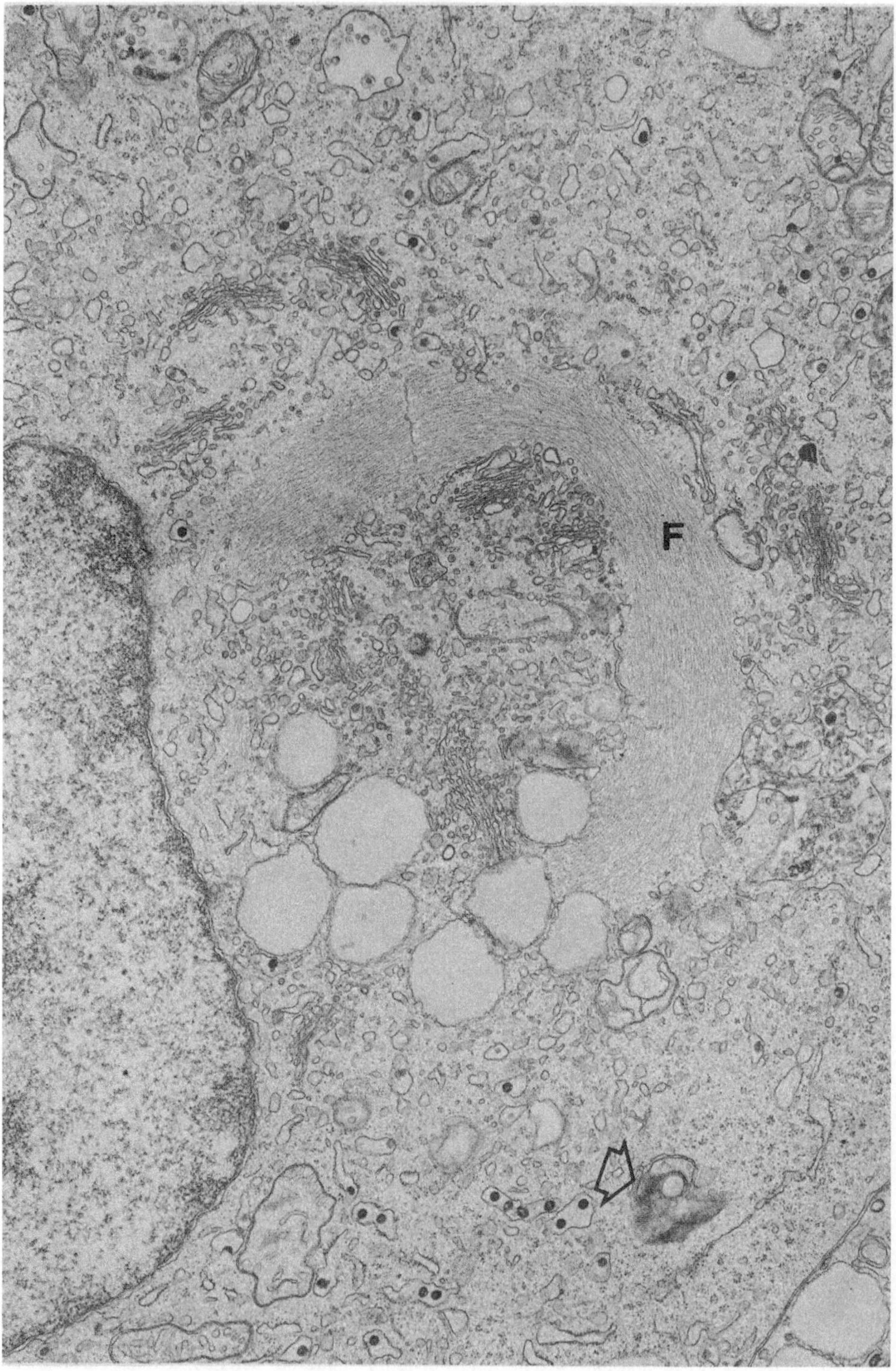

Figure 32. This N2-A neuroblastoma cell contains one dense body (hollow arrow) and a bundle of filaments (F). × 20,550, reproduced at 85%.

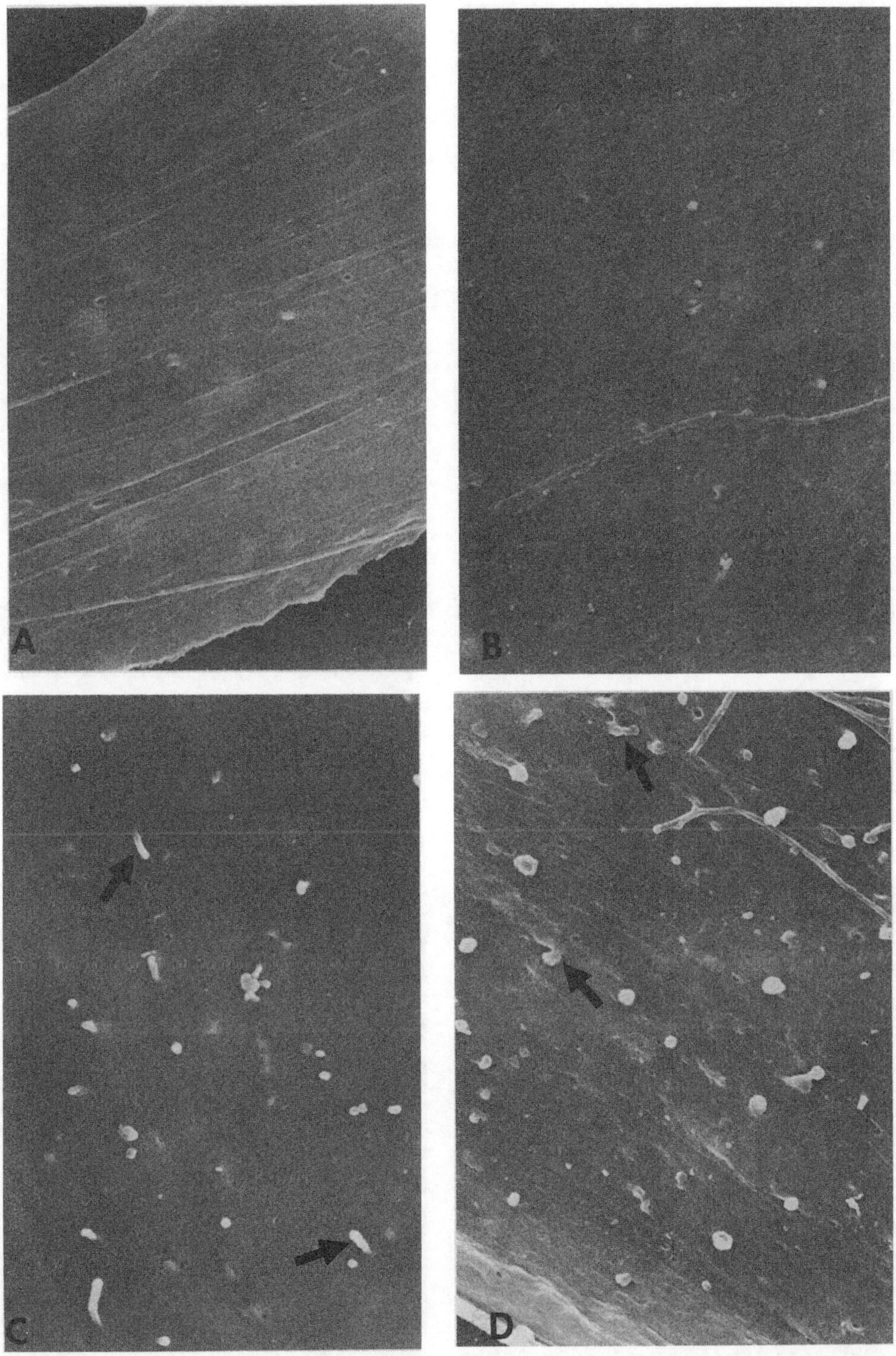

Figure 33. IMR-90 fibroblasts are shown by scanning electron microscopy. Figures 33 A and B were classified as "−" because they had no surface projections (microvilli), while the cells in C and D were classified as "+" since a few projections were present. All micrographs × 10,000, reproduced at 85%. (From Johnson, 1979.)

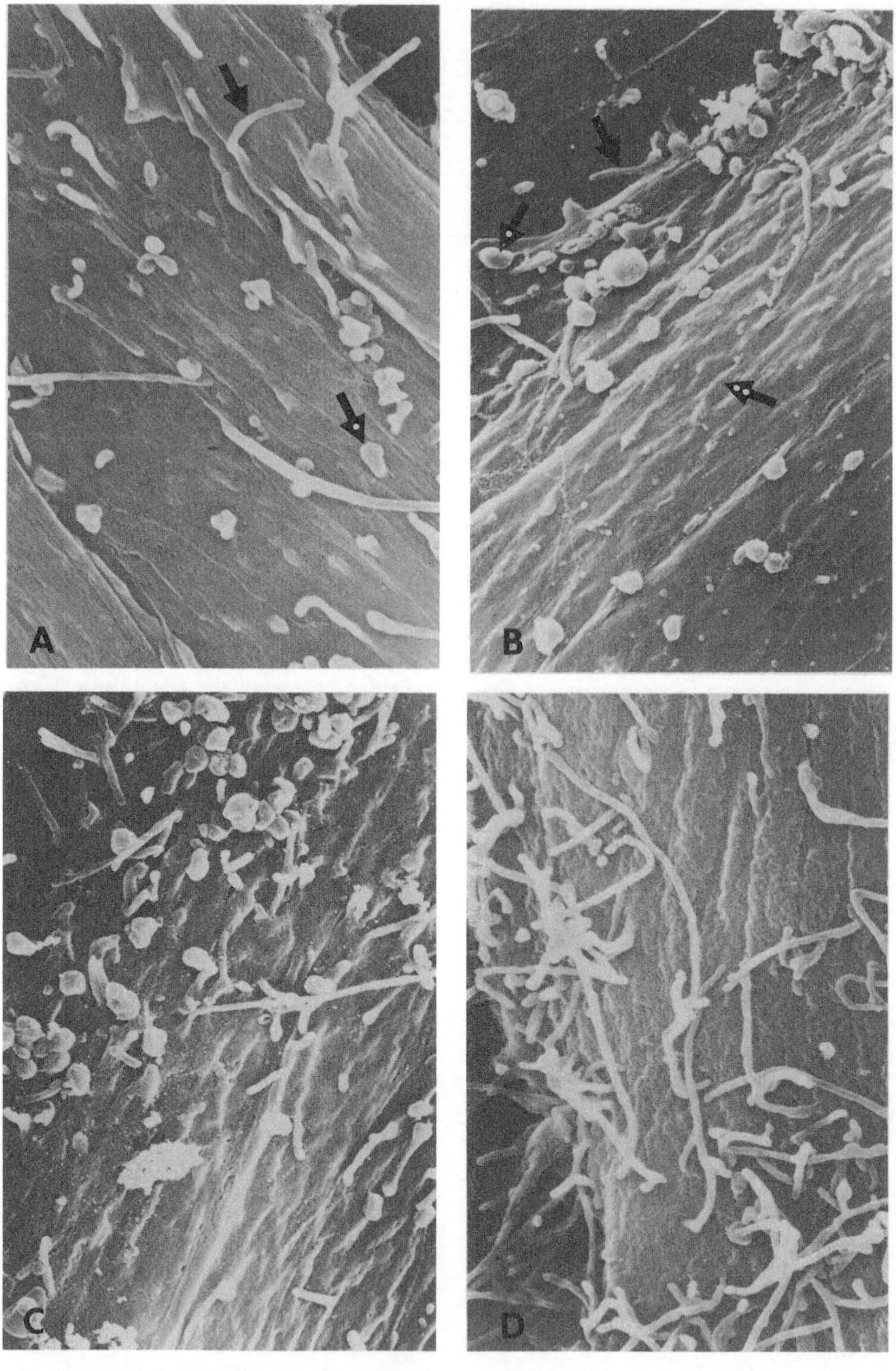

Figure 34. The IMR-90 fibroblasts in 34 A and B were classified as "++", and the cells in C and D were classified as "+++" because of the large numbers of projections on their surface. All micrographs × 10,000, reproduced at 85%. (From Johnson, 1979.)

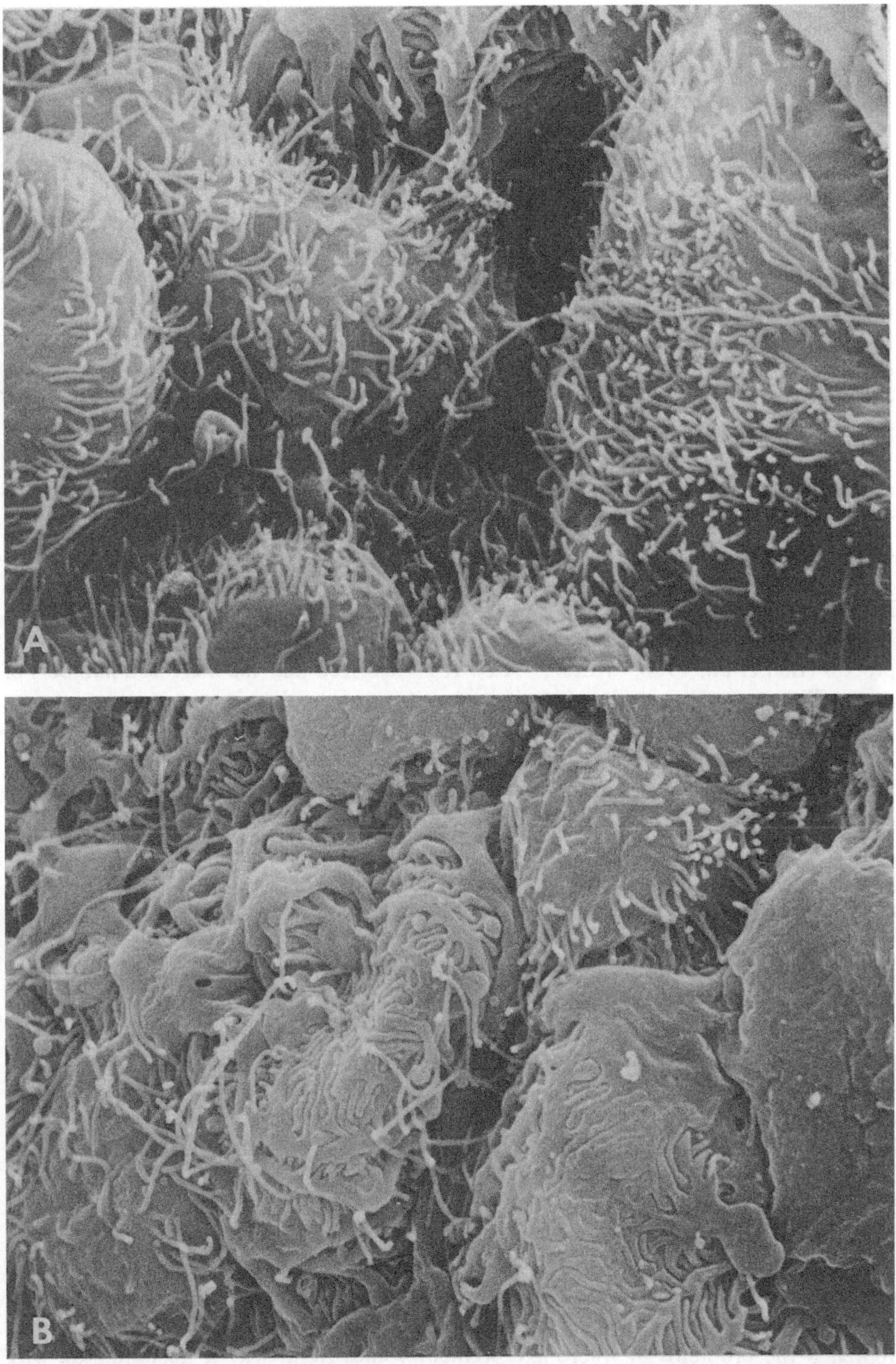

Figure 35. Scanning electron micrographs are shown from mouse kidney glomeruli. In 35 A (24-month-old control animal) there are large numbers of surface projections (microvilli) while in 35 B (24-month-old animal on 4% protein diet begun early in life) the surface projections are reduced. Both micrographs × 7500, reproduced at 85%. (See Johnson and Barrows, 1980.)

The meaning of altered numbers of microvilli on the surface of aging cells is difficult to explain at present. Microvilli are common features on cell surfaces. They may serve to increase membrane transport (Madison *et al.*, 1979), increase surface area (Oder, 1954), entrap glycocalyx (Andrews and Porter, 1973), and act as a site of membrane reserve for cell movement and division (Erickson and Trinkaus, 1976).

It has been generally felt that an inverse relationship exists between the rate of cell division and level of cell differentiation (De Robertis *et al.*, 1975). A recent theory (Walton, 1982) suggests that differentiated cell aging and limited stem cell replicative capacity underlie age-related pathology. Kelley *et al.* (1979) utilized IMR-90 fibroblasts in culture to study gap junctions and their relationships to metabolic cooperation (communication) between neighboring cells in contact with one another. When cell populations were subcultivated until proliferative rates were reduced, formation of gap junctions was delayed, and contact-mediated metabolic cooperation (transfer of [^{3}H]uridine nucleotides between cells) was reduced. It was suggested that gap junctions are implicated as low-resistance pathways for cell-to-cell electrical coupling and as passageways for direct intercellular exchange of low-molecular-weight metabolites (Kelley *et al.*, 1979).

That cellular aging of rat fibroblasts *in vitro* represents a process of differentiation has been suggested (Kontermann and Bayreuther, 1979), and, indeed, the data from several laboratories indicate the importance of state of cell differentiation in the aging process (Bell *et al.*, 1978; Walton, 1982). It is theorized that the increased numbers of microvilli seen on aging cells *in vivo* and *in vitro* may be due, in part, to a decrement in cellular intercommunication which accompanies an increased state of differentiation. Such loss of communication between cells could result in attenuated ability to maintain tissue, organ, and organism homeostasis, theirin precipitating aging, pathology and death.

5. CONCLUSION

In spite of the fact that cells grown in a bottle, flask, dish, or tube are living in an abnormal environment, there is a surprising number of similarities of age-related structural changes *in vitro* to those changes seen *in vivo*. These changes include a loss of granular endoplasmic reticulum and mitochondria, and increases in age pigment and Golgi. Age pigment appears to accumulate at a faster rate *in vitro* probably because the cells lack some of the protective mechanisms present *in vivo*.

The fact that *in vitro* cell populations do not have the normal stimuli from other cell types, hormones, and the variety of milieu components in the interstitial fluid, has been the foundation for criticism of *in vitro* studies. Yet, it is this very fact that represents the value of cell culture; it is isolated, thereby eliminating many other variables.

Some argue that changes seen with increasing population doublings do not represent aging. This is probably true. Yet some of the changes detected *in vitro* seem to mimic those observed *in vivo* and, thus, represent alternate avenues from which we can approach the problems of where age pigment comes from, why it accumulates, why granular endoplasmic reticulum is lost, why mitochondria are lost, etc.

In vivo aging studies are replete with criticisms as well. One of the most salient of these criticisms concerns the use of rodents and generalizing the data obtained from them to human aging. The paired helical filaments seen in the brains of human patients with Alzheimer's disease are not found in rodents. Indeed, even monkey brain filament pairs have a different periodicity. These kinds of differences are outweighed by the fact that many structural changes discerned in rodents *resemble* those seen in humans, plus one of the most significant points of all: rodent populations for an experiment are genetically similar. Add this to the fact that rats and mice grow old in 2½ years, compared to thirty for monkeys and 75 for humans, and one can see why rodent models for gerontology will never disappear.

One might notice that there is a great deal of discrepancy in results between laboratories for both *in vivo* and *in vitro* studies. This can probably be accounted for, in part, by uncontrolled, perhaps undiscovered, variables. Suppose an experimenter holds factor "A" constant and varies factor "B." He does not control factor "C" because he is unaware it exists or does not know how important it is. His results are gathered on a day when factor "C" is at a high point. A second laboratory performs a similar experiment and, again, does not control factor "C." The day they collect their data, factor "C" is at a low point. The results are different. Solution? Control as many factors as possible. Cell culture experiments may represent an excellent way of doing this adjunct to *in vivo* studies. However, we still have a long way to go, because not only do we have the undiscovered variable to contend with, but some of the known ones which we tend to dismiss, e.g., the ways in which the animals are fed or handled, substrates on which the cells are grown, concentration of the serum in the culture media, cell plating density, etc. In essence, there is bad news and good news concerning our knowledge of cell aging. The bad news is that we are still on square one. The good news is that we know it.

REFERENCES

Andrew, W., 1941, Cytological changes in senility in the trigemial ganglion, spinal cord and brain of the mouse, *J. Anat.* **75**:406.

Andrew, W., 1955, Amitotic division in senile tissues as a probable means of self-preservation of cells, *J. Gerontol.* **10**:1.

Andrew, W., 1956, Structural alterations with aging in the nervous system, *J. Chronic Diseases* **3**:575.

Andrew, W., 1959, The reality of age differences in nervous tissue, *J. Gerontol.* **14**:259.

Andrew, W., 1961, An electron microscope study on types of age change in nerve cells, with particular reference to chromophilia and to accumulation of pigment in man and laboratory animals, *J. Gerontol.* **16**:388.

Andrew, W., 1962, An electron microscope study of age changes in the liver of the mouse, *Am. J. Anat.* **110**:1.

Andrew, W., 1964, Changes in the nucleus with advancing age of the organism, *Adv. Gerontol. Res.* **1**:87.

Andrew, W., 1971, *The Anatomy of Aging in Man and Animals*, pp. 222–237, Grune and Stratton, New York.

Andrews, P. M., and Porter, K. R., 1973, The ultrastructural morphology and possible functional significance of mesothelial microvilli, *Anat. Rec.* **177**:409.

Aschenbrenner, J. E., 1979, Ultrastructure of the adenohypophysis, in: *Aging in Nonhuman Primates* (D. Bowden, ed.) Van Nostrand Reinhold, New York.

Bakerman, S., 1969, *Aging Life Process*, Charles C. Thomas, Springfield.

Barr-Nea, L., and Wolman, M., 1977, Lipid pigment formation in cultures of chick hepatocytes. Effects of aging and of oxidizable agents, *Virchows Arch Cell. Pathol.* **23**:79.

Barrows, C. H., Jr., Roeder, L. M., and Falzone, J. A., 1962, Effect of age on the activities of enzymes and the concentration of nucleic acids in the tissues of female wild rats, *J. Gerontol.* **14:**144.

Bell, E., Marek, L. F. Levinstone, D. S., Merrill, C., Sher, S., Young, I., and Eden, M., 1978, Loss of division potential in vitro: Aging or differentiation? *Science* **202:**1158.

Bemiller, P. M., and Lee, L. H., 1976, Nucleolar changes in senescing WI-38 cells, *Mech. Ageing Dev.* **8:**417.

Bemiller, P. M., and Miller, J. E., 1979, Cytological changes in senescing WI-38 cells: A statistical analysis, *Mech. Ageing Dev.* **10:**1.

Bilder, G., and Denckla, W. D., 1977, Restoration of ability to reject xenograph and clear carbon after hypophysectomy of adult rats, *Mech. Ageing Dev.* **6:**153.

Bogart, B. I., 1970, The effect of aging on the rat submandibular gland: An ultrastructural, cytochemical and biochemical study, *J. Morph.* **130:**337.

Bondareff, W., 1957, Genesis of intracellular pigment in spinal ganglia of senile rats. An electron microscope study, *J. Gerontol.* **12:**364.

Bondareff, W., 1959, Morphology of the aging nervous system, in: *Handbook of Aging and the Individual* (J. E. Birren, ed.), pp. 136–173, University of Chicago Press, Chicago.

Bowman, P. D., Meek, R. L., and Daniel, C. W., 1976, Decreased synthesis of nucleolar RNA in aging human cells in vitro, *Exp. Cell Res.* **101:**434.

Brizzee, K. R., and Johnson, F. A., 1970, Depth distribution of lipofuscin pigment in cerebral cortex of rat, *Acta Neuropathol.* **16:**205.

Brizzee, K. R., and Ordy, J. M., 1981, Cellular features, regional accumulation, and prospects of modification of age pigment in mammals, in: *Age Pigments* (R. S. Sohal, ed.), pp. 101–154, Elsevier, New York.

Brock, M. A., and Hay, R. J., 1971, Comparative ultrastructure of chick fibroblasts *in vitro* at early and late stages during their growth span, *J. Ultrastruct. Res.* **36:**291.

Cammermeyer, J., 1963, Cytological manifestations of aging in rabbit and chinchilla brains, *J. Gerontol.* **18:**41.

Chaconas, E., and Finch, C. E., 1973, The effect of aging on RNA/DNA ratios in brain regions of the C57BL/6J male mouse, *J. Neurochem.* **21:**1469.

Collier, R., 1943, Über den Feinbau der Epiphysis von Nagetreran und die Frage seiner funktionellen Veränderungen, *Z. Zelforsch.* **33:**51.

Collins, V. P., and Brunk, U., 1976, Characterization of residual bodies formed in phase II cultivated human glial cells, *Mech. Ageing Dev.* **5:**193.

Cristofalo, V. J., and Kritchevsky, D., 1966, Respiration and glycolysis in the human diploid cell strain WI-38, *J. Cell. Comp. Physiol.* **67:**125.

Denckla, W. D., 1974, Role of pituitary and thyroid glands in the decline of minimal 0_2 consumption with age, *J. Clin. Invest.* **53:**572.

De Robertis, E., and Pellegrino de Iraldi, A., 1961, Plurivesicular secretory processes and nerve endings in the pineal gland of the rat, *J. Biophys. Biochem. Cytol.* **10:**361.

De Robertis, E. D., Saez, F. A., and De Robertis, M. F., 1975, *Cell Biology,* W. B. Saunders, Philadelphia.

Dimitrova, Z., 1901, Recherches sur la structure de la glande pineale chez auelques mammiferes, *Nevraxe* **2:**259.

Dolley, D. H., 1917, The recovery from depression in the Purkinje cell and the decline to senility of depression with the histogenesis of abnormal pigmentation, *J. Comp. Neurol.* **28:**465.

Everitt, A. V., and Burgess, J. A., 1976, *Hypothalamus, Pituitary and Aging,* Charles C. Thomas, Springfield.

Everitt, A. V., Seedsman, N. J., and Jones, F., 1980, The effects of hypophysectomy and continuous food restriction, begun at ages 70 and 400 days, on collagen aging, proteinurea, incidence of pathology and longevity in the male rat, *Mech. Age. Devl.* **12:**161.

Erickson, C. A., and Trinkaus, J. P., 1976, Microvilli and blebs as sources of reserve surface membrane during cell spreading, *Exp. Cell Res.* **99:**375.

Fahim, M. A., and Robbins, N., 1982, Ultrastructural studies of young and old mouse neuromuscular junctions, *J. Neurocytol.* **11:**641.

Feldman, M. L., and Peters, A., 1972, Intracellular rods and sheets in rat cochlear nucleus, *J. Neurocytol.* **1:**109.

Field, E. J., and Peat, A., 1971, Intranuclear inclusions in neurons and glia: a study in the aging mouse, *Gerontologia* **17**:129.

Garg, B. D., Kourounakis, P., and Tuchweber, B., 1979, Fine structural changes in the liver of young and old rats as influenced by microsomal enzyme inducers, *Gerontology* **25**:314.

Gatenby, J. B., and Moussa, T. A., 1950, The sympathetic ganglion cell, with Sudan black and the Zernicke microscope, *J. R. Microsc. Soc.* **70**:342.

Goldstein, S., and Trieman, G., 1974, Glucose consumption by early and late passage human fibroblasts during growth and stationary phase, *Experientia* **31**:177.

Greski, E. W., Brennan, M., and Azmitia, E., 1982, Age-related changes in EGF and protease in submandibular glands of C57BL/6J mice, *Exp. Aging Res.* **8**:87.

Hasan, M., and Glees, P., 1973, Ultrastructural age changes in hippocampal neurons, synapses and neuroglia, *Exp. Gerontol.* **8**:75.

Hayflick, L., 1977, The cellular basis for biological aging, in: *Handbook of the Biology of Aging* (C. E. Finch and L. Hayflick, eds.), pp. 159–186, Van Nostrand Reinhold, New York.

Herbener, G. H., 1976, A morphometric study of age-dependent changes in mitochondrial populations in mouse liver and heart, *J. Gerontol.* **31**:8.

Herrlinger, H., Anzil, A. P., and Blinzinger, K., 1974, Intramitochondrial crystalloid inclusions in neuronal processes of human cerebral cortex, *J. Neurocytol.* **3**:719.

Hodge, C. F., 1894, Changes in ganglion cells from birth to senile death. Observations on man and honey bee, *J. Physiol.* **17**:129.

Hyden, H., 1967, RNA in brain cells, in: *The Neurosciences, A Study Program* (G. C. Quarton, T. Melnechuk, and F. O. Schmitt, eds.), pp. 248–266, The Rockefeller Press, New York.

Johnson, J. E., Jr., 1979, Fine structure of IMR-90 cells in culture as examined by scanning and transmission electron microscopy, *Mech. Ageing Dev.* **10**:405.

Johnson, J. E., Jr., 1980a, Fine structural studies in the aging neuroendrocrine axis: Observations on the pituitary and adrenal, *Soc. Neuroscience Abstr.* 280.

Johnson, J. E., Jr., 1980b, Fine structural alterations in the aging rat pineal gland, *Exp. Aging Res.* **6**:189.

Johnson, J. E., Jr., 1980c, Fine structural effects of hypophysectomy on the aging rat pineal gland, *Gerontology* **26**:206.

Johnson, J. E., Jr., and Barrows, C. H., 1980, Effects of age and dietary restriction on the kidney glomeruli of mice: Observations by scanning electron microscopy, *Anat. Rec.* **196**:145.

Johnson, J. E., Jr., and Cutler, R. G., 1980, Effects of hypophysectomy on age-related changes in the rat kidney glomerulus: Observations by scanning and transmission electron microscopy, *Mech. Age. Devel.* **13**:63.

Johnson, J. E., Jr., and Miquel, J., 1974, Fine structural changes in the lateral vestibular nucleus of aging rats, *Mech. Aging Dev.* **3**:203.

Johnson, J. E., Jr. and Miquel, J., 1978, Comparative electron microscopic studies of various cell types in the aging mammal, *Age* **1**:78.

Johnson, J. E., Jr., Mehler, W. R., and Miquel, J., 1975, A fine structural study of degenerative changes in the dorsal column nuclei of aging mice. Lack of protection by vitamin E, *J. Gerontol.* **30**:395.

Johnson, J. E., Jr., Mehler, W. R., and Oyama, J., 1976, The effects of centrifugation on the morphology of the lateral vestibular nucleus in the rat. A light and electron microscopic study, *Brain Res.* **106**:205.

Johnson, J. E., Jr., Miquel, J., Binnard, R., and Economos, A. C., 1978, Effects of antioxidants on the liver of aging mice, *Age* **1**:162.

Kadenbach, E., 1968, Transfer of proteins from microsomes into mitochondria. Biosynthesis of cytochrome C, in: *Biochemical Aspects of the Biogenesis of Mitochondria* (E. C. Slater, J. J. Tager, S. Papa, and E. Quagliarello, eds.), pp. 415–429, Bari, Adriatica Editrice.

Kelley, R. O., Vogel, K. G., Crissman, H. A., Lujan, C. J., and Skipper, B. E., 1979, Development of the aging cell surface. Reduction of gap junction-mediated metabolic cooperation with progressive subcultivation of human embryo fibroblasts, *Exp. Cell Res.* **119**:127.

Kevorkian, J., and Wessel, W., 1959, So-called "nuclear pellets" (Kernkugeln) of pinealocytes, *Arch. Pathol.* **68**:513.

Kim, S. U., 1975, Brain hypoxia studies in mouse central nervous system cultures, I. Sequential cellular changes, *Lab. Invest.* **33**:658.

Knox, C. A., Yates, R. D., and Chen, I., 1980, Brain aging in normotensive and hypertensive strains of rats. II. Ultrastructural changes in neurons and glia, *Acta Neuropathol.* **52**:7.

Kontermann, K., and Bayreuther, K., 1979, The cellular aging of rat fibroblasts *in vitro* is a differentiation process, *Gerontology* **25**:261.

Kuhlenbeck, H., 1954, Some histologic age changes in the rat's brain and their relationship to comparable changes in the human brain, *Confin. Neurol.* **14**:329.

Lehniger, A. L., 1962, Water uptake and extrusion by mitochondria in relation to oxidative phosphorylation, *Physiol. Rev.* **42**:467.

Leto, S., Kokkonen, G. C., and Barrows, C. H., 1976, Dietary protein, life span, and biochemical variables in female mice, *J. Gerontol.* **31**:144.

Lipetz, J., and Cristofalo, V. J., 1972, Ultrastructural changes accompanying the aging of human diploid cells in culture, *J. Ultrastr. Res.* **39**:43.

London, E. D., Nespor, S., Moore, L., Mahone, P., and Rapoport, S. I., 1979, Age-dependent changes in local cerebral glucose utilization, *Age* **2**:131.

Madison, L. D., Porter, B. B., Torres, A. R., and Shelton, E., 1979, Regulation of surface topography of mouse peritoneal cells, Formation of microvilli and vesiculated pits on omental mesothelial cells by serum and other proteins, *J. Cell Biol.* **82**:783.

Massie, H. R., Colacicco, J. R., and Williams, T. R., 1981, Loss of mitochondrial DNA with aging in the swedish C strain of Drosophila melanogaster, *Age* **4**:42.

McCay, C. M., Maynard, L. A., Sperling, G., and Barnes, L. L., 1939, Retarded growth, life span, ultimate body size, and age changes in the albino rat after feeding diets restricted in calories, *J. Nutr.* **18**:1.

McMartin, D. N., and Schedlbauer, L. M., 1975, Incorporation of (^{14}C)-Leucine into protein and tubulin by brain slices from young and old mice, *J. Gerontol.* **30**:132.

Medvedev, Zh. A., 1964, The nucleic acids in development and aging, *Adv. Geront. Res.* **1**:181.

Meier, C., 1974, Ultrastructural study of the diminution of lipofuscin in nervous tissue-favourable influence of meclofenoxate, *Ann. Anesthesiol. Francaise* **15**:579.

Menzies, R. A., Mishra, R. K., and Gold, P. H., 1972, The turnover of ribosomes and soluble RNA variety of tissues of young adult and aged rats, *Mech. Aging Dev.* **1**:117.

Miquel, J., 1971, Aging of male Drosophila melanogaster: Histological, histochemical and ultrastructural observations, *Adv. Gerontol. Res.* **3**:39.

Miquel, J., Oro, J., Bensch, K., and Johnson, J. E., Jr., 1977, Lipofuscin: Fine structural and biochemical studies, in: *Free Radicals in Biology* (W. A. Pryor, ed.), Academic Press, New York.

Miquel, J., Ludgren, P. R., and Johnson, J. E., Jr., 1978, Spectrophotofluorometric and electron microscopic study of lipofuscin accumulation in the testis of aging mice, *J. Gerontol.* **33**:5.

Miquel, J., Johnson, J. E., Jr., and Cervos-Navarro, J., 1983, Comparison of CNS aging in humans and experimental animals, in: *Brain Aging: Neuropathology and Neuropharmacology* (J. Cervos-Navarro and H. I. Sarkander, eds.), pp. 231–258, Raven Press, New York.

Munnell, J. F., and Getty, R., 1968, Nuclear lobulation and amitotic division associated with increasing cell size in the aging canine myocardium, *J. Gerontol.* **23**:363.

Nandy, K., 1968, Further studies on the effects of centrophenoxine on the lipofuscin pigment in the neurons of senile guinea pigs, *J. Gerontol.* **23**:83.

Nandy, K., and Bourne, G. H., 1966, Effect of centrophenoxine on the lipofuscin pigment in the neurones of senile guinea pigs, *Nature* **210**:313.

Nandy, K., Baste, C., and Schneider, F. H., 1978, Further studies on the effects of centrophenoxine on lipofuscin pigment in neuroblastoma cells in culture: An electron microscope study, *Exp. Gerontol.* **13**:311.

Nosal, G., 1979, Neuronal involution during ageing. Ultrastructural study in the rat cerebellum, *Mech. Ageing Dev.* **10**:295.

Novikoff, A. B., 1967, Lysosomes in nerve cells, in: *The Neuron* (H. Hyden, ed.), pp. 319–377, Elsevier, Amsterdam.

Oder, D. L., 1954, Observations of the rat mesothelium with the electron and phase microscopes, *Am. J. Anat.* **95**:433.

Ogrodnik, J. P., Wulff, J. H., and Cutler, R. G., 1975, Altered protein hypothesis of mammalian aging processes. II. Discrimination ratio of methionine *vs* ethionine in the synthesis of ribosomal protein and RNA of C57BL/6J mouse liver, *Exp. Gerontol.* **10**:119.

Orgel, L., 1963, The maintenance of the accuracy of protein synthesis and its relevancy to aging, *Proc. Natl. Acad. Sci. USA* **49**:517.

Palay, S. L., and Palade, G. E., 1955, The fine structure of neurons, *J. Biophys. Biochem. Cytol.* **1**:69.

Quay, W. B., 1965, Histological structure and cytology of the pineal organ in birds and mammals, *Prog. Brain Res.* **10**:49.

Rees, S., and Cragg, B., 1983, Is silica involved in neuritic (senile) plaque formation? *Acta Neuropathol.* **59**:31.

Ringborg, U., 1966, Composition and content of RNA in neurons of rat hippocampus at different ages, *Brain Res.* **2**:296.

Sahgal, V., Subramani, V., Hughes, R., Shah, A., and Singh, H., 1979, On the pathogenesis of mitochondrial myopathies. An experimental study, *Acta Neuropathol.* **46**:177.

Samis, H. V., Erk, F. C., and Baird, M. B., 1971, Senescence in Drosophila. I. Sex differences in nucleic acid, protein and glycogen levels as a function of age, *Exp. Gerontol.* **6**:9.

Samorajski, T., and Ordy, J. M., 1967, The histochemistry and ultrastructure of lipid pigment in the adrenal glands of aging mice, *J. Gerontol.* **22**:253.

Samorajski, T., and Rolsten, C., 1973, Age and regional differences in the chemical composition of brains of mice, monkeys and humans, *Progr. Brain Res.* **40**:253.

Samorajski, T., Frieda, R. L., and Ordy, J. M., 1971, Age differences in the ultrastructure of axons in the pyramidal tract of the mouse, *J. Gerontol.* **26**:542.

Schmucker, D. L., Mooney, J. S., and Jones, A. L., 1978, Stereological analysis of hepatic fine structure in the Fischer-344 rat. Influence of sublobular location and age, *J. Cell Biol.* **78**:319.

Schneider, E. L., and Shorr, S., 1975, Alteration in cellular RNA's during the *in vitro* life span of cultured human diploid fibroblasts. I. Increased mRNA, rRNA and tRNA content in late passage WI-38 cells, *Cell* **6**:179.

Seite, R., Mei, N., and Vuillet-Luciani, J., 1973, Effect of electrical stimulation on nuclear microfilaments and microtubules of sympathetic neurons submitted to cycloheximide, *Brain Res.* **50**:419.

Shamoto, M., 1968, Age differences in the ultrastructure of hepatic cells of thyroxine-treated rats, *J. Gerontol.* **23**:1.

Shirey, T. L., and Sobel, H., 1972, Compositional and transcriptional properties of chromatins isolated from cardiac muscle of young mature and old dogs, *Exp. Gerontol.* **7**:15.

Sohal, R. S., and Allison, V. F., 1971, Age related changes in the fine structure of the flight muscle of the house fly, *Exp. Gerontol.* **6**:167.

Sosa, J. M., 1952, Aging of neurofibrils, *J. Gerontol.* **7**:191.

Spoerri, P. E., and Glees, P., 1973, Neuronal aging in cultures: An electron microscopic study, *Exp. Gerontol.* **8**:259.

Spoerri, P. E., and Glees, P., 1974, The effects of dimethylaminoethyl-p-chlorophenoxyacetate on spinal ganglia neurons and satellite cells in culture. Mitochondrial changes in the aging neurons. An electron microscope study, *Mech. Ageing Dev.* **3**:131.

Spoerri, P. E., Glees, P., and El Ghazzawi, E., 1974, Accumulation of lipofuscin in the myocardium of senile guinea pigs: Dissolution and removal of lipofuscin following dimethylaminoethyl-p-chlorophenoxyacetate administration. An electron microscope study, *Mech. Ageing Dev.* **3**:311.

Steiness, I. B., 1957, Vibratory perception in normal subjects; a biothesiometric study, *Acta Med. Scand.* **158**:315.

Strehler, B. L., 1964, On the histochemistry and ultrastructure of age pigment, *Adv. Gerontol. Res.* **1**:343.

Strehler, B. L., 1966, Code degeneracy and the ageing process: A molecular genetic theory of aging, *Proc. 7th Congr. Gerontol. Vienna,* 177.

Strehler, B. L., Hirsch, G., Gussek, D., Johnson, R., and Bick, M., 1971, The codon restriction theory of aging and development, *J. Theoret. Biol.* **33**:429.

Sundararaman, V., and Cummings, D. J., 1976, Morphological changes in aging cell lines of Paramecium aurelia. II. Macronuclear alterations, *Mech. Ageing Dev.* **5**:325.

Tandler, B., and Hoppel, C. L., 1972, *Mitochondria,* Academic Press, New York.

Thompson, J. M., London, E. D., and Johnson, J. E., Jr. 1982, Ultrastructural, functional and biochemical characteristics of mouse and human neuroblastoma cell lines, *Neuroscience* **7**:1807.

Timiras, P. S., and Vernadakis, A., 1972, Structural, biochemical and functional aging of the nervous system, in *Developmental Physiology and Aging* (P. S. Timiras, ed.), The Macmillan Co., New York.

Vanneste, J., and vandenbosch de Aguilar, Ph., 1981, Mitochondrial alterations in the spinal ganglion neurons in ageing rats, *Acta Neuropathol.* **54**:83.

Vaughan, W. J., and Calvin, M., 1977, Electrophoretic analysis of brain proteins from young adult and aged mice, *Gerontology* **23**:110.

Vaughan, D. W., and Vincent, J. M., 1979, Ultrastructure of neurons in the auditory cortex of ageing rats: A morphometric study, *J. Neurocytol.* **8**:215.

Walton, J., 1982, The role of limited cell replicative capacity in pathological age change. *A review, Mech. Age. Devel.* **19**:217.

Wayner, M. J., Wulff, V. J., and Piekielniak, M., 1962, Ribonucleic acid content of tissues of rats of various ages, *J. Gerontol.* **17**:455.

Weibel, E. R., 1969, Stereological principles for morphometry in electron microscope cytology, *Int. Rev. Cytol.* **26**:235.

Weiss, J., and Lansing, A. I., 1953, Age changes in the fine structure of anterior pituitary of the mouse, *Proc. Soc. Exp. Biol. Med.* **82**:460.

Wilcox, H. H., 1959, Structural changes in the nervous system related to the process of aging, in *The Process of Aging in the Nervous System* (J. E. Birren, H. A. Imus, and W. F. Windle, eds.), pp. 16–23, Charles C. Thomas Springfield.

Wilson, P. D., and Franks, L. M., 1975, The effect of age on mitochondrial ultrastructure, *Gerontologia* **21**:81.

Wolfe, D. E., 1965, The epiphyseal cell: An electron microscopic study of its intercellular relationships and intracellular morphology in the pineal body of the albino rat, *Prog. Brain Res.* **10**:332.

Zs.-Nagy, V., Bertoni-Freddari, C., Zs.-Nagy, I., Pieri, C., and Giuli, C., 1977, Alterations in the numerical density of perichromatin granules in different tissues during ageing and cell differentiation, *Gerontology* **23**:267.

3

New Concepts in Regulation of the Lifespan of Human Diploid Fibroblasts in Vitro

THOMAS B. POOL and JOHN D. METTER

1. INTRODUCTION

It was the initial observations of Swim and Parker (1957) on the longevity of human fibroblasts in cell culture, later confirmed and extended by Hayflick and Moorehead (1961) and Hayflick (1965), that have given rise to the discipline of cytogeronotology. That human embryonic fibroblasts demonstrate a finite lifespan *in vitro* has been taken as evidence for, and developed as a model system for, aging at the cellular level by a multitude of investigators following the pioneering work of Hayflick. Possibly no other single subdiscipline of cellular biology is more rampant with hypotheses offered to explain a given phenomenon than is cytogerontology. Although no single explanation has been uniformly derived or universally accepted regarding the mechanism of cellular senescence, the significance of the eventual answer has already been considered theoretically with respect to pathological implications (Goldstein *et al.*, 1975; Timiras, 1975; Martin, 1979) and societal implications (Strehler, 1975). Some investigators question the extrapolation of data obtained from the classical *in vitro* observations to the *in vivo* situation (see Orgel, 1973), while others question the phenomenon as being representative of cellular aging at all (see Bell *et al.*, 1978). Whether the phenomenon (1) truly represents the normal sequence of events during cellular senescence, (2) is not aging

THOMAS B. POOL • Department of Anatomy, The University of Texas Health Science Center at San Antonio, San Antonio, Texas 78284. **JOHN D. METTER** • Department of Pathology, The University of Texas Health Science Center at San Antonio, San Antonio, Texas 78284.

but in reality is cellular differentiation, or (3) is simply an artifact of the conditions imposed upon cells *in vitro* will no doubt continue to be the subject of academic debate for years to come. That normal human diploid cells lose their ability to proliferate upon continuous subcultivation is not disputed. This phenomenon provides a valuable model system for investigations concerned with characterizing proliferative and nonproliferative cells from a uniform source in hopes of understanding the regulatory mechanisms involved in the control of cell proliferation. Once we understand these regulatory mechanisms, we will be capable of addressing the more theoretical implications of the observed phenomena in a valid and profitable manner.

It is our purpose in this chapter to review the cultured human diploid fibroblast system as amodel for cellular senescence and, in particular, to examine those events that lead to the loss of proliferative capacity in cultured human diploid fibroblasts (HDF).

2. THE DECLINE IN PROLIFERATIVE CAPACITY OF HUMAN DIPLOID FIBROBLASTS IN CULTURE

HDF in cell culture grow predominantly as anchorage-dependent monolayers at subconfluent density. These cells seem to prefer to be in contact with one another and will readily multilayer to maintain contacts, even at moderate cell density (Figs. 1 and 2). The major part of the *in vitro* lifespan of HDF consists of a period characterized by rapid proliferation and short (weekly) time periods between subcultivation. This period, referred to as phase II (Hayflick and Moorhead, 1961), is also characterized by a gradual decline in the capacity of cells to incorporate [^{3}H]thymidine ([^{3}H]-TdR) into nuclear DNA. An abrupt decline in labeling index signals the onset of crisis or phase III (Fig. 3). This pattern of decline in proliferative capacity is clearly seen in HDF derived from fetal lung, but is not as well defined in cultures derived from adult skin, where a more linear decline in labeling index is apparent.

The following characteristics of HDF grown *in vitro* have been developed over the past 20 years as a rationale for the adaptation of this system as a model for cell aging studies:

1. HDF demonstrate a finite lifespan in culture that is constant, statistically, for different cultures derived from a single source (Swim and Parker, 1957; Hayflick and Moorhead, 1961; Hayflick, 1965).
2. The cultured lifespan of HDF tends to be intimately related, if not inversely related, to the age of the donor from which the cells were derived (Macieria-Coelho, 1970; Martin *et al.,* 1970; Schenider and Mitsui, 1976).
3. Lifespan is measured in terms of proliferative capacity of the cells, although there are a number of ways that this parameter is expressed (population doubling level, cell doublings, specific growth rate, labeling index, cloning efficiency, etc.) (Dell'Orco *et al.,* 1973; Smith *et al.,* 1978).
4. A permanent loss in division potential is both equivalent to and defines "senescent" cell populations.

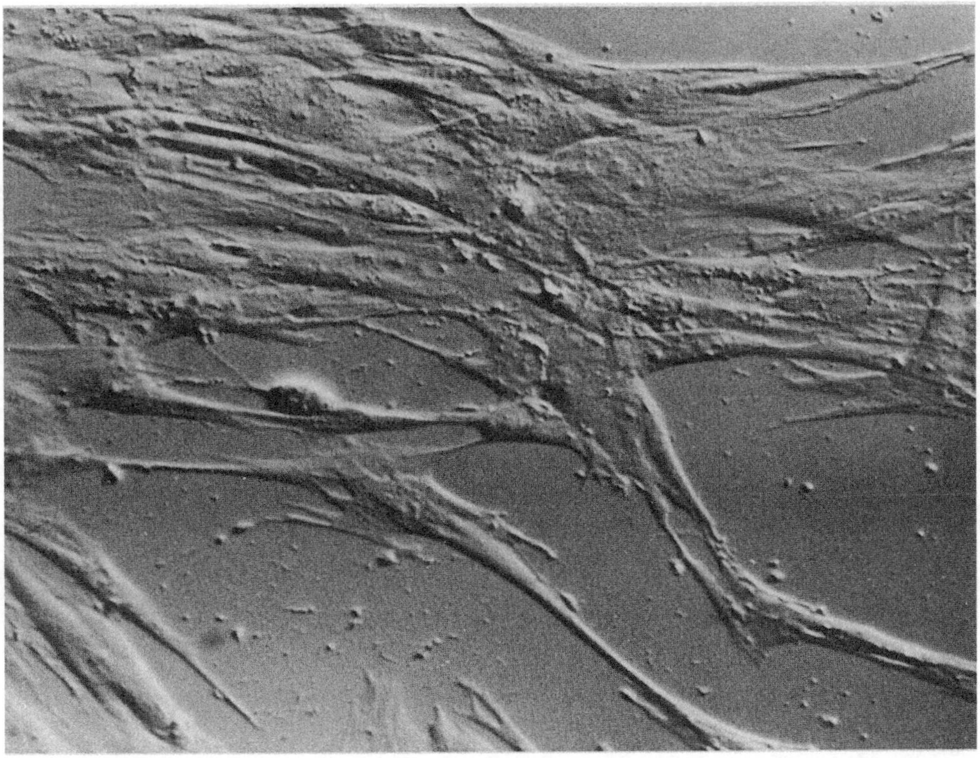

Figure 1. Photomicrograph of midpassage human diploid fibroblasts using Nomarski differential interference microscopy. Note the degree of contacts maintained between cells even at subconfluent density. Rounded cell at left center of field is in mitosis. × 290, reproduced at 88%.

It should be emphasized that an underlying assumption with this model is that loss of growth capacity represents a phenotypic expression of cellular senescence; furthermore, most current research in cell aging is directed towards identifying the mechanisms that control this process. Another assumption then follows—an understanding of the loss of division potential equals an understanding of cellular senescence. Objectivity dictates that we constantly admit these assumptions and periodically reevaluate them as new information surfaces; however, convincing data exist in the literature supporting a direct relationship between *in vivo* lifespan and proliferative capacity *in vitro*. As mentioned above, a direct relationship between donor age and *in vitro* lifespan has been demonstrated. Additionally, fibroblasts derived from patients with aging diseases, such as progeria, Werner's syndrome, and others, tend to have a reduced *in vitro* lifespan when compared to normal cells derived from donors of equivalent age (Goldstein, 1969; Fujiwara *et al.*, 1977). Finally, evidence has been presented (Martin *et al.*, 1980) that demonstrates directly the genetic determination of *in vitro* lifespan, measured in terms of proliferative capacity. In this work, proliferating hybrid and parental tetraploid

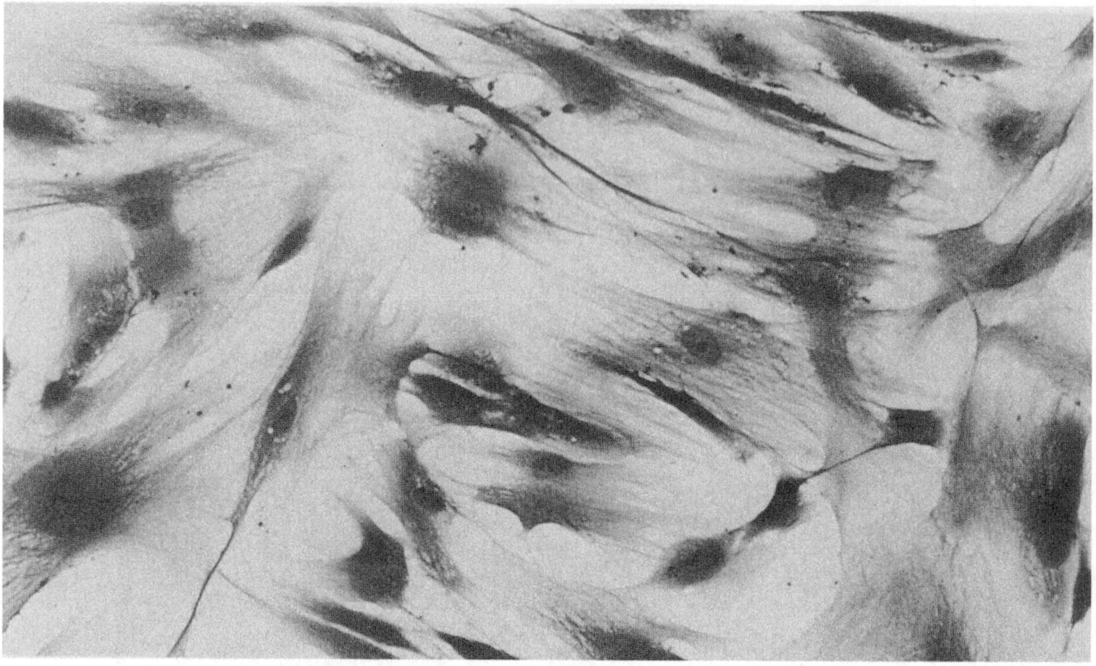

Figure 2. Midpassage human diploid fibroblasts stained with Coomassie blue to reveal cytoskeletal system. This preparation demonstrates the heterogeneity both in cell size and in the degree to which cells spread on the culture substrate. ×240, reproduced at 88%.

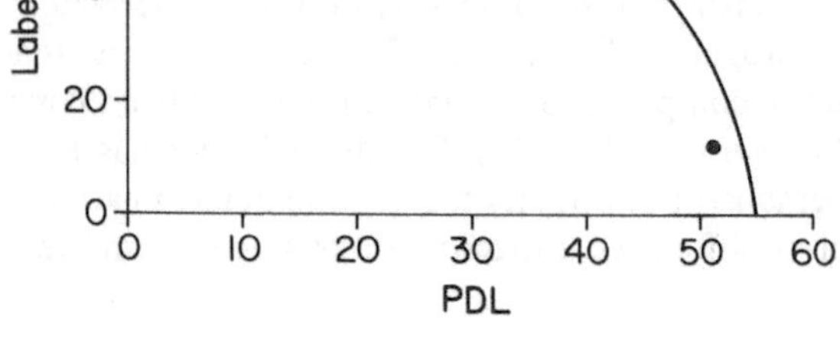

Figure 3. Decline in proliferative capacity of IMR-90 fibroblasts during serial subcultivation. Autoradiographs of cultures exposed to [^{3}H]thymidine for 24 h were used to calculate the percentage of cells with a labeled nucleus (labeling index). From Pool *et al.* (1982).

human cells were isolated and measured as to cultured lifespan. Crosses between short-lived strains, produced by polyethylene glycol-mediated cell fusion, yielded short-lived progeny whereas crosses between long-lived strains resulted in long-lived progeny. The compelling results were obtained when short-lived strains were crossed with long-lived strains to yield hybrids with *in vitro* lifespans intermediate to those of the parentals. Similar findings have been reported recently by Muggleton-Harris and Aroian (1982). These investigators fused individual skin-derived fibroblasts from a neonatal and an aged donor using inactivated Sendai virus as a fusogen. By following the replicative pattern of hybrids, the authors demonstrated that an intermediate mode of replication between that of the two parental cell lines occurred. Another group has used fractionation procedures to produce subpopulations of fibroblasts, from midpassage cultures, that are highly enriched for small proliferating cells; however, replating this enriched population failed to produce cultures with lifespans greater than those of parallel control, unfractioned cultures (Mitsui *et al.*, 1980). Other evidence in support of genetic regulation of *in vitro* lifespan will be considered in a separate section of this chapter.

The loss of division potential in human diploid fibroblasts (Phase III, Hayflick and Moorhead, 1961) with extended time *in vitro* is also seen in other cell types *in vivo* with increasing age. Bucher *et al.* (1964) demonstrated that regenerating hypatocytes in adult rats are delayed in the onset of DNA synthesis as measured by thymidine-2^{14}C incorporation when compared to regenerating hepatocytes of weanling rats. It is, however, the human diploid cell line, WI-38, that has been most carefully studied with respect to the decline in division potential during aging.

Macieira-Coelho *et al.* (1966) examined the cell cycle of WI-38 cells using autoradiographic procedures in both phase II (actively growing) and phase III cultures. It was seen that after the 40th passage, the cultures became heterogeneous with respect to the length of the division cycle. Also, the fraction of cells capable of incorporating labeled precursor into DNA following pulse treatment decreased. It was concluded that the increase in generation time was due to prolonged G_1 and G_2 periods of the cell cycle. Additionally, the rate of RNA synthesis declined in late-passage cells, yet all late-passage cells retained the ability to synthesize RNA in a pattern identical to that of phase II cells. It appears then that the decline in proliferative capacity associated with phase III cells is not due to a generalized decline in nucleic acid metabolism but is specific to DNA synthesis. Klevecz and Kapp (1973) have shown that DNA synthesis occurs in three temporally distinct bursts during S phase in WI-38 cells and suggest that cellular senescence results from a loss in temporal organization of cellular biosynthetic and degradative processes. The cell cycle time of the rapidly proliferating cells is apparently uniform whether measured at passage 25 or passage 50 in WI-38 cells (Kapp and Klevecz, 1976). The decline in the percentage of cells able to incorporate [^{3}H]-TdR into nuclear DNA in aging cultures of WI-38 cells has been correlated both with calendar time (Cristofalo and Sharf, 1973) and with the number of population doublings (Dell'Orco *et al.*, 1973; Cristofalo and Sharf, 1973). The rate of decline varies for different populations measured depending upon the overall lifespan of the culture but is seen to decline linearly regardless of the specific population measured if culture age is normalized and expressed as "percent of lifespan completed."

Where, then, are the postreplicative fibroblasts arrested with respect to the cell cycle? Yanishevsky and co-workers (1974) employed cytophotometric procedures to measure the DNA content of nondividing WI-38 cells propagated serially. The DNA content of these cells was seen to be diploid; thus, they concluded that the nonproliferating fibroblasts were arrested in G_1. Schneider and Fowlkes (1976) have similarly observed an increase in the proportion of cells in G_1 in senescent cultures of WI-38 cells by using flow cytofluorometric techniques with single cells.

3. MORPHOLOGICAL CHANGES IN HDF DURING THE IN VITRO LIFESPAN

In this section we will review the work of other investigators who have examined structural changes that occur in HDF throughout their *in vitro* lifespan. We will then conclude this section by summarizing some of our own work on morphological alterations in HDF, with particular reference to both proliferating and nonproliferating subpopulations throughout their *in vitro* lifespan.

3.1. Cell Volume Changes

A discrete increase in modal cell volume has been reported to accompany the decline in proliferative capacity in senescent human diploid cells. Simons (1967) first described a broadening of cell diameter distributions in phase III cultures of adult human skin fibroblasts when compared to phase II cultures. He also demonstrated in this study that changes in ploidy level do not alter the heterogeneity seen in cell sizes. Other investigators have also reported an increase in cell sizes in late-passage versus early-passage cultures of human fibroblasts (Cristofalo and Kritchevsky, 1969; Macieira-Coelho and Ponten, 1969; Greenberg *et al.*, 1977). Bowman *et al.* (1975) questioned the relationship between cell size and the capacity of cells to synthesize DNA in cultures of WI-38 cells and in cultures of SV40 transformed WI-38 cells. Although the mean cell volume of mitotic cells remained constant at 4300 μm;3 regardless of culture age, those cells that did not incorporate [^{3}H]-TdR and remained in monolayer attained sizes in excess of 5000 μm^3. It is of particular interest that the SV40 transformed WI-38 cells did not show this increase in cell volume; these cells are considered immortal in the sense that there is no decline in proliferative capacity with continuous subcultivation. These authors therefore concluded that the increase in volume occurring as cells enter phase III was associated with the loss in division potential. During the next year, Mitsui and Schenider (1976) contributed work demonstrating that cell populations of fetal skin fibroblasts possessing trisomy 21 show a shift to larger cell sizes when compared to rapidly proliferating control populations at the same passage level. Since fibroblasts with trisomy 21 replicate more slowly than do genetically normal fibroblasts, this observation was taken as support of the relationship between division capacity and cell size. In this same study the authors experimentally manipulated proliferative rates by (1) reducing the serum concentration from 10% to 1%, (2) decreasing the incubation temperature from 37 °C to 33 °C, and (3) inhibiting DNA synthesis with hydroxyurea in cultures of WI-38 cells. In each instance, a shift to larger cell volumes was seen when the experimental cultures were compared to controls. It therefore is becoming increasingly evident that (1) an inverse relationship exists between proliferative capacity and cell volume in cultured human diploid fibroblasts and that (2) senescent cultures of these cells contain two distinct subpopulations of cells—smaller, proliferating fibroblasts and larger postmitotic cells.

Several groups of investigators have capitalized upon the shift toward increasing cell volumes in senescent cultures to physically separate nonproliferating subpopulations from rapidly dividing ones. Bowman and Daniel (1974) and Bowman, Meek, and Daniel (1975) were able to separate cells which had not incorporated [^{3}H]-TdR in a previous continuous labeling period according to size by rate velocity sedimentation in a linear Ficoll gradient. Mitsui and Schneider (1976a; 1976b) have used separation methods at unit gravity to recover (a) cell population fractions enriched for small cell volumes and rapidly replicating cells and (b) cell population fractions enriched for both large cell volumes and slow and/or nonreplicating cells. Both groups of investigators, therefore, have further confirmed the relationship between cell size and mitotic capacity in human diploid fibroblasts. Although the separation techniques described by both laboratories are relatively inexpensive and simple, few current investigators have employed these tools in cytogerontologic investigations. From the foregoing review of the literature, the possibility emerges that any study, regardless if the approach is morphological, biochemical, or biophysical, that in design considers senescent cultures to consist of a homogeneous population of cells is, at least in part, in error. Whatever characterizations we make of late-passage cultures in comparison to early-passage cultures must account for the presence of subpopulations differing distinctly in their sizes and capacities to replicate.

3.2. Nuclear Size

In addition to experimentally demonstrating the shift toward increasing cell volumes in slow and/or nonreplicating fibroblasts, Mitsui and Schneider (1976c) have shown that nuclear size increases as proliferative potential deceases in WI-38 cells. By use of [^{3}H]-TdR autoradiography, these workers showed that (1) there is a shift to larger cell nuclei as a function of *in vitro* passage, (2) regardless of passage level, the rapidly replicating cell subpopulations have smaller nuclear sizes than comparable subpopulations of slow or nonreplicating cells, and (3) there is a marked shift in nuclear size distribution of the slow or nonreplicating subpopulation with increasing passage level while the size distributions of nuclei in rapidly replicating cells remains constant. That this phenomenon occurs in normal human fibroblasts derived from various aged donors (8, 40, and 84 years) in addition to fibroblasts of embryonic origin has been reported by Lee *et al.* (1978). In all three cell lines, the increase in mean nuclear area of cells was statistically significant during aging *in vitro*. Also, an interesting relationship between the increase in mean nuclear area during *in vitro* aging and the age of the donor was noted: the older the donor, the faster was the beginning of mean nuclear area increases with successive doublings and the larger was the mean nuclear area when measured at the 10th doubling from the end of doubling activity *in vitro*. In addition to increasing in size, nuclei of senescent cells tend to become highly lobulated (Lipetz and Cristofalo, 1972; Brandes *et al.*, 1972; Brock and Hay, 1971; Munnell and Getty, 1968), yet the significance of this is not fully understood.

3.3. Nucleoli

Bemiller and Lee (1976) have examined changes in the area, dry mass, and morphology of nucleoli in aging WI-38 cells. Their results indicate that there is a dramatic increase in the fraction of cells containing one large nucleolus per nucleus in aging cultures. Mean nucleolar dry mass and mean nucleolar area were also seen to increase significantly in sensecent cells. This appears to correlate well with the decreased synthesis of nucleolar RNA measured in aging human cells *in vitro* by Bowman, Meek, and Daniel (1976b).

3.4. Golgi Apparatus and Lysosomes

Increases in both the amount of Golgi apparati and the number of lysosomes have been reported in senescent human diploid cells (Robbins *et al.*, 1970; Lipetz and Cristofalo, 1972). The increase in the number of lysosomes and residual bodies seen in senescent fibroblasts occurs shortly after culture seeding and precedes the marked decrease seen in proliferative capacity in aging cultures. These observations have been used to support a proposed causal relationship between lysosomal enlargement and cellular senescence (Lipetz and Cristofalo, 1972).

3.5. Mitochondria

Age-related changes in the structure of mitochondria and in the extent of cytoplasmic microfilaments (Brock and Hay, 1971; Lipetz and Cristofalo, 1972; Johnson, 1979) have also been reported for fibroblasts grown in cell culture. It should be emphasized that although several of the aforementioned studies have employed some type of quantitative methods in describing morphological characteristics of sensecent cells, none have considered senescent cell cultures as containing two subpopulations of cells; the quantitation is therefore representative of more than one cell type. Only those studies described previously that have demonstrated nuclear size increases with culture age have addressed themselves to the question of whether or not the organelle belonged to a cell capable of division. Great difficulty therefore arises in trying to causally relate observed changes in organelle systems to cellular senescence when the quantitative measures so carefully made actually constitute a large mean representing two physiologically distinct cell types. It is certainly conceivable that causal relationships can be drawn if the presence of these two subpopulations in senescent cultures is considered in experimental design.

3.6. Structural Changes and the Loss of Proliferative Capacity—Mechanistic Considerations

If one is to understand cellular senescence, which is defined as the permanent loss of division potential, then one must understand what events direct cells from the cycling compartment to the postmitotic compartment. Since ultrastructural features have been used to characterize senescent cells, we have asked the following questions: (1) Do postmitotic cells from cultures of human diploid fibroblasts differ quantitatively from actively proliferating fibroblasts with respect to the area fractions of the various organelle systems (Golgi, mitochondria, lysosomes, nuclei, nucleoli, etc.)? (2) If so, are these quantitative changes postponed until crisis or do they accrue in cells slowly as more and more cells enter the postmitotic fractions? To answer these question, early-, mid-, and late-passage cultures (population doubling levels 12, 35, and 51, respectively) of IMR-90 fibroblasts were exposed to [^{3}H]-TdR for 48 h prior to fixation *in situ* for morphometric analysis of autoradiographs, both at the light microscopic and electron microscopic levels, with an image analyzer. This allowed us to compare quantitatively the area fraction of subcellular components from both the cycling (labeled) and postmitotic (unlabeled) fractions (Pool *et al.*, 1982). These experiments were based upon the simplest kinetic model that could account for fibroblast senescence *in vitro* (Fig. 4). This approach considers cultures of HDF as consisting of two subpopulations: (1) those cells that are actively traversing the cell cycle and (2) those cells that have withdrawn permanently from the cycling fraction and, thus, constitute the postmitotic or senescent fraction.

The data from this work are summarized in Tables I–III and in Fig. 5. Briefly, we demonstrated that mid- and late-passage cultures of IMR-90 fibroblasts contain an increased number of cells that are both larger and fail to incorporate [^{3}H]-TdR, but that the increase in cell size is secondary to the loss of division potential. No other significant morphological changes were seen to accompany the loss of division potential in IMR-90 fibroblasts. These findings do not mean that certain morphologic criteria cannot be used to characterize senescent cells *in vitro*. Such features as modal cell volume and lysosomal content on a per cell basis may well be characteristic features of late-passage cells. We do conclude, however, that these changes are not related mechanistically to the loss of division potential.

We have also examined other qualitative changes in HDF morphology that have been associated with senescent cells (Pool *et al.*, 1981). As mentioned earlier in this chapter, several investigators have noted an increase in nuclear lobulation associated with a loss of division potential in HDF. Basler *et al.* (1979) have noted that altered nuclear structure of senescent fibroblasts *in vitro* correlates with a decline in DNA synthetic capacity. Munnell and Getty (1968) speculated that nuclear lobulations seen in aging canine myocardial cells *in vivo* were a compensatory attempt by enlarging cells to maintain a given nucleocytoplasmic ratio. We have examined nuclear morphology in

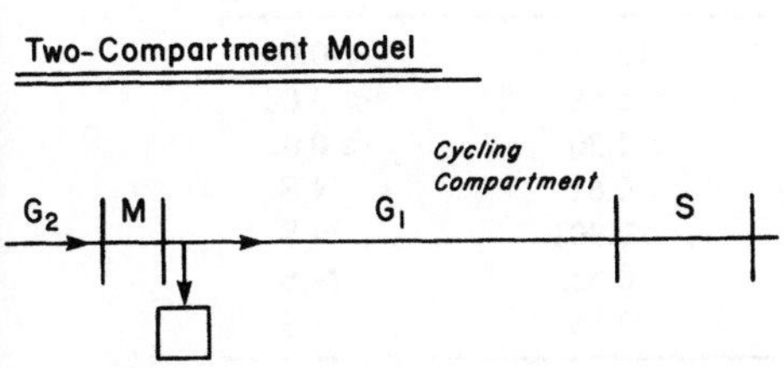

Figure 4. Theoretical consideration of kinetic activity of human fibroblasts *in vitro*. Cells leave the cycling compartment in early G_1 to become permanently postmitotic or senescent.

Table I. Quantitative Characterization of IMR-90 Fibroblasts as a Function of Proliferative Status and Population Doubling Level[a]

	PDL 12		PDL 35		PDL 51	
Parameter (μm^2)	Labeled	Unlabeled	Labeled	Unlabeled	Labeled	Unlabeled
Relative cell area	45.89 ± 1.9	50.52 ± 2.1	66.85 ± 2.5	107.62 ± 5.2	74.22 ± 2.6	113.58 ± 4.3
Nuclear area	16.99 ± 0.5	18.08 ± 0.5	22.00 ± 0.5	24.94 ± 0.9	25.59 ± 0.7	30.69 ± 1.2
Nucleolar area	1.76 ± 0.1	1.76 ± 0.1	1.96 ± 0.1	3.21 ± 0.1	2.03 ± 0.1	3.59 ± 0.3
Nucleoli/μm^2 nucleus	0.008 ± 0.005	0.091 ± 0.006	0.084 ± 0.005	0.108 ± 0.006	0.069 ± 0.004	0.082 ± 0.006
n	40	36	50	30	45	90

[a]From Pool *et al.* (1982).

Table II. Quantitative Changes in Organelle Areal Density as a Function of Proliferative Status and Population Doubling Level in IMR-90 Fibroblasts ($\bar{X}$ ± S.E.M.)[a]

	PDL 12		PDL 35		PDL 51	
Parameter (μm^2)	Labeled	Unlabeled	Labeled	Unlabeled	Labeled	Unlabeled
Mitochondria	0.027 ± 0.002	0.037 ± 0.002	0.061 ± 0.004	0.038 ± 0.002	0.051 ± 0.002	0.052 ± 0.002
Golgi	0.057 ± 0.006	0.046 ± 0.004	0.033 ± 0.005	0.045 ± 0.004	0.056 ± 0.005	0.063 ± 0.006
Lysosomes	0.021 ± 0.002	0.061 ± 0.007	0.041 ± 0.001	0.024 ± 0.001	0.016 ± 0.002	0.042 ± 0.003
n	18	16	17	15	21	21

[a]From Pool *et al.* (1982).

two strains of HDF (Pool *et al.*, 1981) both with phase contrast optics (living cells) and with electron microscopy. We do not detect nuclear lobulation in living cells and do not see a difference in nuclear morphology at the ultrastructural level between cells that can and cannot incorporate [^{3}H]-TdR if the cells are well spread and fixed *in situ* (Fig. 6,7). We do see pronounced nuclear lobulations in cells that have been rounded by enzymatic treatment or have otherwise been perturbed to alter normal cytoskeleton–membrane–culture-substrate interactions (Fig. 8).

Why have most investigators seen pronounced ultrastructural changes in senescent HDF? At least two possibilities come to mind. First, Kelley *et al.* (1980) have shown

Table III. F Statistics from One-Way Analysis of Variance of Labeled and Unlabeled IMR-90 Fibroblasts at Early-, Mid-, and Late-Passage Levels[a]

Parameter	*F* Value	Probability
Relative cell area	12.11	< 0.001
Nuclear area	6.88	< 0.001
Nucleolar area	2.30	< 0.05
Nucleolar area/μm^2 nucleus	0.62	N.S.
Mitochondrial area	2.90	N.S.
Golgi area	0.55	N.S.
Lysosomal area	3.86	N.S.

[a]From Pool *et al.* (1982).

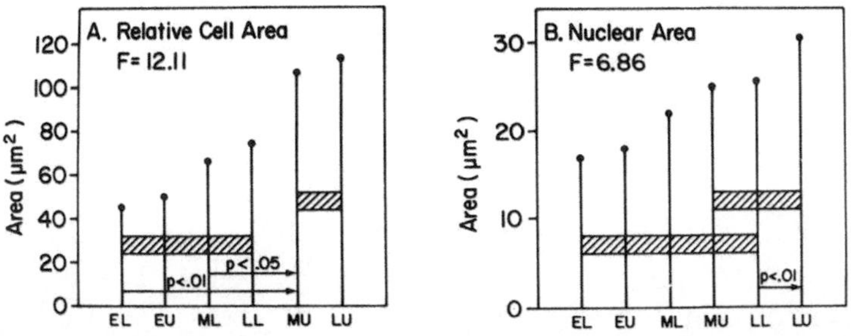

Figure 5. Summary of statistical differences from Student Newman–Keuls multiple comparison procedure. Horizontal cross-hatched bar indicates groups with mean values that are not significantly different. Arrows indicate probability levels between groups that are significantly different. For this purpose, group mean values were ranked in ascending order from left to right: E, M, L—early, mid-, late passage; U or L designates labeled or unlabeled. (A) Statistical groupings by relative cell area. (B) Statistical groupings by nuclear area. From Pool *et al.* (1982).

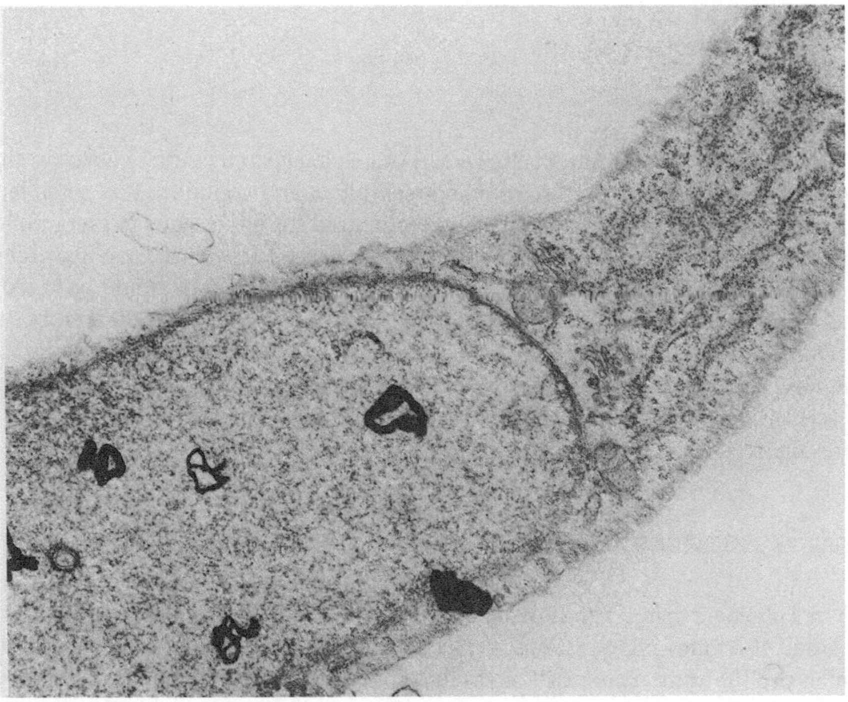

Figure 6. High-resolution autoradiogram of late-passage (PDL 51) IMR-90 fibroblast grown in the presence of [^{3}H]thymidine for 48 h prior to fixation. From Pool *et al.* (1981). ×20,800, reproduced at 95%.

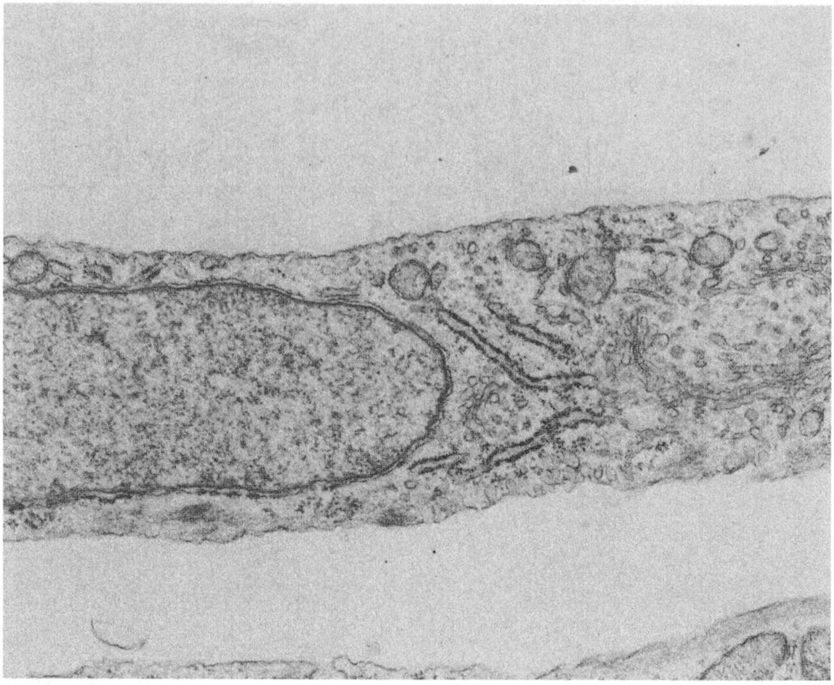

Figure 7. See Fig. 6.

that late-passage IMR-90 fibroblasts spread more slowly on a culture substrate than do early-passage cells due to a retardation in assembly of actin bundles. It is possible, then, that if both early- and late-passage cells are prepared for microscopy too soon following subcultivation, the late-passage cells will not have spread as well as early-passage cells; under these conditions, a distinct difference in nuclear morphology would be noted. Secondly, Johnson (1979) has indicated that the ultrastructure of IMR-90 fibroblasts can be altered drastically if optimal culture conditions, such as pH of the medium, are not maintained prior to preparation of the cells for electron microscopy. We have confirmed this observation and, therefore, have adopted strict protocols for the maintenance of all cultures deemed for ultrastructural analysis.

4. KINETIC COMPARTMENTS OF HYMAN DIPLOID FIBROBLASTS IN VITRO

Since morphologic criteria fail to shed light upon the mechanism of growth arrest, we have asked the following questions: (1) Is there an innate, genetically determined rate at which proliferating cells exit the cycling fraction to enter the postmitotic fraction? (2) Once a cell enters the postmitotic fraction, does it remain viable during subcultivation procedures so that it gets passed to a new flask? If these statements are true, then the declining labeling index characteristic of mass cultures at late passage may

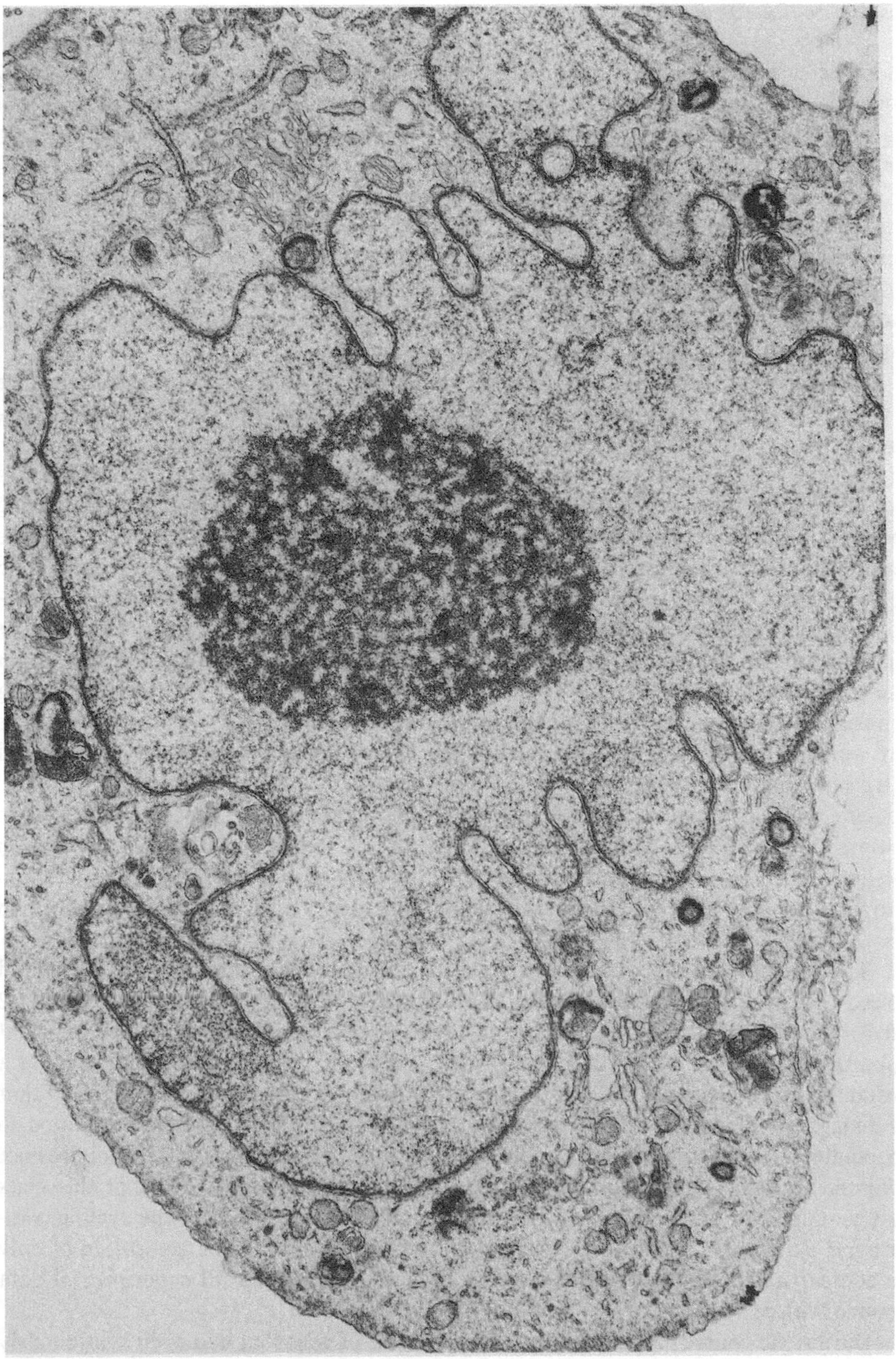

Figure 8. Incompletely spread IMR-90 fibroblast at PDL 51. Note nuclear lobulations and cortical network of microfilaments. From Pool *et al.* (1981). ×22,000.

not reflect a loss of division potential at the cellular level, but may more accurately reflect the subsequent accumulation of the postmitotic fraction over a long series of subcultivations. In other words, the fraction of cycling cells that incorporate [^{3}H]-TdR could remain stable over a long series of subcultivations whereas the unlabeled, postmitotic fraction would increase in size. The net effect would be (a) that a decline in the labeling index over the *in vitro* lifespan of a cell strain could be demonstrated without there being an actual reduction in the proliferative potential of the cycling fraction, and (b) that cultures would eventually reach a point where the majority of cells would be postmitotic, due to accumulation. This condition would be indistinguishable from the classical description of phase III or terminal senescence.

The method of Rao (1981) was used in experiments designed to measured the rate at which cells leave the cycling fraction to enter the postmitotic fraction per cell generation (Pool, 1981). Cells in exponential growth were exposed for 4 h to cytochalasin B (1 μg/ml) in order to generate binucleated cells by inhibiting cytokinesis. This was done to identify those cells that passed through division, and thus were in the cycling fraction, during the treatment period. The cytochalasin B was then removed; the cells were rinsed with two changes of saline and were refed with growth medium containing [^{3}H]-TdR (0.1 μCi/ml). The cells were incubated for 24 h prior to fixation for autoradiography; binucleated cells were then scored as to whether they contained labeled or unlabeled nuclei. Longer incubations in medium containing [^{3}H]-TdR (36 h) did not alter the results. Rao (1981) has even incubated with label for 72 h without changing his results with WI-38 cells. This experimental design allowed us to ask: Of those cells that were in the cycling fraction during the cytochalasin B treatment (binucleated cells), how many will progress to S phase of the next cell cycle? This design assumes that cytochalasin B treatment for 4 h does not affect other cellular processes that govern G_1 transition. The work of Fournier and Pardee (1975) indicates that this assumption is valid. The design also assumes that binucleated cells were generated by an inhibition of cytokinesis and not by random cell fusion. Separate control experiments confirmed this.

The results from this experiment were in agreement with those of Rao (1981) and showed that 53% of binucleates failed to incorporate label when measured at PDL 22 and 35 (Pool, 1981). That the values for each class of binucleates remained constant supports the idea that cells retire from the cycling to the postmitotic compartment at a constant rate. However, if the exit rate remains this high, then calculations reveal that not enough proliferating cells would exist at the end of a single cultivation period to repopulate a new flask. Clearly the binucleate class with both unlabeled nuclei does not represent a homogeneous class of permanently arrested cells. We feel that this class must contain some transiently growth-arrested nuclei that can reenter the cycling compartment. It appears, then, that considering cultures of HDF to be comprised of only two compartments (cycling and senescent) does not account for all experimental data obtained with HDF.

During the course of analyzing autoradiographs of cells that had been continuously exposed to [^{3}H]-TdR for 72 h, it became apparent that the number of silver grains overlying a labeled nucleus fell into one of two distinct classes: (1) lightly labeled nuclei (25–100 grains above background) and (2) heavily labeled nuclei (totally covered nucleus). The percentages of each of the classes were scored for both IMR-90 and GM 730 cells and are summarized in Table IV.

Table IV. Labeling Pattern Following Continuous Exposure to [^{3}H]-TdR (72 h)

	Total L.I.	Heavy	Light	% Light of total L.I.	Total cells
IMR-90					
A	12.4	9.0	3.4	27	1000
B	11.5	9.1	2.4	21	1000
C	14.9	11.5	3.4	23	1000
D	12.8	10.5	2.3	18	1000
GM730	54.6	48.7	5.9	11	600

Lightly labeled cells constituted from 18% to 27% of the total labeled cells using IMR-90 fibroblasts and slightly over 10% of all labeled GM 730 cells. What do these lightly labeled cells represent? A few possibilities are as follows: (1) Cells that were in the last part of S phase when label was added and progressed to M, divided and entered the postmitotic state. That as many as 27% of all labeled IMR-90 cells were lightly labeled seems to exclude this possibility. (2) Light labeling represents repair synthesis. The specific activity of the label used (1–2 Ci/mmol) and the amount used (0.1 μCi/ml) should not result in 25–100 reduced grains within the exposure time used (5 days at 4°C) simply due to repair synthesis. (3) Lightly labeled cells may be cells that have only undergone one round of synthesis (one strand labeled during semiconservative replication) versus cells that have undergone several rounds of replication (all strands radioactive). If cells only underwent one round of DNA synthesis during 3 days continuous exposure, then they have either reentered the cycle from a transient state of growth arrest or have exited the growth fraction following a single round of DNA synthesis. For the same reasons stated above in possibility 1, the former conclusion is much more probable than the latter. Results similar to these have been reported by Mitsui *et al.* (1980).

In another type of experiment, we have asked: Do permanently growth arrested cells differ in the types of proteins that they synthesize as compared to cells induced to enter transient quiescence by serum deprivation? To answer this, fetal human fibroblasts were treated with bromodeoxyuridine (BUdR) for 3 days. Hoeschst 33258 was then added to the medium and the cells were exposed to fluorescent light for 30 min. Cells were refed with regular growth medium and returned to the incubator. This procedure eliminates the cycling fraction (Burmer and Norwood, 1980) and yields high quantities of postmitotic cells (99–100% unlabeled following 24-h exposure to [^{3}H]-TdR) in our laboratory. Parallel flasks were treated with growth restriction medium (0.5% fetal calf serum) for 3 days to induce transient quiescence. [^{3}H]leucine was added to each group for 24 h prior to the isolation of soluble cell proteins. Samples were electrophoresed in one dimension on SDS-gradient gels (7%–12%) and the resulting gels were fluorographed. Although many polypeptides comigrated in the two samples, at least six bands were quantitatively different between these two states of growth arrest. This experiment, however simple, does illustrate that permanent growth arrest and transient growth arrest are distinct physiological states. Fluorographs from this experiment are shown in Fig. 9.

That human diploid fibroblasts can enter into a state of transient growth arrest has

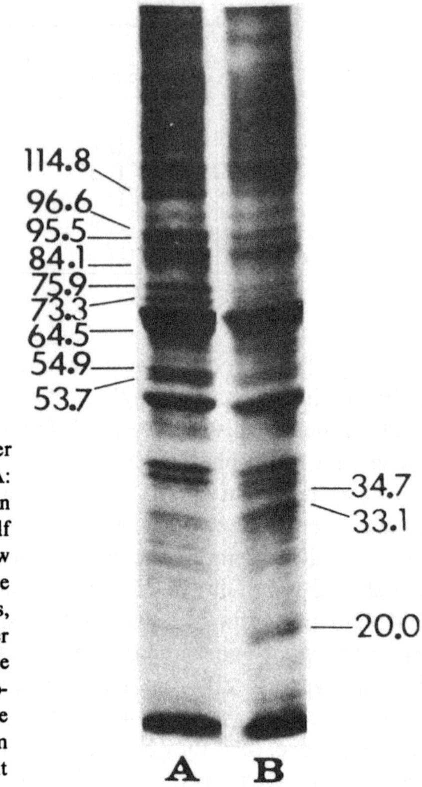

Figure 9. Fluorograph of cellular proteins synthesized over a 24-h period and separated on SDS gradient gels. Lane A: Proteins that have incorporated [^{3}H]leucine from human fibroblasts grown in medium containing 0.5% fetal calf serum. Cells that are growth restricted in this manner show quantitative increases in the radioactivity of 9 proteins. Lane B: Proteins synthesized by postmitotic human fibroblasts, produced by the BrdU/Hoeschst 33258 method of Burmer and Norwood (1980). Less than 1% of these cells are capable of incorporating [^{3}H]thymidine into nuclear DNA. Two proteins (20 kd and 34.7 kd) have incorporated more [^{3}H]leucine than those of equivalent molecular weight from serum-starved cells (lane A). One protein 33.1 kd, is present only in postmitotic cells. From Pool (1982).

been demonstrated directly by Sarber *et al.* (1981). It was clearly shown that both IMR-90 fibroblasts and foreskin fibroblasts of either low or high population doubling levels cease to divide within four days when incorporated into collagen lattices *in vitro.* Cells of both strains grow to confluence when plated on plastic surfaces. That the growth arrest induced by cultivation in collagen lattices is reversible is demonstrated directlv when cells are permitted to leave the lattice and populate a plastic substrate.

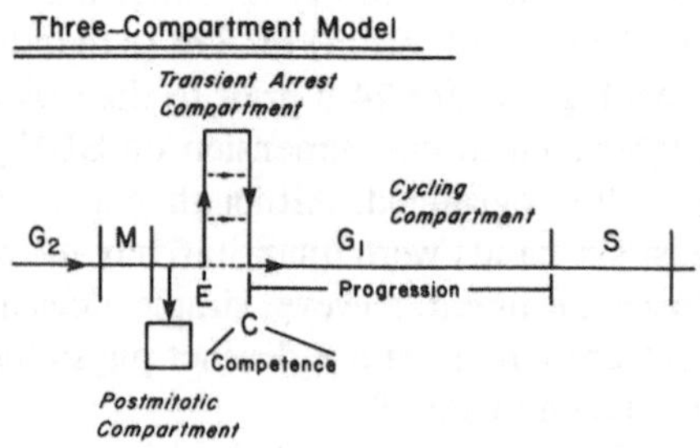

Figure 10. Current concept of kinetic compartments of human diploid fibroblasts *in vitro.* This view depicts a compartment of growth arrest in G_1 that is populated by cells capable of reentering the cycling compartment. E indicates the entry point while the small arrows within the transient compartment indicate different states of transient growth arrest. In this model, the entry points into the transiently arrested and postmitotic compartments are distinct and, presumably, regulated by different processes.

From this study alone, we must become aware that meaningful data that will lend themselves to unambiguous interpretation regarding the regulation of cellular senescence *in vitro* can only be had from experimental designs that account for the presence of at least two distinct compartments of growth arrest. The demonstration that human fibroblasts can enter into a reversibly growth arrested state *in vitro,* when placed into a three-dimensional matrix brings the cultured human fibroblast model of cellular senescence an order of magnitude closer to paralleling the behavior of these cells *in vivo.*

We are currently viewing the HDF system as consisting of at least three kinetic compartments (Fig. 10) and feel that fibroblasts are guided towards various states of growth arrest by different biochemical mechanisms.

5. GENETIC REGULATION OF THE LOSS OF PROLIFERATIVE CAPACITY

We have summarized evidence in the previous section that transiently growth arrested cells coexist with permanently arrested or senescent cells *in vitro.* In this section we will focus upon the genetic regulation of entry into the postmitotic or senescent state. Our review of cell fusion studies and those using other genetic approaches will be far from comprehensive; instead, our aim is to acquaint the reader with experimental approaches that appear to be particularly fruitful in advancing our understanding of the permanent loss of proliferative capacity in human fibroblasts.

Studies by Littlefield (1973) and by Norwood *et al.* (1974) involving cell fusion of senescent with young cells showed dominance of the senescent phenotype. Heterodikaryons formed from young cells fused to senescent cells were incapable of further division or DNA synthesis. Homodikaryons formed by fusion of senescent to senescent cells were also incapable of further division, providing evidence against the hypothesis that cellular senescence is caused by the accumulation of mutations, by a random mechanism, within genes necessary for continued mitotic cycling. If this were the case one might expect to find complementation in the senescent–senescent fusions resulting in renewed growth.

The nature of this senescent phenotypic dominance was explored further in a series of experiments by Wright and Hayflick (1975) involving the fusion of anucleate cytoplasts with whole cells. Cytoplasts were generated by centrifugation of cytochalasin B treated cells, which caused removal of the nuclei. The fusion of young cytoplasts to senescent whole cells resulted in hybrids which were incapable of further division, whereas the fusion of senescent cytoplasts to young cells resulted in hybrids which did not differ significantly from the unfused young cells in proliferative potential. These results suggest that the dominant senescent phenotype is a property of the nucleus rather than due to changes in the cytoplasm or cell membrane, since both hybrids are composed of a mixture of sensecent and young cell cytoplasms and cell membranes.

Norwood *et al.* (1975) showed that dominance of the sensecent phenotype was abolished when senescent fibroblasts were fused to HeLa cells or to SV-40 transformed human fibroblasts. The resulting heterodikaryons showed incorporation of tritiated thymidine by the senescent cell nuclei, analogous to the results of similar experiments by Harris (1966) in which fusion of three types of terminally differentiated cells to tumor cells of unlimited growth potential resulted in reactivation of DNA synthesis by the postmitotic differentiated cell nuclei. Norwood used this evidence to support his hypothesis that senescent fibroblasts are really terminally differentiated cells. He proposed that the senescent cell nuclei produce a substance which can block mitotic cycling of young fibroblasts, but which can be overridden by factors present in transformed cells.

Stein and Yanishevsky (1979) fused senescent human diploid fibroblasts to cells from two "partially transformed" cell lines. One of these cell lines was derived from a humgn glioblastoma and the other from rabbit kidney. Both were immortal in tissue culture, but displayed contact growth inhibition and anchorage dependence for growth. The heterodikaryons resulting from the fusion of senescent fibroblasts to these "partially transformed" cells showed no reactivation of DNA synthesis in the senescent nuclei. To the contrary, it was shown by fusing synchronized cells that the senescent cell could inhibit mitotic cycling of the glioblastoma or rabbit kidney cells at a point during G_1 phase. If fused during S phase, the cycling nucleus would continue to synthesize DNA. To explain the discrepant results from fusion of senescent cells to HeLa or SV40 transformed cells on the one hand, or to "partially transformed" cells on the other, Stein proposed that transformed cells could arise by at least two different mechanisms. They could arise by loss of the ability to produce a "senescence factor" which blocks mitotic cycling in G_1 phase, as with the glioblastoma and kidney cell lines. When these cell types are fused to senescent fibroblasts the senescence factor produced by the fibroblasts causes cessation of mitotic cycling by a G_1 block. HeLa cells and SV40 transformed fibroblasts, in contrast, are presumed to arise by the presence of a dominant transforming factor which promotes DNA synthesis even in the presence of senescence factor, and hence can reactivate DNA synthesis by senescent fibroblast nuclei in hybrids.

These studies indicate that the phenomenon of cellular senescence in human diploid fibroblasts appears to be mediated by changes in the nucleus. Whatever these changes are, they appear to cause changes in the cytoplasm, since the message is conveyed to the other nucleus in a fusion hybrid. Numerous biochemical studies have searched for the putative "senescence" and "transforming" factors. Recent work (Norwood, personal communication) has shown that the level of DNA-polymerase in a given cell correlates well with the ability of that cell to induce DNA synthesis in a fusion hybrid.

The nonhistone or acidic group of nuclear proteins have been shown by two-dimensional electrophoresis to be composed of several hundred different species and are tissue specific. In contrast, the histones are not tissue specific and are composed of only five different species. Whereas the histones are felt to play a role in DNA packaging, the nonhistone proteins are felt to be important in gene regulation. Comparison of these nuclear proteins in senescent versus young cells might provide additional clues as to the mechanism of cellular senescence.

Stein (1975) investigated the acidic DNA-binding proteins in young and senescent fibroblasts, assuming that this group of proteins might be important in the regulation of replication. The most striking difference was the markedly increased level of a protein, which preferentially bound to single-stranded DNA, in senescent cells as compared to stationary phase young cells. This same protein was also increased in replicating phase young cells. The significance of this protein is not known; however, the marked difference in the amount present in senescent as compared to contact-inhibited cells underscores the difference between the two growth-arrested states.

Sakagami *et al.* (1979) studied the nuclear acidic proteins of senescent fibroblasts by high-resolution two-dimensional electrophoresis. Differences in seven major nonhistone nuclear proteins were found in late-passage as compared to early-passage confluent cultures. Four protein species were present in young cells but not senescent cells, two species were present in senescent cells but not young cells, and one protein showed a shift in isoelectric point possibly representing a change in phosphorylation. Further fractionation of these proteins will be necessary to determine their intracellular location and function, and what if any bearing they may have on production of the senescent phenotype.

Another approach has been to look for differences in chromatin structure and DNA sequence arrangement between young and senescent cells. Using alkaline sucrose-gradient centrifugation of DNA, Icard and Beaupain (1980) showed that with the onset of senescence there was a shift to smaller size single-stranded DNA in both human and mouse fibroblasts. This suggests the presence of single-stranded nicks or hydrolysis sensitive sites in senescent cell DNA. With the establishment of immortal cell lines in the mouse fibroblast system, there was a return to the high-molecular-weight single-stranded DNA. When caffeine was administered to cells during growth crisis there was death of the cultures within four population doublings, whereas continued growth occurred if the caffeine was given during early passage, or after establishment of an immortal cell line in the case of mouse fibroblasts. The effects of caffeine are complex, but studies have shown that caffeine can enhance the rate of DNA synthesis and increase the number of active replicons, thus possibly leading to single strand breaks if DNA replication occurs before DNA repair processes are completed (Painter, 1980; Cleaver, 1981). These data raise the possibility that the production of the senescent phenotype involves a mechanism which includes the production of single strand breaks.

Strehler *et al.* (1979) and Strehler and Chang (1979) used a probe against ribosomal DNA and found that in human myocardium and brain tissue there was a progressive decrease in hybridizable DNA with increasing donor age. Shmookler Reis and Goldstein (1980) have used a probe to the 340 base pair repetitive sequence generated by *Eco*RI digestion of human genomic DNA, to quantitate the number of hybridizable sequences present in the DNA of human fibroblasts at early and late population doubling levels. They found a significant decrease in hybridizable sequences in the late-passage cells. These two studies suggest that the senescent cellular phenotype is associated with a decrease in repetitive sequence DNA. The implications of these findings at this time are not clear. Both ribosomal and *Eco*RI repetitive sequences have been shown to cluster at a few specific chromosomal locations (Manuelidis, 1978). This suggests that they are arranged tandemly and raises the possibility that a molecular clock involving sequential removal of tandem repeats may be involved in the generation of a reproducible finite replicative lifespan for human, diploid fibroblasts. Roberts and Axel (1982) have shown, however, that repetitive arrays, which are localized to specific chromosomal regions, may have a local polytene rather than tandem arrangement, and that extensive revision of the entire array can occur following a single division cycle. Due to the heterogeneity of the cell populations in which these studies demonstrated a reduction of hybridizable repetitive DNA with increased donor age or increased passage level, it is possible that this reduction is an all-or-none phenomenon for individual cells, rather than a progressive deletion mechanism constituting a molecular clock.

Recent work by Calabretta *et al.* (1982) and Robberson *et al.* (1982) have demonstrated that genomic rearrangement may be involved in the production of the senescent phenotype. Following the observation that Alu-family repetitive sequences are present in human circular extrachromosomal nuclear DNA, and the finding of extrachromosomal circular DNA as intermediate forms in bacterial transposition, it was proposed that this repetitive family may be involved in genomic rearrangements. To test this hypothesis a clone rich in Alu-repeat sequences was selected from a human genomic library. A probe was generated to a unique sequence fragment present between multiple Alu repeats in this clone, on the assumption that this segment may have arrived at this site by a transposition event. Using this probe it was shown that hybridizable restriction fragment length polymorphisms existed in the DNA digests of different tissue types from the same individual indicating different genomic arrangements in these tissues. Furthermore, it was shown that a unique circular DNA species containing complementary sequences appeared in late-passage human diploid fibroblasts which was not detectable in early-passage cells. These findings suggest that genomic rearrangement may be important in the production of the senescent phenotype.

6. CONCLUSIONS AND FUTURE DIRECTIONS

The human diploid fibroblast model has been studied exhaustively over the past two decades in hopes of identifying the underlying basis of aging. We have yet not found a single answer to the question of what causes aging. We are, however, beginning to appreciate the complexities of the HDF model and should be in a better position to use this system to further our understanding of the cellular processes involved with aging. It is clear that populations of HDF in culture are heterogenous with respect to size, morphology, proliferative status, and proliferative potential. We now realize that HDF *in vitro* behave more like their counterparts *in vivo* when grown in a three-dimensional matrix. Further, we know that other growth-arrested states exist in cultures of HDF that are not permanent and are, no doubt, mediated by mechanisms totally different from those that produce the senescent or terminally differentiated state. Future studies designed to identify these various transiently growth-arrested states, and subsequent investigations aimed at elaborating the regulatory processes that guide cells into and out of growth arrest, will be essential to advancing our understanding of cellular senescence.

From the data summarized in this chapter, it is apparent that the production of the senescent phenotype in HDF is under nuclear control. That this process is akin to terminal differentiation is suggested by the cell fusion studies (Martin *et al.*, 1974; 1975; Loutermann and Bayreuther, 1979), by several morphologic and biochemical studies of diploid fibroblast clones, and, most recently, by studies of nuclear extrachromosomal DNA. For instance, the translocation of genetic material has been demonstrated as the mechanism of yeast mating type switch (Klar *et al.*, 1981) and the mechanism of surface glycoprotein antigenic variation in *Trypanosoma brucei* (Vander Ploeg *et al.*, 1982). These systems represent well-characterized models of enkaryotic cell differentiation that underscore the importance of the identification of new species of extrachromosomal circular DNA in late-passae HDF (Calabretta *et al.*, 1982; Robberson *et al.*, 1982). A model of transposition has been proposed that involves single strand breaks

(Harshey and Bukhari, 1981). Perhaps this is related to the caffeine sensitivity observed in late-passage HDF (Painter, 1980; Cleaver, 1981). The study of genomic rearrangements appears to be a fertile area of future endeavor, both for unifying seemingly diverse observations made with HDF and for elucidating the mechanism of cellular differentiation and senescence.

ACKNOWLEDGMENTS

This work was supported in part by grant No. AG02212 from the National Institute on Aging to Thomas B. Pool. We wish to thank Teri O. Heitman for technical assistance during the course of this project and Mrs. Judy Kelley for help in preparing this chapter.

REFERENCES

Basler, J. W., David, J. D., and Agris, P. F., 1979, Deteriorating collagen synthesis and cell ultrastructure accompanying senscence of human normal and Werner's syndrome fibroblast cell strains, *Exp. Cell Res.* **118**:73.

Bell, E., Marek, L. F., Levinstone, D. S., Merril, C., Sher, S., Young, I. T., and Eden, M., 1978, Loss of division potential *in vitro:* Aging or differentiation? *Science* **202**:1158.

Bemiller, P. M., and Lee, L., 1976, Nucleolar changes in senescing WI-38 cells, *Mech. Ageing Dev.* **8**:417.

Bowman, P. D., and Daniel, C. W., 1974, Separation of nondividing from dividing cells in cultures of aging WI-38 cells, *In Vitro* **9**:352.

Bowman, P. D., Meek, R. L., and Daniel, C. W., 1975, Aging of human fibroblasts *in vitro.* Correlations between DNA synthetic ability and cell size, *Exp. Cell Res.* **93**:184.

Bowman, P. D., Meek, R. L., and Daniel, C. W., 1976, Decreased synthesis of nucleolar RNA in aging human cells *in vitro, Exp. Cell Res.* **101**:434.

Brandes, D., Murphy, D. G., Anton, E. B., and Barnard, S., 1972, Ultrastructural and cytochemical changes in cultured human lung cells, *J. Ultrastruc. Res.* **39**:465.

Brock, M. A., and Hay, R. J., 1971, Comparative ultrastructure of chick fibroblasts *in vitro* at early and late stages during their growth span, *J. Ultrastruct. Res.* **36**:291.

Bucher, N. L. R., Swaffield, M. N., and DiGroia, J. F., 1964, The influence of age upon the incorporation of thymidine-2C^{14} into the DNA of regenerating rat liver, *Cancer Res.* **24**:509.

Burmer, G. C., and Norwood, T. H., 1980, Selective elimination of proliferating cells in human diploid cell cultures by treatment with BrdU, 33258 Hoechst and visible light, *Mech. Ageing Dev.* **12**:151.

Calabretta, B., Robberson, D. L., Barrera-Saldana, H. A., Lambrou, T. P., and Saunders, G. F., 1982, Genomic instability in a region of human DNA enriched in Alu repeat sequences, *Nature* **296**:219.

Cleaver, J. E., 1981, Inhibition of DNA replication by hydroxyurea and caffeine in ultraviolet-irradiated human fibroblast cell line, *Mutation Res.* **82**:159.

Cristofalo, V. J., and Kritchevsky, D., 1969, Cell size and nucleic acid content in the diploid human cell line WI-38 during aging, *Med. Exp.* **19**:313.

Cristofalo, V. J., and Sharf, B. B., 1973, Cellular senescence and DNA synthesis, *Exp. Cell Res.* **76**:419.

Dell'Orco, R. T., Mertens, J. G., and Kruse, Jr., P. F., 1973, Doubling potential, calendar time, and senescence of human diploid cells in culture, *Exp. Cell Res.* **77**:356.

Fujiwara, Y., Higashikawa, T., and Tatsumi, M., 1977, A retarded rate of DNA replication and normal level of DNA repair in Werner's syndrome fibroblasts in culture, *J. Cell Physiol.* **92**:365.

Greenberg, S. B., Grove, G. L., and Cristofalo, V. J., 1977, Cell size in aging monolayer cultures, *In Vitro* **13**:297.

Goldstein, S., 1969, Lifespan of cultured cells in progeria, *Lancet* **1**:424.

Goldstein, S., Niewiarowski, S., Singal, D. P., 1975, Pathological implications of cell aging *in vitro, Fed. Proc.* **34**:56.

Harris, H., Watkins, J. F., Ford, C. E., and Schoefl, G. I., 1966, Artificial heterokaryons of animal cells from different species, *J. Cell Sci.* **1**:1.

Harshey, R. M., and Bukhari, A. I., 1981, A mechanism of DNA transposition, *Proc. Natl. Acad. Sci. USA* **78**:1090.

Hayflick, L., 1965. The limited *in vitro* lifetime of human diploid cell strains. *Exp. Cell Res.* **37**:614.

Hayflick, L., and Moorhead, P. S., 1961, The serial cultivation of human diploid cell strains, *Exp. Cell Res.* **25**:585.

Icard, C., and Beaupain, R., 1980, Spontaneous structural changes in DNA during fibroblast aging and the establishment process *in vitro, Mech. Ageing Dev.* **14**:81.

Johnson, J. E., Jr., 1979, Fine structure of IMR-90 cells in culture as examined by scanning and transmission electron microscopy, *Mech. Ageing Dev.* **10**:405.

Kapp, L. N., and Klevecz, R. R., 1976, The cell cycle of low passage and high passage human diploid fibroblasts, *Exp. Cell Res.* **101**:154.

Kelley, R. O., Trotter, J. A., Marek, L. F., Perdue, B. D., and Taylor, C. B., 1980, Variation in cytoskeletal assembly during spreading of progressively subcultivated human embryo fibroblasts (IMR-90), *Mech. Ageing. Dev.* **13**:127.

Klar, A. J. S., Strathern, J. N., Broach, J. R., and Hicks, J. B., 1981, Regulation of transcription in expressed and unexpressed mating type cassettes of yeast, *Nature* **289**:239.

Klevecz, R. R., and Kapp, L. N., 1973, Intermittent DNA synthesis and periodic expression of enzyme activity in the cell cycle of WI-38, *J. Cell Biol.* **58**:564.

Kouterman, K., and Bayreuther, K., 1979, The cellular aging of rat fibroblasts *in vitro* is a differentiation process, *Gerontology* **25**:261.

Lee, Song-Chiav, Bemiller, P. M., Bemiller, J. N., and Pappelis, A. J., 1978, Nuclear area changes in senescing human diploid fibroblasts, *Mech. Ageing Dev.* **7**:417.

Lipetz, J., and Cristofalo, V. J., 1972, Ultrastructural changes accompanying the aging of human diploid cells in culture, *J. Ultrastruct. Res.* **39**:43.

Littlefield, J. W., 1973, Attempted hybridizations with senescent human fibroblasts, *J. Cell. Physiol.* **82**:129.

Macieira-Coelho, A., 1970, The decreased growth potential *in vitro* of human fibroblasts of adult origin, in: *Aging in Cell and Tissue Culture* (E. Holeckova and V. J. Cristofalo, eds.), Plenum Press, New York.

Macieira-Coelho, A., and Ponten, J., 1969, Analogy in growth between late passage human embryonic and early passage human adult fibroblasts, *J. Cell Biol.* **43**:374.

Macieira-Coelho, A., Ponten, J., and Phillipson, L., 1966, The division cycle and RNA-synthesis in diploid human cells at different passage levels *in vitro, Exp. Cell Res.* **43**:673.

Manuelidis, Laura, 1978, Chromosomal localization of complex and simple repeated human DNAs, *Chromosoma* **66**:23.

Martin, G. M., 1979, Proliferate homeostatis and its age-related aberrations, *Mech. Ageing Dev.* **9**:385.

Martin, G. M., Sprague, C. A., and Epstein, C. J., 1970, Replicative life-span of cultivated human cells. Effects of donor's age, tissue and genotype, *Lab Invest.* **23**:86.

Martin, G. M., Sprague, C. A., Norwood, T. H., and Pendergrass, W. R., 1974, Clonal selection, attenuation and differentiation in an *in vitro* model of hyperplasia, *Am. J. Path.* **74**:137.

Martin, G. M., Sprague, C. A., Norwood, T. H., Pendergrass, W. R., Bornstein, P., Hoehn, H., and Arend, W. P., 1975, Do hyperplastoid cell lines "differentiate themselves to death"? *Adv. Exp. Med. Biol.* **53**:67.

Martin, G. M., Hoehn, H., and Bryant, E. M., 1980, Evidence against somatic mutation as a mechanism of clonal senescence, in *Aging Phenomena, Advances in Experimental Medicine and Biology, Vol. 129* (K. Oota, T. Makinodan, M. Iriki, and L. S. Baker, eds.), pp. 139–145, Plenum Press,New York.

Mitsui, Y., and Schneider, E. L., 1976a, Increased nuclear sizes in senescent human diploid fibroblast cultures, *Exp. Cell Res.* **100**:147.

Mitsui, Y., and Schneider, E. L., 1976b, Relationship between cell replication and volume in senescent human diploid fibroblasts, *Mech. Ageing Dev.* **5**:45.

Mitsui, Y., and Schneider, E. L., 1976c, Characterization of fractionated human diploid fibroblast cell populations, *Exp. Cell Res.* **103**:23.
Mitsui, Y., Matsuoka, K., Aizawa, S., and Noda, K., 1980, New approaches to characterization of aging human fibroblasts at individual cell level, "Aging Phenomena," *Adv. Exp. Med. Biol.* **129**:5–23.
Muggleton-Harris, A. L., and Aroian, M. A., 1982, Replicative potential of individual cell hybrids derived from young and old donor human skin fibroblasts, *Somat. Cell Genet.* **8**:41.
Munnell, J. F., and Getty, R., 1968, Nuclear lobulation and amitotic division associated with increasing cell size in the aging canine myocardium, *J. Gerontol.* **23**:363.
Norwood, T. H., Pendergrass, W. R., Sprague, C. A., and Martin, G. M., 1974, Dominance of the senescent phenotype in heterokaryons between replicative and post-replicative human fibroblast-like cells, *Proc. Natl. Acad. Sci. USA* **72**:2231.
Norwood, T. H., Pendergrass, W. R., and Martin, G. M., 1975, Reinitiation of DNA synthesis in senescent human fibroblasts upon fusion with cells of unlimited growth potential, *J. Cell Biol.* **64**:551.
Orgel, L. E., 1973, Ageing of clones of mammalian cells, *Nature* **243**:441.
Painter, R. B., 1980, Effect of caffeine on DNA synthesis in irradiated and unirradiated mammalian cells, *J. Mol. Biol.* **143**:289.
Pool, T. B., 1981, Rate measurements of retirement from the cell cycle by human fibroblasts in vitro, *J. Cell Biol.* **91**:11a.
Pool, T. B., 1982, A comparison of protein synthesis between serum deprived and postmitotic human diploid fibroblasts, *Cell Biol. Int. Rep.* **6**:658.
Pool, T. B., Heitman, T. O., and Buck, M. A., 1981, Changes in nuclear shape and mitochondrial structure do not accompany the loss of division potential in human fibroblasts in vitro, *Am. J.Anat.* **162**:369.
Pool, T. B., Heitman, T. O., and Buck, M. A., 1982, Quantitative morphological analysis of proliferating and nonproliferating subpopulations of IMR-90 fibroblasts during aging in vitro, *Am. J. Anat.* **164**:255.
Rao, M. V. N., 1981, Cell population heterogeneity in the inducibility of DNA synthesis in human diploid fibroblasts, *J. Cell Biol.* **89**:194.
Robberson, D. L., Calabretta, B., Somasundaran, T., Barrera-Saldana, H. A., Lambrou, T. P., Yen, N., Brock, D. L., Peterson, C. A., and Saunders, G. F., 1982, Genome rearrangements and extrachromosomal circular duplex DNAs in human cells, in: *Perspectives on Genes and the Molecular Biology of Cancer, 35th Annual Symposium on Fundamental Cancer Research,* The University of Texas M. D. Anderson Hospital and Tumor Institute at Houston, Program and Abstracts of Papers, pp. 5–6.
Robbins, E., Levine, E. M., and Eagle, H., 1970, Morphologic changes accompanying senescence of cultured human diploid cells, *J. Exp. Med.* **131**:1211.
Roberts, J. M., and Axel, R., 1982, Gene amplification and gene correction in somatic cells, *Cell* **29**:109.
Sakagami, H., Mitsui, Y., Murota, S., and Yamada, M., 1979, Two-dimensional electrophoretic analysis of nuclear acidic proteins in senescent human diploid cells, *Cell Struct. Func.* **4**:215.
Sarber, R., Hull, B., Merril, C., Saranno, T., and Bell, E., 1981, Regulation of proliferation of fibroblasts of low and high population doubling levels grown in collagen lattices, *Mech. Ageing Dev.* **17**:107.
Schneider, E. L., and Fowlkes, B. J., 1976, Measurement of DNA content and cell volume in senescent human fibroblasts utilizing flow multiparameter single cell analysis, *Exp. Cell Res.* **98**:298.
Schneider, E. L., and Mitsui, Y., 1976, The relationship between *in vitro* cellular aging and *in vivo* human age, *Proc. Natl. Acad. Sci. USA* **73**:3584.
Shmookler Reis, R. J., and Goldstein, S., 1980, Loss of reiterated DNA sequences during serial passage of human diploid fibroblasts, *Cell* **21**:739.
Simons, J. W. I. M., 1967, The use of frequency distributions of cell diameters to characterize cell populations in tissue culture, *Exp. Cell Res.* **45**:336.
Smith, J. R., Pereira-Smith, O. M., and Schenider, E. L., 1978, Colony size distributions as a measure of in vivo and in vitro aging, *Proc. Natl. Acad. Sci. USA* **75**:1353.
Stein, G. H., 1975, DNA-binding proteins in young and senescent normal human fibroblasts, *Exp. Cell Res.* **90**:237.
Stein, G. H., and Yanishevsky, R. M., 1979, Entry into S phase is inhibited in two immortal cell lines fused to senescent human diploid cells, *Exp. Cell Res.* **120**:155.
Strehler, B. L., 1975, Implications of aging research for society, *Fed. Proc.* **34**:5.

Strehler, B. L., and Chang, M., 1979, Loss of hybridizable ribosomal DNA from human post-mitotic tissues during ageing: II. Age dependent loss in human cerebral cortex-hippocampal and somatosensory cortex comparison, *Mech. Ageing Dev.* **11:**379.

Strehler, B. L., Chang, M., and Johnson, L. K., 1979, Loss of hybridizable ribosomal DNA from human post-mitotic tissues during ageing: I. Age dependent loss in human myocardium, *Mech. Ageing Dev.* **11:**371.

Swim, E. H., and Parker, R. F., 1957, Culture characteristics of human fibroblasts propagated serially, *Am. J. Hyg.* **66:**235.

Timiras, P. S., 1975, Aging of homeostatic control systems, *Fed. Proc.* **34:**81.

Van der Ploeg, L. H. T., Bernards, A., Rijsewijk, A. M., and Borst, P., 1982, Characterization of the DNA duplication-transposition that controls the expression of two genes for variant surface glycoproteins in Trypanosoma brucei, *Nucleic Acids Res.* **10:**593.

Wright, W. E., and Hayflick, L., 1975, Nuclear control of cellular aging demonstrated by hybridization of anucleated and whole cultured normal human fibroblasts, *Exp. Cell Res.* **96:**113.

Yanishevsky, R., Mendelsohn, M. L., Mayall, B. H., and Cristofalo, V. J., 1974, Proliferative capacity and DNA content of aging human diploid cells in culture: A cytophotometric and autoradiographic analysis, *J. Cell. Physiol.* **84:**615.

4

Aging at the Neuromuscular Junction

DEAN O. SMITH and JULIE L. ROSENHEIMER

1. INTRODUCTION

Aging is manifest in many ways. Some are subtle while others are quite pronounced. In the nervous system, aging is particularly evident, for there are distinct behavior changes. Impaired motor coordination and reduced physical strength are among the most apparent and widespread attributes of senescence. For example, the rates and the magnitudes of various reflex responses, including simple movements of a finger or a foot, generally decrease. This is associated with a progressive decline in muscle strength after about 25 years of age (Åstrand and Rodahl, 1977). Furthermore, the rate of decrease in strength is greater for the leg and trunk muscles than for the arm muscles.

These changes in behavior presumably reflect changes in the structure and the function of the nervous system at the cellular and the subcellular level. The underlying mechanisms are being studied in both the central and the peripheral nervous systems. Although many significant changes probably occur within the rich neurochemical milieu of the brain, aging has been investigated most profitably at the neuromuscular junction. At this peripheral synapse, it has been possible to examine age-related changes in a uniform population of cells from a physiological, biochemical, and morphological point of view. Salient findings obtained from each of these different approaches will be presented. Because they have been studied most extensively, particular attention will on age-related architectural changes.

2. PHYSIOLOGICAL CHANGES

Physiological changes in both the pre- and the postsynaptic components of the neuromuscular junction during aging have been the subject of several investigations. Exam-

DEAN O. SMITH and JULIE L. ROSENHEIMER • Department of Physiology, University of Wisconsin, Madison, Wisconsin 53706.

Table I. Age-Related Physiological Changes at the Diaphragm Neuromuscular Junction of Rats

Property	Age (months) 10	28
Resting membrane potential (mV)	−81	−77
Miniature end-plate potentials (number/s)	1.2	2.0
End-plate potentials		
Mg^{2+}-block	No significant difference	
Curare-block		Greater synaptic depression
Mean quantal content	75	116
K^+-diffusion from junction (cm^2/s)	2.2×10^{-5}	1.2×10^{-5}

ples of differences found in the diaphragm are summarized in Table I, which is derived from data presented by Smith (1979, 1982a). There is often considerable variability, however, between results obtained from this and other preparations which warrant more detailed presentation.

2.1. Membrane Properties

Basic membrane properties change during senescence. Using intracellular recording techniques, several investigators (Gutmann *et al.*, 1971; Vyskočil, 1974; Frolkis *et al.*, 1976; Smith, 1979) have measured the resting potential in muscle fibers obtained from young adult and senescent rats. As in diaphragm, the resting potential generally declined by 3 to 4 mV with age.

Ionic changes could readily explain the depolarizing shift of the membrane potential. This could involve increased permeability to ions involved in establishment of membrane potential. For example, intracellular Na^+ and K^+ content have been reported to increase and to decrease, respectively, in aged animals (Frolkis *et al.*, 1976). However, it is questionable whether such small shifts seriously affect postsynaptic neuromuscular function in large-diameter muscle fibers. In small cells, which have a larger surface-to-volume ratio, permeability changes might attain greater significance.

In related studies, membrane resistance and capacitance were not found to change significantly in experiments using Fischer 344 rats (Smith, 1979). In contrast, a slight decrease in membrane resistance was reported in old albino rats by Frolkis *et al.* (1976). This latter result is stated in their text, although the data appear to show a slight increase in membrane resistance.

2.2. Action Potentials

Pronounced changes in action potential shape and time course have been reported, although inconsistencies mar the data. The threshold depolarization from the resting potential for spike initiation in rat gastrocnemius-gracilis increased with age (Frolkis *et al.*, 1976) but did not change with age in diaphragm fibers (Vyskočil, 1974). Also, the duration of the action potential was found to increase during senescence in gastrocnemius-gracilis fibers (Frolkis *et al.*, 1976), to decrease in diaphragm fibers (Vyskočil, 1974), and to remain unchanged in the levator ani muscle (Gutmann *et al.*, 1971). Frolkis *et al.* (1976) also cite increased relative and absolute refractory periods and prolonged after-depolarizations. In two studies (Vyskočil, 1974; Frolkis *et al.*, 1976) a significant decrease was observed in the amplitude of the action potentials in diaphragm and gastrocnemius-gracilis fibers; however, in a third study, no change was reported in the levator ani muscle (Gutmann *et al.*, 1971). Thus, variations among preparations preclude any generalizations.

Action potential conduction velocity apparently becomes slower in the axons of senescent nervous systems. This has been shown in peripheral nerves of both humans (Norris *et al.*, 1952; Campbell, McComas, and Petito, 1973) and rats (Birren and Wall, 1956; Retzlaff and Fontaine, 1965). Lower conduction velocities have also been measured in senescent skeletal muscle fibers (Vyskočil, 1974). Interpretation of these results is complicated by the observation that prior exercise can increase nerve conduction velocity (Retzlaff and Fontaine, 1965). It is conceivable that at least part of the age-associated decline in propagation speed may be associated with reduced activity.

Action potential propagation failure occurs more readily in old than in young animals. That is, the frequency of repetitive stimulation at which action potential conduction first blocks is consistently lower in aged rats (Smith, 1979). This is illustrated in Fig. 1. Two consequences of presynaptic conduction block have been reported which are consistent with these data. Namely, experiments have demonstrated an onset of "Wedensky inhibition" at abnormally low frequencies of nerve stimulation in aged rats (Frolkis *et al.*, 1976). Also, Campbell *et al.* (1973) recorded a reduction of motor unit action potential amplitude and twitch tension with no change in the number of nerve fibers in old humans; this suggests that the α-motoneuron action potential may not have invaded every branch, resulting in a functional reduction of motor unit size.

Enhanced conduction failure could be explained by age-related differences in the extent of K^+ accumulation near the axon during repetitive nerve stimulation. Increasing

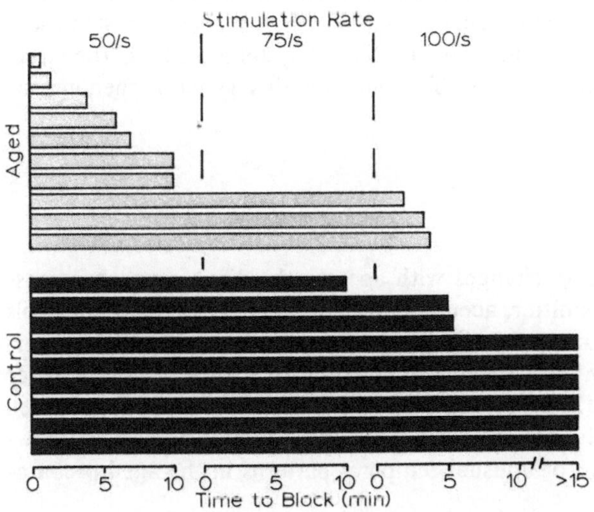

Figure 1. Nerve stimulation parameters required to produce the first occurrence of intermittent conduction block in aged and control rats. The occurrence of conduction block was monitored in two ways: end-plate potentials (EPPs) were recorded intracellularly from a single muscle fiber and extracellularly from several muscle fibers in the region of the intracellular electrode. Intermittent failures of EPPs in either record were taken to indicate presynaptic conduction block. Stimulation was delivered first at 50 Hz; if failure did not occur, it was increased to 75 Hz and then to 100 Hz. (From Smith, 1979.)

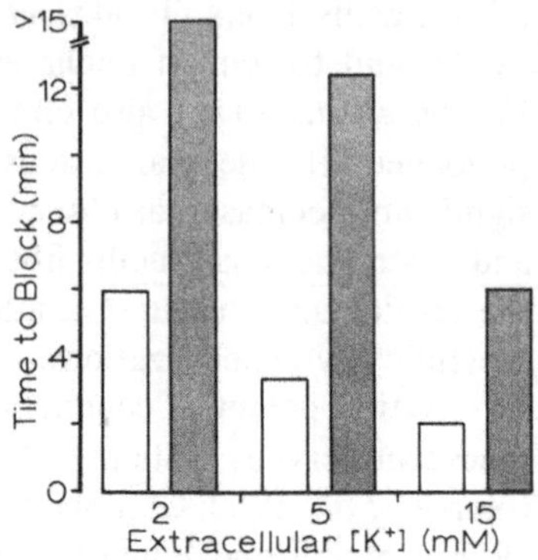

Figure 2. Length of time during which the nerve must be stimulated repetitively to produce conduction block as a function of the concentration of K^+ in the extracellular bath solution. Stimulation was at 100 Hz. Initial block was observed in 5 mM [K^+]. The bath was then flushed with 2 mM [K^+]; after 20 min, the duration of stimulation required for conduction block was measured again. This was repeated with 15 mM [K^+]. The electrodes with which end-plate potentials were recorded were not moved during the change of solutions. The clear and the solid bars represent data obtained from the aged and the control rats, respectively. (From Smith, 1979.)

the concentration of K^+ in the physiological saline bath reduced the length of time during which stimulation of the nerve at a given rate was required to attain conduction block. Conversely, if the concentration of extracellular K^+ was decreased, the time to block increased at a given rate of stimulation. This is illustrated using examples from both age groups in Fig. 2.

The role of K^+ was explored more specifically by measuring the activity of extracellular K^+ near the presynaptic axon terminal during repetitive stimulation using K^+-selective microelectrodes (Smith, 1982a). Accumulation of K^+ was observed in young and aged rats. The kinetics were considerably faster than those for conduction failure in both age groups, indicating that some other mechanism probably underlies the blocking phenomenon.

There was, however, significantly greater K^+ buildup at the neuromuscular junction of the aged than the young rats. This was attributed to an age-related drop in the diffusion coefficient for K^+ (Smith, 1982a). Values of the diffusion coefficient were estimated to be 2.2×10^{-5} and 1.2×10^{-5} cm^2/s for rats aged 10 and 28 months, respectively. This restricted diffusion of K^+ may be associated with more general age-related changes in diffusion through extracellular space. This could affect not only small ions in the nervous system but also other molecules in different systems, such as the circulatory system. Further studies might profitably focus on this general phenomenon because of its practical relevance.

2.3. Synaptic Transmission

Spontaneous transmitter release changes with age. At the diaphragm neuromuscular junction, leakage of the transmitter, acetylcholine (ACh), from cytoplasmic pools is about 4 times greater in old (28 mos) rats (Smith, 1983). Moreover, the rate of occurrence of miniature end-plate potentials (MEPPs) in aged rats is about 1.75 times faster than in young (10 mos) animals (Smith, 1979). The distribution of intervals between the times of occurrence of successive MEPPs was exponential for both age groups, however, and there were no indications of unusual temporal patterns in the aged preparations. Conversely, in the levator ani muscle (Gutmann *et al.*, 1971; Vyskočil and Gutmann, 1972), the MEPP rate in rats aged 30 months was considerably slower than in rats aged 2 months. However, it should be recognized that there are also differences in the MEPP rates of 2- and 10-month animals (Kelly, 1978). Consequently, the results of these different studies cannot be compared critically because the age groups do not correspond.

End-plate potentials (EPPs) in response to stimulation of the phrenic nerve have been analyzed in the diaphragm muscle of aged (28 months) rats (Smith, 1979). Muscle action potentials and the resulting contractions were precluded by adding either curare (*d*-tubocurarine) or elevating extracellular Mg^{2+} levels. In general, the amount of either curare or Mg^{2+} required to block contractions within 20 min was less for the old rats than for young (10 months) animals. These differences in sensitivity must be taken into account in any study of synaptic transmission which utilizes exogenously applied drugs, such as curare, or alterations in ion concentrations.

The average number of quanta released per action potential, and the associated EPPs, are greater at the diaphragm of aged rats (Smith, 1983). However, the time course of the EPPs in the aged and the control animals did not differ significantly. Furthermore, in the Mg^{2+}-blocked preparations there were no significant age-related differences in the extent of synaptic facilitation and in posttetanic potentiation.

In curare-blocked preparations, however, synaptic depression was more pronounced in the aged animals (Smith, 1979). This is illustrated in Fig. 3. This age-related difference has been attributed to more rapid depletion of the releasable store of transmitter in the older rats (Smith, 1983). Similarly, more extensive synaptic depression has been reported in the hippocampus of aged rats (Landfield *et al.*, 1978), suggesting that this may be a general phenomenon during senescence.

Enhanced conduction failure and synaptic depression during aging could result in trophic deficiencies which are associated with changes in pre- or postsynaptic structures. Direct evidence of these associations is not available. However, biochemical and morphologic studies provide indirect indications of such interactions.

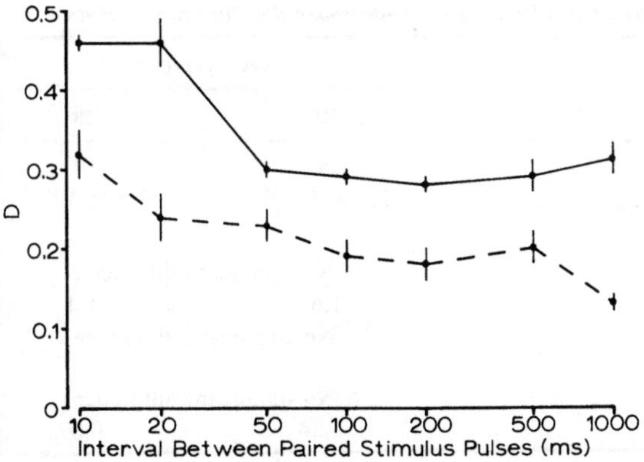

Figure 3. Enhanced synaptic depression during senescence. Synaptic depression, *D*, is calculated as the fractional decrease in the amplitude of successive end-plate potentials. The solid and the dashed lines represent data obtained from aged and control rats, respectively. The averages (±s.d.) of nine values obtained from five different animals are shown for each age group. (From Smith, 1979.)

3. BIOCHEMICAL CORRELATES

Biochemical studies at the neuromuscular junction have focused primarily on ACh. Results obtained from the rat diaphragm are summarized in Table II. In general, these results complement the physiological data.

3.1. Acetylcholine (ACh)

The amount of ACh in the presynaptic terminals of the rat phrenic nerve has been assayed in nonstimulated and stimulated tissue (Smith, 1983). In both cases, the amount of ACh was less in the aged animals. In nonstimulated tissue, ACh levels were 29 and 17 fmol/end plate in the young (10 months) and the old (28 months) preparations, respectively. The corresponding values following stimulation (1000 impulses at 3 Hz) were 27 and 13 fmol/end plate. Qualitatively similar results have been observed at the iris of aged chickens (Marchi *et al.*, 1980).

In the presence of hemicholinium (HC-3), which blocks uptake of choline (a substrate in ACh synthesis), ACh levels in nonstimulated and stimulated tissue from young rats dropped to 18 and 15 fmol/end plate, respectively. In contrast, there were no significant changes in the older animals. Thus, sensitivity to HC-3 is markedly less in the aged preparations.

During stimulation, greater amounts of ACh were released into the bath by the 28-month animals. As shown in Table II, 22% more ACh per end plate was released by the older rats (Smith, 1983). Judging from these results, the fraction of the resting ACh content which is released during nerve stimulation increased with age. During stimulation (1000 impulses delivered at 3 Hz), 10.92% and 22.76% of the ACh assayed in the nonstimulated tissue was released in the 10- and the 28-month animals, respectively. These values represent an increase of 108%.

Table II. Age-Related Biochemical Changes at the Diaphragm Neuromuscular Junction of Rats

Property	Age (months)	
	10	28
ACh content (fmol/end plate)	29	17
ACh released per action potential (mol/end plate)	3.2×10^{-18}	3.9×10^{-18}
Choline uptake by phrenic nerve terminal		
K_m	No significant difference	
V_{max} (nmol/4 min/end plate)	1.6	1.4
ChAT activity	No significant difference	
AChE activity		
Innervated	No significant difference	
Noninnervated V_{max} (nmol/min/mg protein)	0.04	0.09

3.2. Choline Acetyltransferase (ChAT)

Reduced amounts of ACh in the older animals may be due to lower activity of the enzyme choline acetyltransferase (ChAT) which catalyzes synthesis of ACh from the substrates choline and acetyl-CoA. In the phrenic nerve, there are no significant age-related changes in ChAT activity (Tuček and Gutmann, 1973). However, in the peroneal and the sciatic nerves, ChAT activity decreases during senescence (Tuček and Gutmann, 1973; Frolkis *et al.*, 1976). It has been suggested that maintenance of ChAT activity in phrenic nerve terminals is due to the continuous activity of the diaphragm muscle and its associated motor nerves (Tuček and Gutmann, 1973).

3.3. Choline

Although ChAT activity is not reduced in the phrenic nerve, availability of the substrate choline appears to decline with age. Specifically, choline uptake by phrenic nerve terminals has been assayed in rats aged 10 and 28 months. In both age groups, K_m values of the high-affinity uptake system was about 1.5 μM (Table II). However, the values of V_{max} were 1.6 and 1.4 nmol/4 min/fiber for the young and the aged rats, respectively. This age-related difference probably does not represent a limiting step in ACh synthesis, for these values of V_{max} are considerably greater than choline levels utilized in ACh synthesis. At most, these results may indicate that there are fewer transport sites in the membrane of the aged animals, but the effectiveness of each site is not necessarily changed during aging. In general, the rate-limiting step in ACh synthesis in nerve terminals of aged animals is not known.

3.4. Acetylcholinesterase (AChE)

In some preparations, there may be age-related changes in the rate of hydrolysis of ACh by acetylcholinesterease (AChE). Frolkis *et al.* (1976) report a significant decrease in old age of true and false cholinesterase activity. However, they do not document which muscle was studied or which muscle region was assayed. In the diaphragm, both junctional and extrajunctional AChE were assayed (Smith, 1983); myotendonous components were excluded. The activity of AChE was not significantly different in the innervated tissue obtained from 10- and 28-month-old rats. However, the activity in noninnervated muscle was 2.4 times greater in the older rats (Smith, 1983; cf. Smith, 1982b). This is associated with an age-related decrease in the ratio of AChE in innervated/noninnervated tissue. Different results may be expected in muscles, such as soleus, for example, which may be used less intensively in older individuals.

Collectively, these physiological and biochemical data demonstrate age-related changes in the communication between pre- and postsynaptic elements at the neuromuscular junction. Using morphological techniques at the light-microscope level, correlated structural changes have been studied.

4. MORPHOLOGY

Morphological techniques have been used to investigate numerous aspects of sensecent neuromuscular junctions. These include descriptive studies of changes in muscle fiber properties and end-plate characteristics using the light and the electron microscope. They have also provided information concerning the mechanisms of synapse maintenance and receptor distribution in aged preparations.

4.1. Histology

4.1.1. Muscle Fibers

Numerous age-related changes in muscle morphology have been documented (e.g. Gutmann and Hanzlíková, 1972; Shafiq *et al.,* 1981). Thus, only the most salient features will be noted, and special attention will focus on aspects related to neuromuscular function.

The most evident change is muscle atrophy. This involves primarily decreased muscle mass, which has been associated with both fewer and smaller fibers (Rowe, 1969; Gutmann and Hanzlíková, 1972; Tomonaga, 1977). However, the extent of size reduction depends on muscle type. In five different muscles of aged mice, Rowe (1969) observed a general tendency towards reduced fiber number and increased fiber diameter; only the sternomastoid of male mice was considered to exhibit "atrophy in terms of mean fiber diameter." Indeed, in our laboratory, we have observed a marked reduction in the mass of extensor digitorum longus (EDL) muscles in rats aged from 10 to 31 months. In contrast, we have not seen any meaningful differences between diaphragm muscles of 10- and 28-month-old rats. The density of fibers in the muscle did not change with age (Smith, 1982b). Muscle diameter was about 6% smaller in the aged rats (Smith and Rosenheimer, 1982), although this probably does not indicate any meaningful functional changes. An attractive conclusion is that continuous activity may retard age-related atrophy.

Muscle atrophy present in senile preparations ("senile muscle atrophy") may be quite distinct from atrophy due to denervation. Some changes, such as fewer and smaller muscle fibers, disorganized myofibrils, and increased proliferation of sarcoplasmic reticulum and t-tubules (Gutmann and Hanzlíková, 1972; Tomonaga, 1977) are common to senile, denervation, and disuse atrophy. However, fragmentation of muscle fibers followed by degeneration of the fragments, which is indicative of denervation atrophy, is seldom observed in aged animals (Gutmann and Hanzlíková, 1972; cf. Fahim and Robbins, 1982). Thus, senile muscle atrophy is probably a distinct phenomenon which does not necessarily follow complete disruption of innervation, although it most probably reflects some age-related disturbance of nerve-muscle trophic interaction.

During aging there is also a regression towards more uniform muscle fiber types. For example, the fractions of EDL muscle fibers which are of the type IIB (fast glycolytic) are 73% and 42% in rats aged 3 and 30 months. Conversely, the fractions of soleus muscles which are of the type I (slow oxidative) are 73% and 41% in 3- and 30-month animals (Caccia *et al.,* 1979). Currently, it is unknown how these changes are associated with altered synaptic physiology (cf. Caccia *et al.,* 1979), primarily because these synapses have not been well-studied in senescent animals.

4.1.2. End Plate

To examine the neuromuscular junction with the light microscope, it is first necessary to stain the muscle end plate. A common procedure is to use a cholinesterase stain to define the end-plate region. Using an acetylthiocholine staining procedure, Gutmann and Hanzlíková (1965) studied end-plate morphology of soleus muscle and report an irregular staining pattern in the older animals. However, they do not comment on the consistency of this observation or whether it is diagnostic of an aged preparation.

In our laboratory, we use a bromoindoxylacetate stain to visualize the extent of the postjunctional region (Smith and Rosenheimer, 1982; cf. Pestronk and Drachman, 1978). The cholinesterase-stained end plates are readily distinguishable. Representative examples are shown in Fig. 4a. With this method, diaphragm, soleus, and EDL muscle end plates obtained from 10-month rats are virtually indistinguishable from those obtained from 28-month rats when viewed under the light microscope. A similar conclusion was drawn by Courtney and Steinbach (1981), who viewed rat sternomastoid end plates labeled with a fluorescent derivative of α-bungarotoxin. This labels ACh receptors, however, and not cholinesterase.

The number of end plates per muscle fiber does not change appreciably with age. In diaphragm muscles from rats aged 10 and 28 months, the percentage of fibers exhibiting two end plates was 1.7% and 1.6%, respectively (Smith and Rosenheimer, 1982). More than two end plates were never observed on a single muscle fiber. Also, in soleus and EDL, preliminary studies indicate that there are no age-related differences in the number of end plates. Similar results have been reported for the rat sternomastoid muscle (Courtney and Steinbach, 1981).

End-plate area has been measured using electronic planimetry, and the size of EDL and soleus end plates has been reported to increase significantly with age (Rosenheimer and Smith, 1983). This could represent spreading of junctional sensitivity to ACh. In contrast, there were no age-related changes in diaphragm end-plate area (Smith and Rosenheimer, 1982). End-plate length in aged rats has been found to decrease in soleus (Pestronk *et al.*, 1980) and to remain unchanged in sternomastoid (Courtney and Steinbach, 1981). Judging from the cholinesterase-stained tissue, the shapes of end plates can be irregular, however, so area measurements may provide more meaningful data from which to draw conclusions.

Nerve terminals may also be visualized in the end-plate region. In our laboratory, we have used silver-gold impregnation of the terminals in addition to the cholinesterase stain of the end plate (Smith and Rosenheimer, 1982; cf. Pestronk and Drachman, 1978). This allows the darkly stained axons to be seen clearly against the light-blue cholinesterase-stained end plate, as shown in Fig. 4b. The architecture of the terminal arborization can then be reconstructed by frequently adjusting the plane of focus and tracing the axon using a drawing tube. A sample tracing obtained with this technique is shown in Fig. 5.

Quantitative measurements have been made from these tracings. Results obtained from the diaphragm neuromuscular junction, initially reported in Smith and Rosenheimer (1982), are presented in Table III. In the diaphragm, the average number ($\pm$S.E.) of terminals per end plate was reported (Smith and Rosenheimer, 1982) to be 13 $\pm$ 1 and 18 $\pm$ 1 in 10- and 28-month rats, respectively (Table III). This is associated with

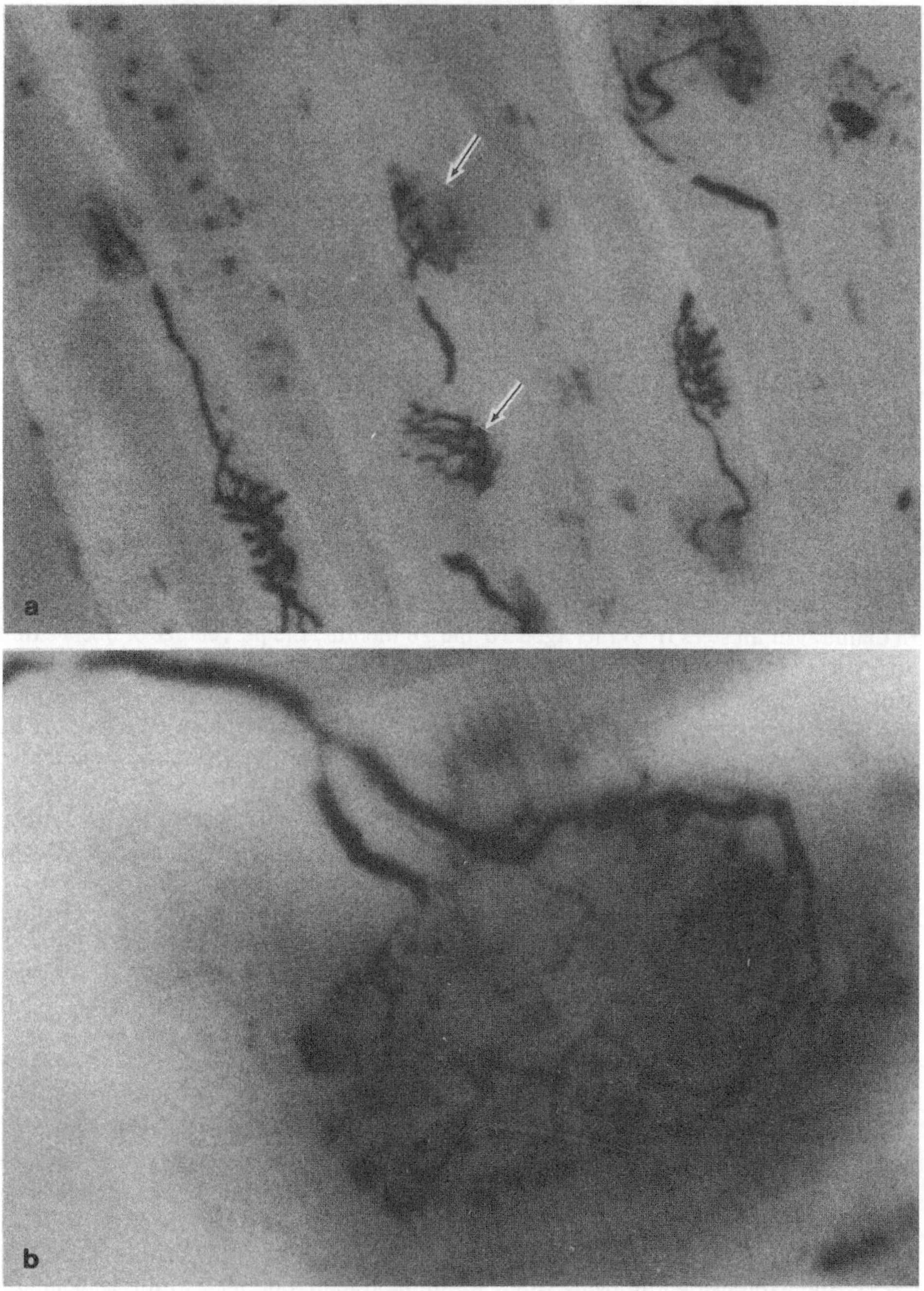

Figure 4. Nerve terminals in the end plates of the rat phrenic nerve-diaphragm muscle. The axon terminals, which appear black due to the silver-gold impregnation, stand out against the underlying end-plate region, which is lighter due to the blue cholinesterase stain. Each illustration is from a 10-month rat. In (a), one end plate appears on each muscle fiber; an example is indicated by the large arrow. Fragments of the main phrenic nerve, which projects across the fibers, are also seen (small arrow). The magnification is 325×. In (b), extensive nerve terminal arborization is shown. Many terminal branches end in small boutons; the magnification is 2050×, reproduced at 90% (from Smith and Rosenheimer, 1982).

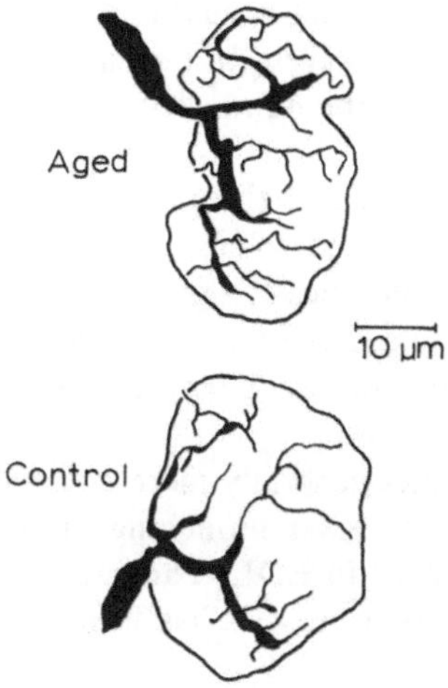

Figure 5. Nerve-terminal arborization in the end plate of the phrenic nerve-diaphragm muscle of the rat. Using a drawing-tube attachment, the perimeter of the cholinesterase-stained region and the axon branches were traced. (From Smith and Rosenheimer, 1982.)

a 52% greater total length of terminal branches per end plate in the aged animals. These differences were found to be statistically significant at the 0.005 level. However, the average length of the terminal branch did not differ by an amount that is significant statistically, indicating that the difference in total length of terminal ramifications is due to the existence of more individual arborizations.

These results in the diaphragm raise an interesting paradox. In the older animals there are more phrenic nerve terminals per end plate, but there is less ACh per end plate. There are two alternative explanations. Either some terminals are nearly devoid of transmitter ("ghosts") or each terminal may have somewhat less ACh. Preliminary studies indicate that the latter explanation is correct (Smith, 1983).

Pronounced age-related differences in the number of nerve terminals within the end-plate region of other muscles have also been reported. The nature of these differences depends upon the specific muscle type. In a related study (Rosenheimer and Smith, 1983), an apparent downward trend in the average number of terminals per end plate in rat EDL, a fast-twitch muscle used in running was observed. In contrast, there was a slight trend toward increased terminal arborization in soleus, a slow-twitch muscle involved in postural maintenance. Pestronk *et al.* (1980) report a decrease in the

Table III. Age-Related Histological Changes at the Diaphragm Neuromuscular Junction of Rats

Property	Age (months) 10	28
End-plate area	No significant difference	
Nerve terminals per end plate	13	18
Degenerating terminals (% of end plates studied)	43%	23%
Sprouting (% of end plates studied)	41%	10%

number of axon branch points within the end-plate region of soleus muscle obtained from 10- and 28-month rats. This discrepancy is currently unresolved. In the cat peroneus digiti quinti (PDQ), there is also an apparent decrease in the number of terminal endings per end plate (Tuffery, 1971).

The number of myelinated motor axon branches projecting into the end plate has also been counted in these preparations. In diaphragm, EDL, and soleus the number increases with age, although the magnitude of the change is statistically significant only in EDL. Likewise, in the cat soleus and PDQ (Tuffery, 1971) and the rat sternomastoid (Courtney and Steinbach, 1981) the number of myelinated axon branches innervating an end plate increased with age.

These data indicate that presynaptic innervation becomes generally more complicated during senescence. This results from increased motor axon branching either within the terminals, as in diaphragm, or more proximally, as in EDL. The cause of such diverse innervation patterns between functionally different muscles has been further examined in related studies of end-plate remodeling.

4.1.3. Degeneration and Sprouting

At the neuromuscular junction, motor nerve terminals undergo a process of renewed growth, or sprouting, in response to interruption of normal innervation of the muscle. The terminal region of the axon will sprout new branches when the axons of other neurons innervating the same muscle are cut (Brown and Ironton, 1978) or when conduction of action potentials supplying the same muscle are cut (Brown and Ironton, 1978) or when conduction of action potentials supplying the same muscle is blocked with tetrodotoxin (Brown and Ironton, 1977). Similar results are also observed when transmitter release is suppressed by administration of botulinum toxin (Duchen, 1970) and when postsynaptic acetylcholine receptors are blocked by curare (Wernig *et al.*, 1980).

Motor nerve terminals may also undergo a process of growth and degeneration in the absence of disrupted innervation, constantly remodeling the structure of the end plate (Barker and Ip, 1965; Tuffery, 1971; Wernig *et al.*, 1980; cf. Cardasis and Padykula, 1981). Although sprouting may be in response to degeneration of existing terminals (Barker and Ip, 1965), the two events may also proceed independently (Tuffery, 1971).

During aging, however, terminal growth and degeneration may become unbalanced, resulting in significant changes in the end-plate architecture. Such an imbalance has been the subject of several investigations (Tuffery, 1971; Smith and Rosenheimer, 1982) which will be considered.

When viewed under the light microscope, a terminal has generally been considered to exhibit degeneration if it is swollen and irregularly shaped, in contrast to the thinner processes associated with normal terminals. Examples are shown in Fig. 6. These identifying criteria, which may manifest silver-stained neurofilament accumulation, have been characterized by previous investigators (e.g., Barker and Ip, 1965).

In the diaphragm muscle, terminal degeneration was seen less frequently in older rats (Smith and Rosenheimer, 1982). In 43% and 23% of the end plates in 10- and 28-month rats, respectively, some evidence of terminal degeneration was observed. Similar results have been observed in EDL, but there is no apparent change in the rate of degen-

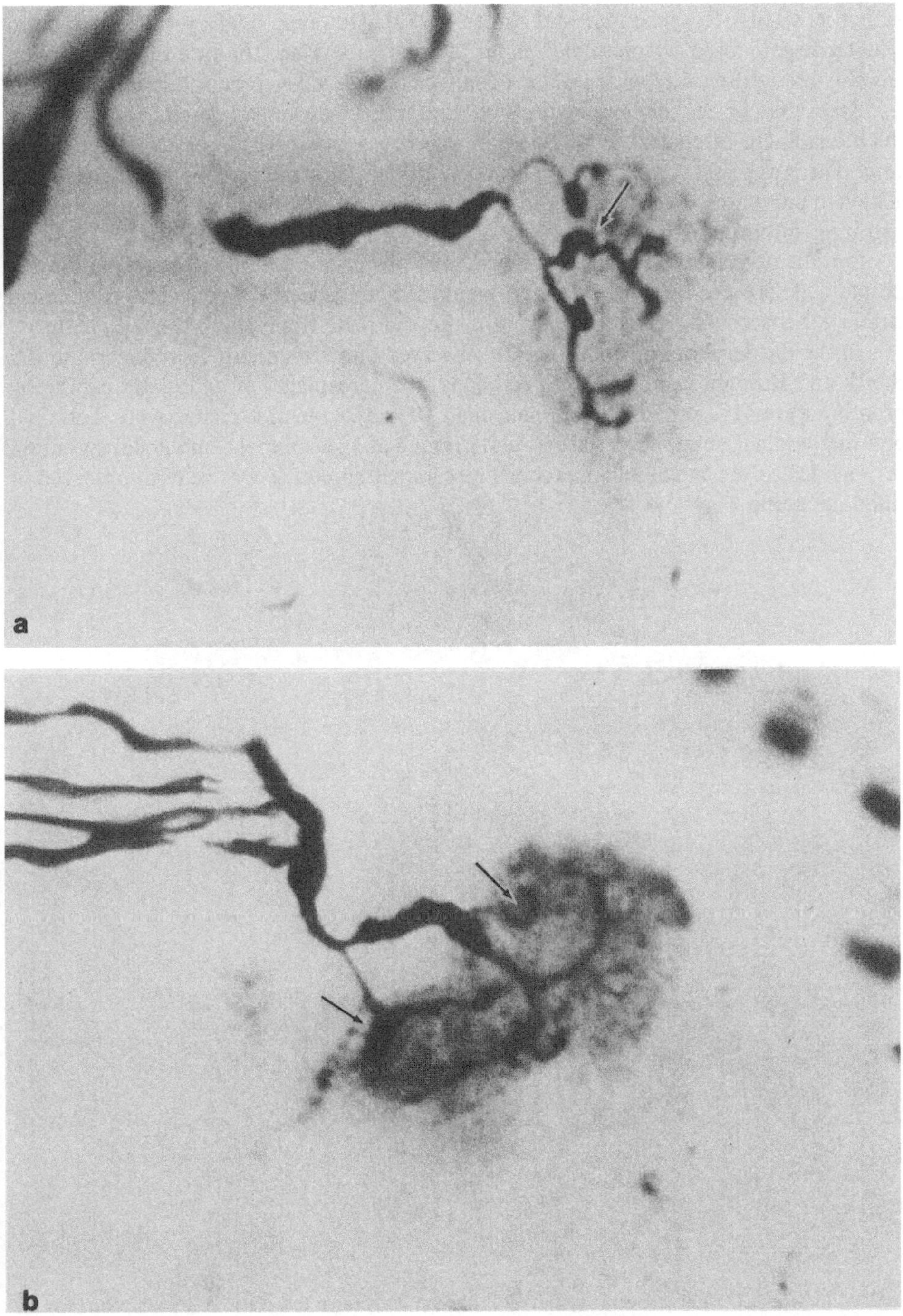

Figure 6. Indications of axon terminal degeneration in the end plate of rat diaphragm. In both (a) and (b), swollen axon terminals (arrows) indicative of degeneration are shown. The magnification in both cases is 2450×, reproduced at 95%. (From Smith and Rosenheimer, 1982.)

eration in soleus (Rosenheimer and Smith, 1983). However, Tuffery (1971) reported reduced degeneration in soleus end plates of aged cats. Thus, the rate of degeneration may fluctuate differently with age between functionally diverse muscles and animals.

Sprouting, or new nerve growth, is characterized by nonmyelinated axon processes which terminate in conical or bulbous growth cones either within or close to the end plate. The morphology of sprouts viewed with the light microscope has been documented in previous studies (Smith and Rosenheimer, 1982). A sample micrograph demonstrating sprouting is shown in Fig. 7.

Sprouts originate from either the node of Ranvier or a terminal branch within the end plate. These are termed nodal and terminal sprouts, respectively. They are called ultraterminal sprouts if they project out of the end plate from which they originate.

In the rat diaphragm, sprouting was observed more frequently in younger animals (Smith and Rosenheimer, 1982). Specifically, the percentages of end plates exhibiting sprouting were 41% and 10% in animals aged 10 and 28 months, respectively. Terminal sprouting was not observed in the end-plate region of aged animals, and nodal sprouting was seen less often in the older rats. Ultraterminal sprouting was seldom observed in either age group.

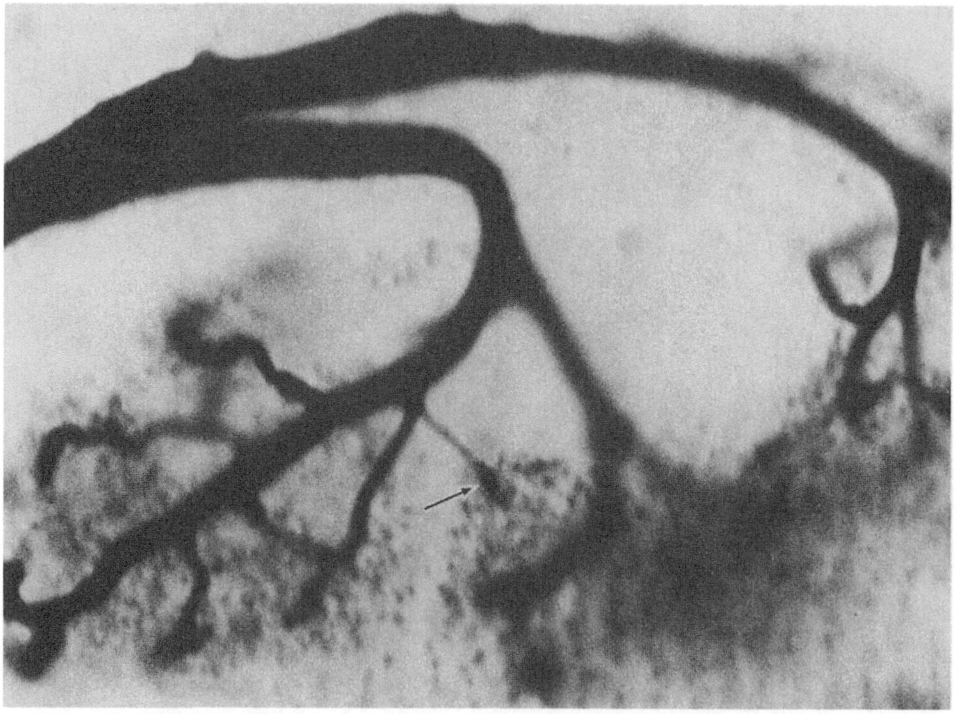

Figure 7. Indications of phrenic nerve sprouting in the rat diaphragm end plate. A growth cone (arrow), which characterizes a sprout, is shown. Because the sprout terminates within the same end plate from which it originated, it is considered a terminal sprout. The magnification is 4100×. (From Smith and Rosenheimer, 1982.)

There was a similar age-related decrease in the rate of sprouting within the rat EDL end plate, while there were no significant changes in the frequency of sprouts in soleus (Rosenheimer and Smith, 1983). Although sprouting within the end plate declined with age in EDL, there was a significant age-related increase in occurrences of ultraterminal sprouting, in which sprouts emanating in the EDL end-plate region projected out of the end plate. An example of ultraterminal sprouting is shown in Fig. 8. In cat soleus and PDQ muscles, there was a pronounced increase in sprouting with age; however, this was associated with a rise in the number of ultraterminal sprouts (Tuffery, 1971).

In young rats (Smith and Rosenheimer, 1982) and cats (Tuffery, 1971), end plates were observed in which there was evidence of both degeneration and sprouting occurring simultaneously. The frequency with which this occurred dropped considerably during aging in both species. Indeed, in diaphragm of 28-month rats, there were no observed cases (Smith and Rosenheimer, 1982). It has been concluded that both degeneration and sprouting are independent processes which may occur less often during senescence.

There is also an age-related decrease in the sprouting response to disruption of normal end-plate innervation (Pestronk *et al.*, 1980). Following "pharmacological

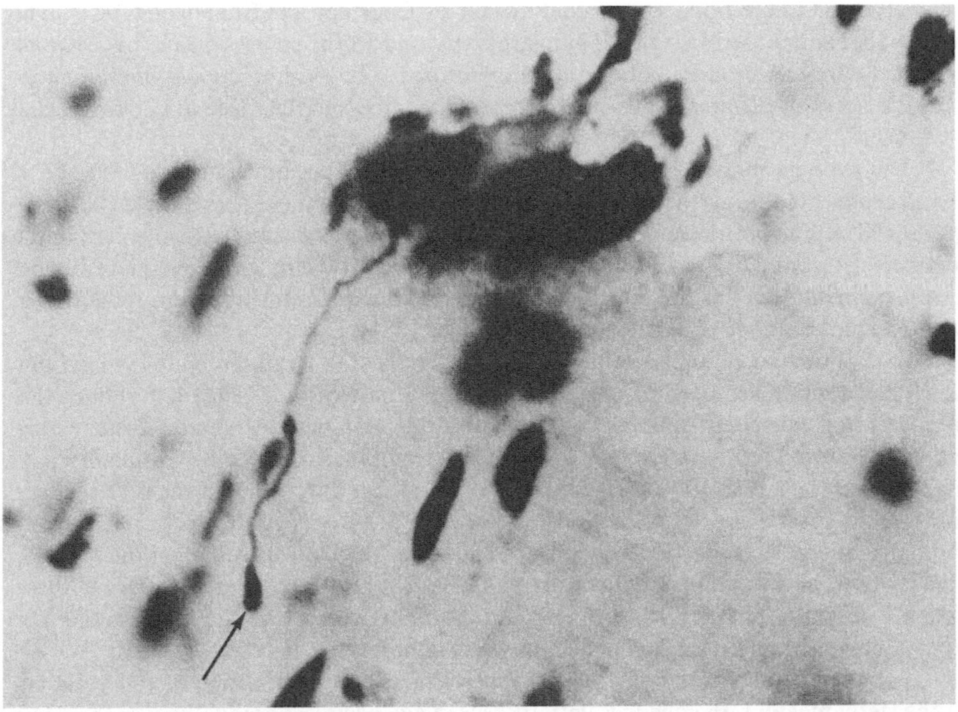

Figure 8. Indications of sprouting in the rat EDL end plate. A growth cone (arrow) can be seen projecting outside the end-plate region. This is considered an ultraterminal sprout. The magnification is 2500×.

denervation" of rat soleus end plates by botulinum toxin, there is a striking increase in end-plate length and terminal arborization in 10-month animals. However, in 28-month rats, botulinum failed to elicit any significant sprouting response. In a related study, Pestronk *et al.* (1980) also showed that axon regeneration following focal nerve crush was much less extensive in 28-month than 10-month rats. Results similar to these have been reported in the central nervous system (Scheff *et al.*, 1978). It may be concluded that during senescence there is a general reduction in an axon's capability to produce new growth either during normal remodeling of the synapse or in response to disrupted innervation.

4.1.4. Receptor Distribution

Functional denervation results in the spreading of sensitivity to ACh beyond the synaptic region. Likewise, the occurrence of increased extrajunctional sensitivity is indicative of disruption of normal synaptic relationships. If the reduced capability to maintain synaptic transmission is tantamount to eventual synapse disruption, extrajunctional sensitivity in aged rats may be increased. In view of the possibility of nonfunctional terminals within the end plate (cf. Smith and Rosenheimer, 1982), it is intriguing to precisely know the receptor distribution.

The number of extrajunctional receptors appears to increase with age. This has been observed in the rat sternomastoid muscle by Courtney and Steinbach (1981), who used a rhodamine derivative of α-bungarotoxin, and in the soleus muscle by Pestronk *et al.* (1980), who used ^{125}I-labeled α-bungarotoxin. The extrajunctional binding probably did not result from the presence of denervated fibers in the muscles (Courtney and Steinbach, 1981).

The average number of junctional ACh receptors appears to decline with age in the sternomastoid (Courtney and Steinbach, 1981) but not the soleus muscle (Pestronk *et al.*, 1980). Furthermore, in the sternomastoid, there was a significant increase in the number of discrete regions of high ACh receptor density within the end-plate region. The "fragmentation" of the receptor-rich area could explain the apparent reduction in total receptor number (Courtney and Steinbach, 1981).

In related studies, supersensitivity to ACh similar to that observed following functional denervation has been reported in aged rats. Frolkis *et al.* (1976) determined this by measuring muscle membrane responses to drugs (ACh, neostigmine, *d*-tubocurarine) administered intra-arterially. Contradictory results are reported by Gutmann (Gutmann and Hanzlíková, 1972), who found no age-related differences in the sensitivity of the muscle membrane to ACh. Furthermore, by applying ACh electrophoretically through a microelectrode, he was unable to find any significant spreading of the receptor area away from the end plate (Gutmann *et al.*, 1971). However, interpretations of these data are difficult, for it is not mentioned if motor innervation was intact or whether the treated fibers exhibited signs of senile muscle atrophy.

The time course of the observed changes in receptor distribution is also variable. In the sternomastoid, the changes with increased age occurred gradually over the adult life of the animal (Courtney and Steinbach, 1981). In soleus, they appeared to occur more abruptly after the age of 18 months (Pestronk *et al.*, 1980). These contradictory findings could conveniently be attributed to differences in the functional uses of the different muscles or the overall activity of the animals. However, definite conclusions are premature.

4.2. Ultrastructure

Ultrastructural studies of the neuromuscular junction in aged preparations have been primarily descriptive. In many cases, anecdotal observations have been reported without any mention of necessary controls. Although conclusions based on such observations may indeed be valid, they should be regarded with appropriate caution. Quantitative morphometric data have also been published. Results obtained from rat diaphragm (Smith and Rosenheimer, 1982) are summarized in Table IV. In general, these data complement results obtained using physiological techniques.

4.2.1. Nerve-Terminal Structure

Within the central nervous system, neurofibrillary tangles, lipofuscin, or other cellular inclusions are associated with senescence. Similar features visible with the electron microscope have been observed in motor nerve terminals obtained from aged animals. As seen in the neuromuscular junctions of the aged mouse (Fahim and Robbins, 1982) and rat (Gutmann and Hanzlíková, 1972), these include more frequent occurrence of lipofuscin deposition, neurotubules, and neurofilaments. Frequent circular myelin figures, illustrated in Fig. 9, have also been reported (cf. Fahim and Robbins, 1982). They may be fixation artifacts associated with polymerization of glutaraldehyde fixative.

In diaphragm tissue obtained from an aged animal it is more common to see neuromuscular junctions which are virtually indistinguishable from those of a younger animal. For comparison, examples obtained from diaphragms of 10- and 28-month rats are shown in Fig. 10. Thus, most distinctions between tissue obtained from young and old animals are based on morphometric analyses.

Significant differences in nerve terminal diameters have seldom been reported. An anecdotal incidence of terminal swelling is documented by Tomonaga (1977), but

Table IV. Age-Related Ultrastructural Changes at the Diaphragm Neuromuscular Junction

Property	Age (months) 10	28
Terminal diameter and length	No significant	
Number of synaptic vesicle (per μm^3)	712	990
Synaptic vesicle volume (nm^3)	66,953	60,774
Synaptic cleft width (nm)	43	51

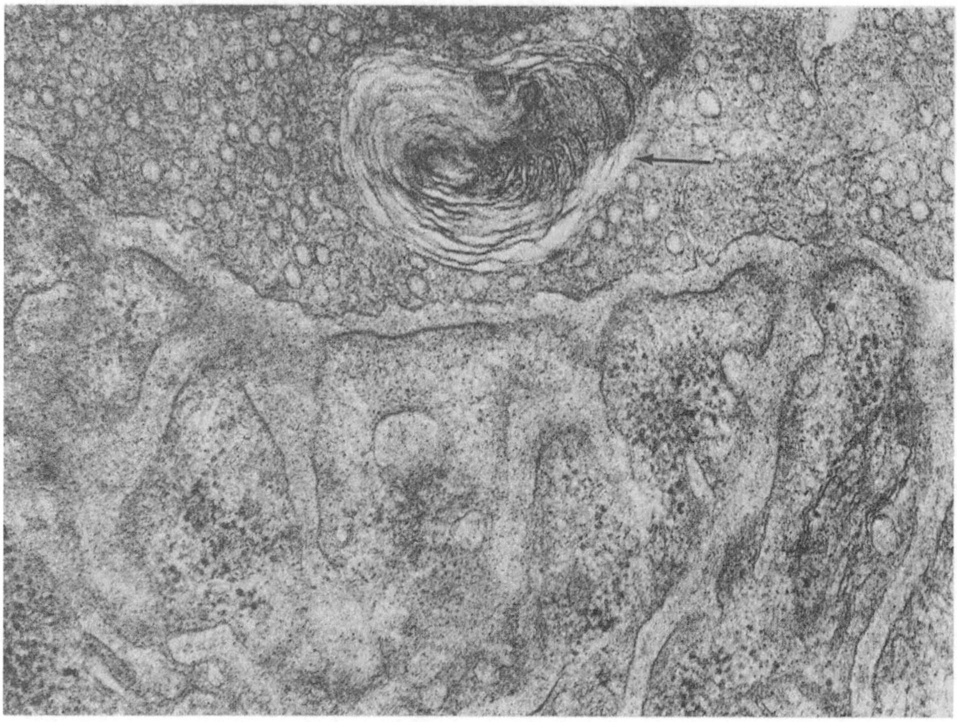

Figure 9. Circular myelin figure in a phrenic nerve terminal of a 28-month rat. These structures (arrow) are more prevalent in aged preparations. However, they are also associated with polymerization of glutaraldehyde fixative. The magnification is 70,200×.

insufficient data are presented to evaluate the case. In the examples shown in Fig. 10, the phrenic nerve terminals are about 2 μm in diameter for both age groups (Smith and Rosenheimer, 1982). Also, there were no age-related differences in axon diameter at sites proximal to the region of insertion into the muscle. Thus, the surface-to-volume ratio of motor nerve terminals does not change appreciably during senescence. Judging from these data, physiological properties which depend on this ratio, such as activity-dependent changes in intracellular ion concentrations, may not change significantly during aging.

Dual terminals have been reported to occur more frequently in older mice (Fahim and Robbins, 1982). An example from the rat diaphragm is shown in Fig. 11. This phenomenon has been attributed to interposition of Schwann cell cytoplasm between the terminals, which may be suggestive of nerve terminal withdrawal (Fahim and Rob-

Figure 10. Synaptic sites in the nerve terminals of the rat phrenic nerve-diaphragm muscle. The synaptic vesicles are concentrated more heavily in the synaptic region overlying the junctional folds of the muscle. Terminals obtained from control and aged rats are shown in (a) and (b), respectively. The magnifications are 23,800× (a) and 28,100× (b). (From Smith and Rosenheimer, 1982.)

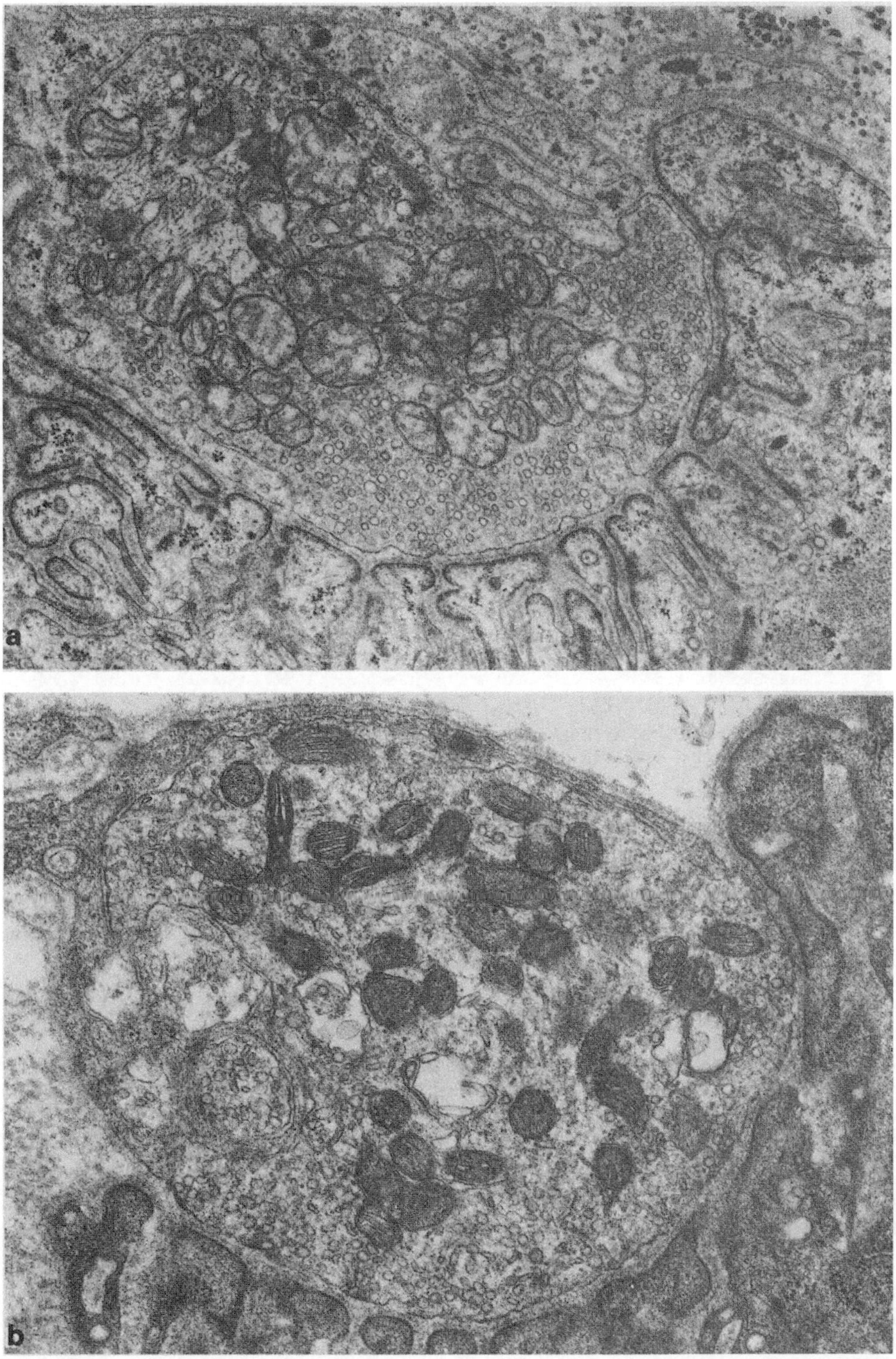
a
b

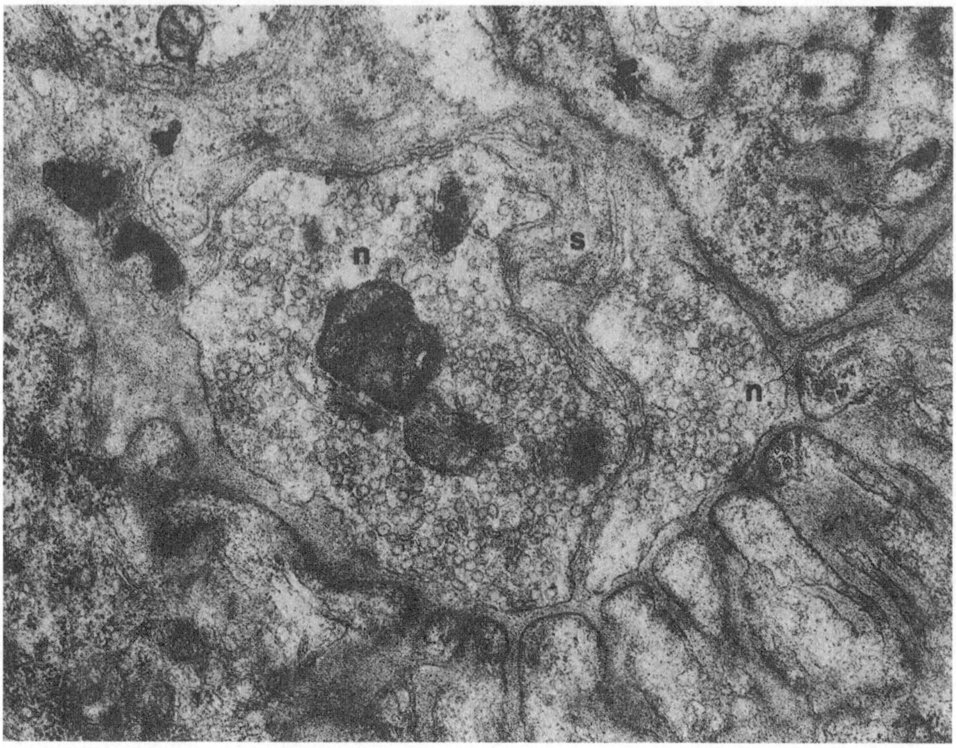

Figure 11. Dual phrenic nerve terminals in the end-plate region of a 28-month rat. The two terminals (n) are separated by Schwann cell (s) tissue. The magnification is 28,600×.

bins, 1982). Alternatively, it may be related to end-plate remodeling, or it may simply be closely coupled branches innervating the end plate. In our laboratory, we have seen dual terminals with about equal frequency in diaphragm preparations from both young and old rats.

4.2.2. Neuromuscular Junction

The junctional folds of aged animals may be more complicated and extensively branched (Gutmann and Hanzlíková, 1972; Tomonaga, 1977; Fahim and Robbins, 1982). This conclusion has been based on observations and morphometric analysis of micrographs similar to Fig. 12. These results must be interpreted with caution, however, for the profile of the junctional folds depends upon tissue orientation during sectioning.

In aged mouse soleus, Fahim and Robbins (1982) have reported postsynaptic folds partially or completely devoid of nerve terminals or Schwann cells (cf. Cardasis and Padykula, 1981). These types of observations require cautious interpretation, for they

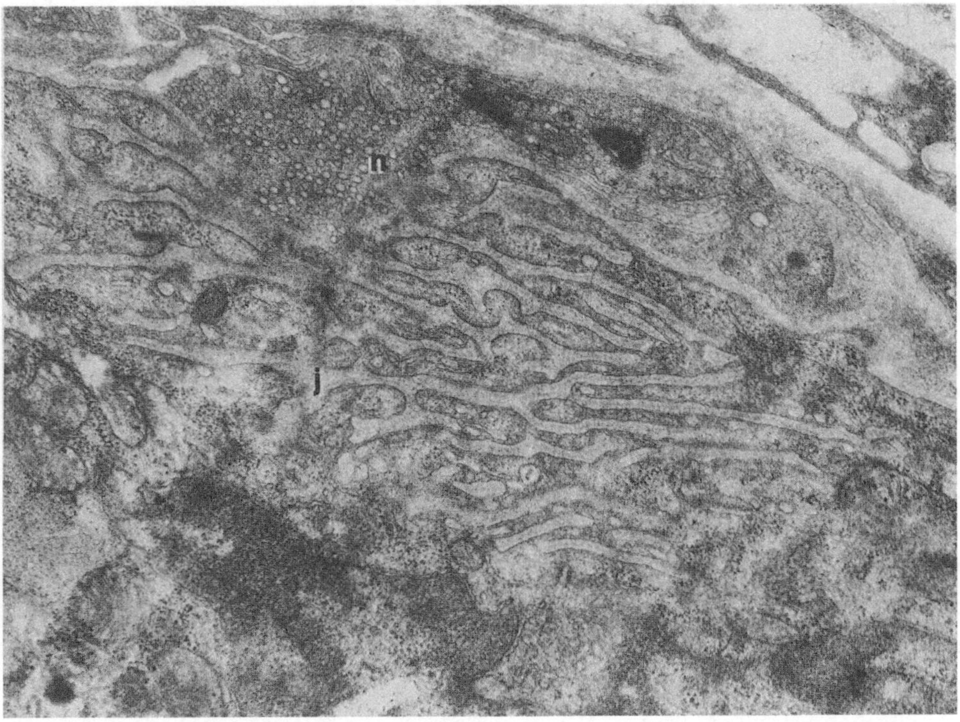

Figure 12. Extensive junctional folding at the diaphragm neuromuscular junction of a 28-month rat. The junctional folds (j) appear to be more extensive than the associated nerve terminal (n); this may be due to the angle of sectioning. The magnification is 36,800×.

may also result from sectioning through junctional folds at an angle that precluded full inclusion of the associated nerve terminal. In our laboratory, this has been observed in diaphragm tissue obtained from rats of all ages; an example is shown in Fig. 13. Likewise, ultrastructural studies of rat levator ani (Gutmann and Hanzlíková, 1972) have also failed to reveal age-related differences in this area. Serial sectioning is required before these conflicting observations can be properly evaluated.

In senescent animals, synaptic clefts are consistently wider (Gutmann and Hanzlíková, 1972; Fahim and Robbins, 1982). For example, in the rat diaphragm we have found the cleft of 28-month rats to be 18% wider than in 10-month animals. These observations indicate that the volume of extracellular space in the immediate vicinity of the neuromuscular junction must increase with age. Larger volumes may attenuate the extent to which extracellular K^+ concentrations rise due to restricted K^+ diffusion away from the terminal region (Smith, 1982a).

Axoplasmic fragmentation with envelopment of the axon terminals by Schwann cells has occasionally been observed in mouse soleus (Fahim and Robbins, 1982) and rat "thigh" (Fujisawa, 1976). This may involve extension of the Schwann cell into the enlarged synaptic clefts. An example obtained from rat diaphragm muscle is shown in

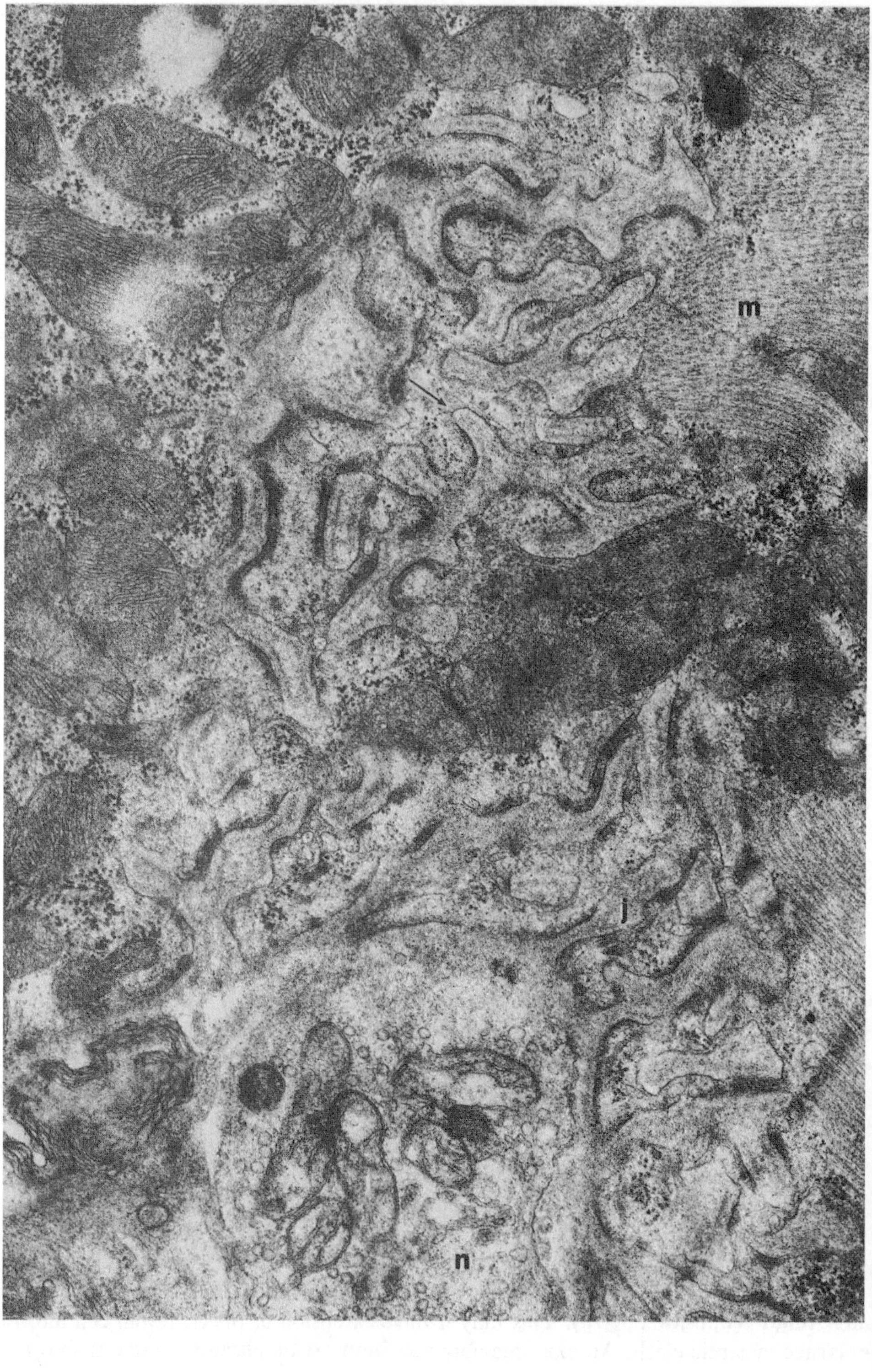
m
j
n

Fig. 14. In such cases, however, further sections should be examined to rule out the possibility that unusual profiles are associated with the angle of sectioning.

4.2.3. Synaptic Vesicles

Synaptic vesicles in the older animals were less regularly shaped. This can be seen in Fig. 10. The significance of this is difficult to interpret, however. It could be due to age-related differences in response to fixatives, or it could be related to the observed drop in transmitter content within the cell.

The number of synaptic vesicles has been found to increase in some preparations. In the diaphragm, there are 39% more vesicles in the terminal junctional region of 28-month rats relative to 10-month animals (Smith and Rosenheimer, 1982). Similar results were obtained from levator ani muscle of aged rats (Gutmann and Hanzlíková, 1972), but in mouse soleus fewer synaptic vesicles were counted (Fahim and Robbins, 1982).

In diaphragm, increased numbers of synaptic vesicles parallel increased numbers of nerve terminals (Smith and Rosenheimer, 1982) during senescence and further exaggerate the paradoxical decrease in end-plate ACh content. As in the nerve terminals, there may be an imbalance between synthesis and degradation. Namely, the rate at which vesicle membrane is synthesized or reclaimed from terminal axolemma may exceed the rate at which it is degraded during senescence. This could explain the rise in vesicle numbers. Until the fate of synaptic vesicle membrane is better understood, however, further speculation may be misdirected.

In the diaphragm, synaptic vesicle size decreased with age (Smith and Rosenheimer, 1982). The vesicle profile diameter and calculated volume in 28-month rats were found to be 3% and 9%, respectively, smaller than in 10-month animals. These slight changes are significant statistically, and they may be associated with the reduced ACh content of the end plates. It is questionable, however, whether volume changes this small would seriously affect synaptic transmission.

5. SUMMARY AND CONCLUSIONS

A composite description of the changes at the neuromuscular junction during aging appears to emerge from these various data. The capability to sustain synaptic transmission decreases, the amount of ACh declines, and the balance between nerve terminal

←

Figure 13. Extensive junctional folding in the apparent absence of an associated phrenic nerve terminal at the diaphragm neuromuscular junction of a 28-month rat. A nerve terminal (n) is seen in one region of junctional folding (j), but at a nearby site of junctional folding (arrow) there are no apparent terminals; this may be due to the angle of sectioning. Surrounding diaphragm muscle (m) is also seen. The magnification is 28,600×.

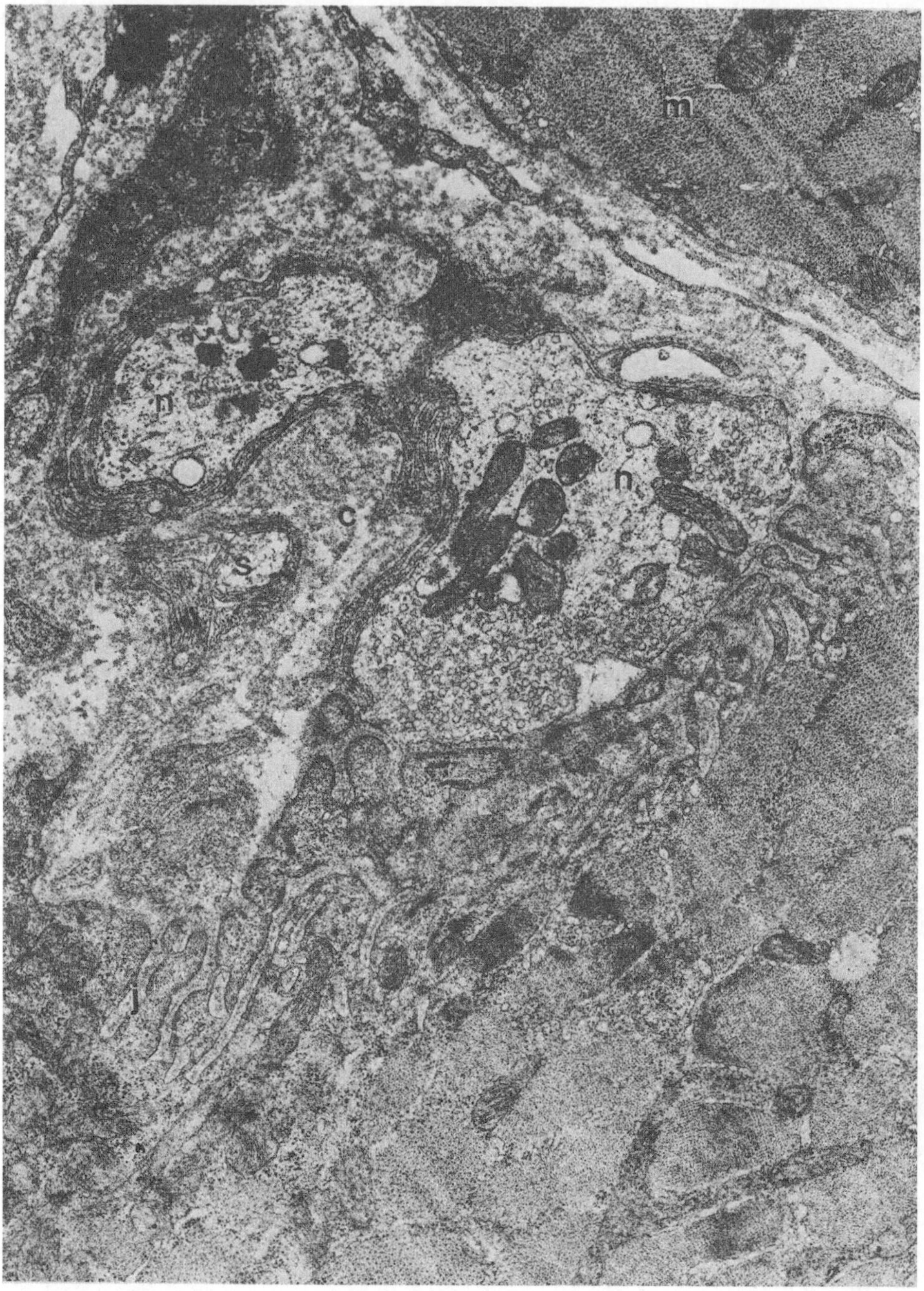

Figure 14. Disrupted diaphragm neuromuscular junction in a 28-month rat. The Schwann cell (s) appears to have become inserted between the junctional folds (j) and a nerve terminal (n). Connective tissue (c) and diaphragm muscle (m) are also seen. The magnification is 15,200×.

growth and degeneration becomes less stable. These changes are associated with altered membrane structure, which is manifest as reduced membrane potentials, lowered choline uptake, and less uniform ACh receptor distribution.

A hallmark of these and most other aging studies is variability. Although many phenomena are associated with senescence, it is not unusual to observe aged preparations which are indistinguishable from younger counterparts. Furthermore, age-related changes at the neuromuscular junction of diaphragm, for example, may not be the same as those of another muscle, such as EDL. Although some variation may be attributed to differences in laboratory techniques or animal strains, it seems quite probable that other factors contribute to the observed variability.

Environmental influences may have a profound effect on the rate at which different preparations exhibit senescence. These may include factors such as diet, social interactions, and exercise. Specifically, the amount of activity at the neuromuscular junction may affect terminal growth (Eldridge, Liebhold, and Steinbach, 1981). Because of its vital function, the diaphragm remains active throughout the life of the animal. In contrast, the soleus and the EDL may be relatively inactive in caged animals. These differences may modulate the rate at which age-related changes in degeneration and sprouting are expressed. Other species-related differences may compound these effects. Thus, variability may be considered a cardinal feature of the aging process.

ACKNOWLEDGMENTS

This work was supported by an NIH Research Career Development Award (NS00380); research performed in our laboratory was supported in part by NIH grant No. AG01572.

REFERENCES

Åstrand, P-O., and Rodahl, K., 1977, *Textbook of Work Physiology: Physiological Bases of Exercise. Second Ed.*, pp. 121–123, McGraw-Hill, New York.

Barker, D., and Ip, M. C., 1965, Sprouting and degeneration of mammalian motor axons in normal and de-afferented skeletal muscle, *Proc. R. Soc. B* **163**:538–554.

Birren, J. E., and Wall, P. D., 1956, Age changes in conduction velocity, refractory period, number of fibers, connective tissue space and blood vessels in sciatic nerve of rats, *J. Comp. Neurol.* **104**:1–16.

Brown, M. C., and Ironton, R., 1977, Motoneurone sprouting induced by prolonged tetrodotoxin block of nerve action potentials, *Nature (London)* **265**:459–461.

Brown, M. C., and Ironton, R., 1978, Sprouting and regression of neuromuscular synapses in partially denervated mammalian muscles, *J. Physiol. (London)* **278**:325–348.

Caccia, M. R., Harris, J. B., and Johnson, M. A., 1979, Morphology and physiology of skeletal muscle in aging rodents, *Muscle Nerve* **2**:202–212.

Campbell, M. J., McComas, A. J., and Petito, F., 1973, Physiology changes in aging muscles, *J. Neurol. Neurosurg. Psychiat.* **36**:174–182.

Cardasis, C. A., and Padykula, H. A., 1981, Ultrastructural evidence indicating reorganization at the neuromuscular junction in the normal rat soleus muscle, *Anat. Rec.* **200**:41–59.

Courtney, J., and Steinbach, J. H., 1981, Age changes in neuromuscular junction morphology and acetylcholine receptor distribution in rat soleus muscle fibres, *J. Physiol. (London)* **320**:435–447.

Duchen, L. W., 1970, Changes in motor innervation and cholinesterase localization induced by botulinum toxin in skeletal muscle of the mouse: Differences between fast and slow muscles, *J. Neurol. Neurosurg. Psychiat.* **33**:40–54.

Eldridge, L., Liebhold, M., and Steinbach, J. H., 1981, Alterations in cat skeletal neuromuscular junctions following prolonged inactivity, *J. Physiol. (London)* **313**:529–545.

Fahim, M. A., and Robbins, N., 1982, Ultrastructural studies of young and old mouse neuromuscular junctions, *J. Neurocytol.* **11**:641–656.

Frolkis, V. V., Martynenko, O. A., and Zamostyan, V. P., 1976, Aging of the neuromuscular apparatus, *Gertontology* **22**:244–279.

Fujisawa, K., 1976, Some observations on the skeletal musculature of aged rats. III. Abnormalities of terminal axons found in motor end-plates, *Exp. Geront.* **11**:43–47.

Gutmann, E., and Hanzlíková, V., 1965, Age changes of motor end-plates in muscle fibers of the rat, *Gerontologia* **11**:12–24.

Gutmann, E., and Hanzlíková, V., 1972, *Age Changes in the Neuromuscular System*, Scientechnica Ltd, Bristol, pp. 21–113.

Gutmann, E., Hanzlíková, V., and Vyskočil, F., 1971, Age change in cross striated muscle of the rat, *J. Physiol. (London)* **219**:331–343.

Kelly, S. S., 1978, The effects of age on neuromuscular transmission, *J. Physiol. (London)* **274**:51–62.

Landfield, P. W., McGaugh, J. L., and Lynch, G., 1978, Impaired synaptic potentiation processes in the hippocampus of aged, memory-deficient rats, *Brain Res.* **150**:85–101.

Marchi, M., Hoffman, D. W., Giacobini, E., and Frederickson, T., 1980, Age dependent changes in choline uptake of the chick iris, *Brain Res.* **195**:423–431.

Norris, A. H., Shock, N. W., and Wagman, I. H., 1953, Age changes in the maximum conduction velocity of motor fibers of human ulnar nerves, *J. Appl. Physiol.* **5**:589–593.

Pestronk, A., and Drachman, D. B., 1978, A new stain for quantitative measurement of sprouting at neuromuscular junctions, *Muscle Nerve* **1**:70–74.

Pestronk, A., Drachman, D. B., and Griffin, J. W., 1980, Effects of aging on nerve sprouting and regeneration, *Expl. Neurol.* **70**:65–82.

Retzlaff, E., and Fontaine, J., 1965, Functional and structural changes in motor neurons with age, in: *Behavior, Aging, and the Nervous System* (A. T. Welford and J. E. Birren, eds.), Charles C. Thomas, Springfield, pp. 340–352.

Rosenheimer, J. L., and Smith, D. O., 1983, Changes in plasticity at the neuromuscular junction during aging, *Proc. Intl. Union Physiol. Sci.* **XV**: 448.

Rowe, R. W. D., 1969, The effect of senility on skeletal muscles in the mouse, *Exp. Geront.* **4**:119–126.

Scheff, S. W., Benardo, L. S., and Cotman, C. W., 1978, Decrease in adrenergic axon sprouting in the senescent rat, *Science* **202**:775–778.

Shafiq, S. A., Lewis, S., Leung, B., and Schutta, H. S., 1981, Fine structure of aging skeletal muscle, in: *Aging and Cell Structure*, Vol. 1 (J. Johnson, Jr., ed.), pp. 333–346, Plenum, New York.

Smith, D. O., 1979, Reduced capabilities of synaptic transmission in aged rats, *Expl. Neurol.* **66**:650–666.

Smith, D. O., 1982a, Restricted diffusion of extracellular potassium at the neuromuscular junction of aged rats, *Brain Res.* **239**:668–673.

Smith, D. O., 1982b, Physiological and structural changes at the neuromuscular junction during aging, in: *The Aging Brain: Cellular and Molecular Mechanisms of Aging in the Nervous System* (E. Giacobini, G. Filogamo, G. Giacobini, and A. Vernadakis, eds.), pp. 123–137, Raven, New York.

Smith, D. O., 1983, Acetylcholine storage, release, and leakage at the neuromuscular junction of mature adult and aged rats, *J. Physiol (London)*, in press.

Smith, D. O., and Rosenheimer, J. L., 1982, Decreased sprouting and degeneration of nerve terminals of active muscles in aged rats, *J. Neurophysiol.* **48**:100–109.

Tomonaga, M., 1977, Histochemical and ultrastructural changes in senile human skeletal muscle, *J. Am. Geriatr. Soc.* **25**:125–131.

Tuček, S., and Gutmann, E., 1973, Choline acetyltransferase activity in muscles of old rats, *Expl. Neur.* **38**:349–360.

Tuffery, A. R., 1971, Growth and degeneration of motor end-plates in normal cat hind limb muscles, *J. Anat.* **110**:221–237.

Vyskočil, F., 1974, Action potentials of the rat diaphragm and their sensitivity to tetrodotoxin during post-natal development of old age, *Pflügers Arch.* **352**:155–163.

Vyskočil, F., and Gutmann, E., 1972, Spontaneous transmitter release from nerve endings and contractile properties in the soleus and diaphragm muscles of senile rats, *Experientia* **28**:280–281.

Wernig, A., Pécot-Dechavassine, M., and Stöver, H., 1980, Sprouting and regression of the nerve at the frog neuromuscular junction in normal conditions and after prolonged paralysis with curare, *J. Neurocytol.* **9**:277–303.

5

Human Oral Tissues as a Model for the Study of Cellular Age Changes by Light Microscopy

SOL BERNICK

1. INTRODUCTION

Age-dependent changes are known to occur in the oral tissues similar to what occurs in other body tissues. Teeth wear down, the degree depending upon the changes in occlusion, the decrease in muscular activity, abnormal taste sensations, excessive bone loss, and delayed wound healing. Detailed morphological studies on age changes to the various components of the oral cavity have been limited in number both in man and experimental animals.

It is not the aim of this presentation to report a detailed analysis of age-related alterations to all the components of the oral cavity. I have elected to describe age changes to specific structures of the oral cavity from human individuals that we have studied. The specimens were obtained at surgery, autopsy, and from the "Willed Body Program" in the Department of Anatomy, School of Medicine, University of Southern California. The structures chosen for study were secured within a period of time after death which allowed adequate fixation for light microscopy but not for ultrastructural analysis. This allowed one to carry out a light microscopic study using various histochemical reactions.

SOL BERNICK • Department of Anatomy and Cell Biology, University of Southern California School of Medicine, Los Angeles, California 90033.

2. TOOTH AND PERIODONTIUM

2.1. Dental Pulp

The dental pulp, like other connective tissue complexes, undergoes age changes. The pulp volume appears to decrease with age, brought about by the continued apposition of dentine (Seltzer and Bender, 1965). The pulpal stroma of erupted functional teeth from young individuals is made of loose connective tissue. It consists of fine collagen fibers, cellular elements, glycosaminoglycans, and specialized cells which include odontoblasts, fibroblasts. macrophages, and undifferentiated mesenchymal cells. Nerves and blood vessels are observed throughout the coronal pulp (Fig. 1a). One of the characteristic changes in the pulp with time is the progressive decrease in surface area of the pulp (Figs. 1b and 1c). The reduction in pulpal area is the result of continued deposition of occlusal dentin, diffuse calcification of pulpal stroma, and dentinal apposition at, and above, the area of furcation. Diffuse pupal calcification is a common finding in 90% of all the teeth obtained from individuals over the age of 50 years (Bernick, 1967). This percentage agrees with the findings of Hill (1949), who noticed that 90% of the aged teeth exhibited diffuse calcification of the pulp. Diffuse calcification begins in the root pulp as small isolated masses (Fig. 2a) and extends into the coronal pulp (Fig. 1c). The calcification may become so extensive as to obliterate the pulp chamber (Fib. 2b). Milch (1966) postulated that apparently increased cross-linkage of collagen with advancing age enhances its tendency for mineralization. It is noticed in this study that the collagen bundles of the vascular and neural sheaths of the young pulps stain blue with the alcian blue–PAS combination, whereas in the old pulps they appeared Schiff-positive. The Schiff-positive fibers were the sites of calcification (Fig. 2c).

Aging adversely alters the vascular and nerve supply to the pulp. There is an extensive branching of both the vascular and neural strucfures in young adult functional teeth. As the pulpal vessels and nerves enter the coronal region, vascular branches course toward the lateral and occlusal surfaces of the pulp. As these branches reach the cuspal horns, a rich subodontoblastic plexus of nerves and blood vessels is formed (Fig. 3a). Nonmyelinated nerve terminals are traced into the odontoblastic zones. Capillary networks are observed within the odontoblastic layer lying adjacent to the predentin (Fig. 3b).

In the aging process, there is a gradual narrowing of the root pulp circumference. The blood vessels and nerves appear to become more prominent in the core of the pulp (Fig. 4a). One of the earliest changes observed in the small arteries and arterioles is the intense Schiff-positive staining of their walls, as this staining reaction gave a "hyalinized" appearance to these vessels (4a), a reaction not observed in the vessels in "young"

Figure 1. (a) Section of a functional tooth from a young adult individual. Note the wide pulp chamber; the stroma consisting of loose connective tissue; and abundant neural and vascular structures. N, nerves; V, vessels. PAS–hematoxylin. ×20. (b) Section of a molar tooth from an individual 60 years of age. Note the narrowing of the pulp chamber as result of continual deposition of dentin; the loose appearance of pulpal tissue; and sparcity of nerves and blood vessels. PAS–hematoxylin. ×20. (c) Section of a molar tooth from an aged female 65 years. Notice the narrowed pulp chamber and the diffuse calcified masses in both the root and coronal pulp. PAS–hematoxylin. ×20.

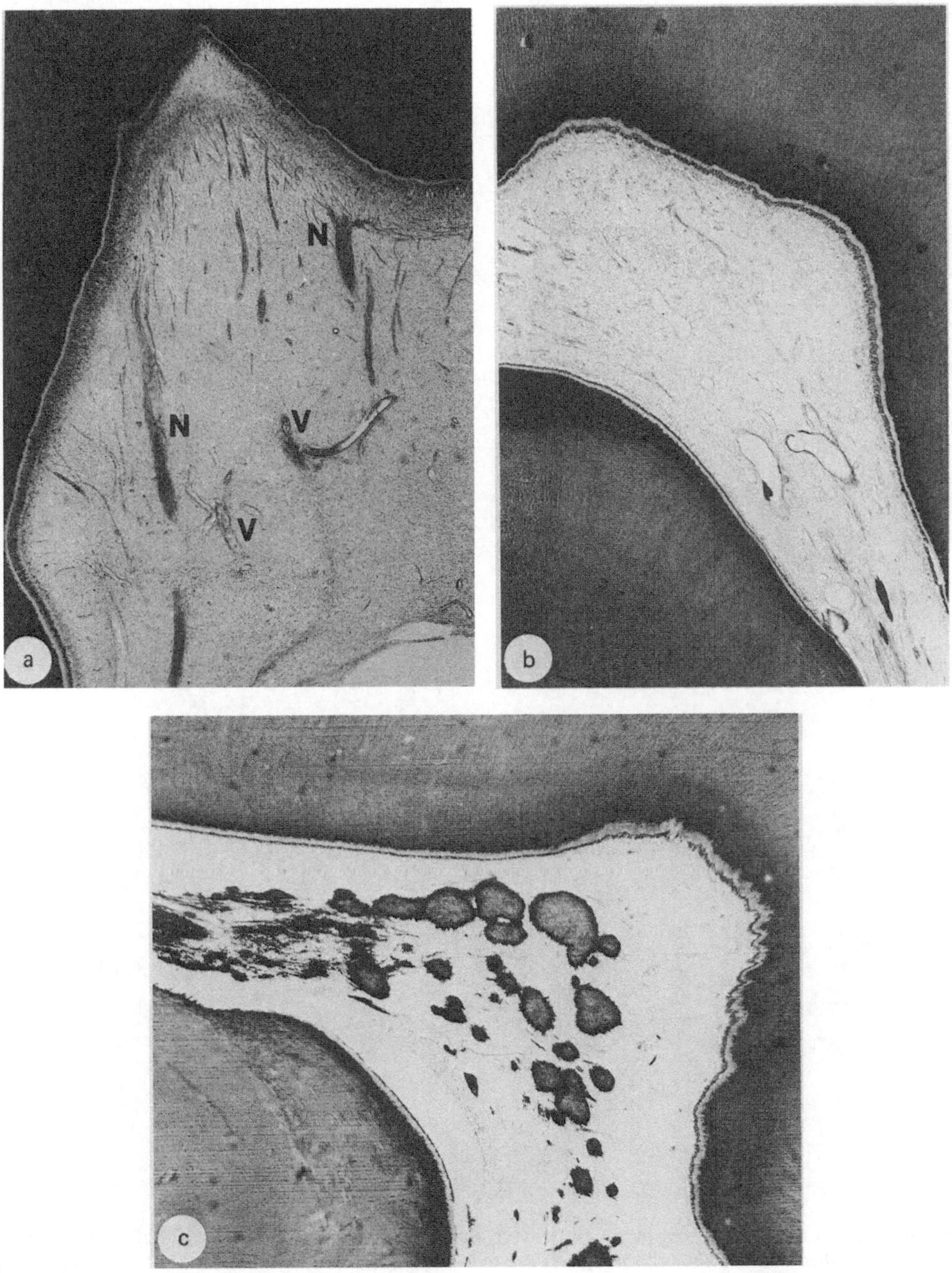
N
N
V
V
a
b
c

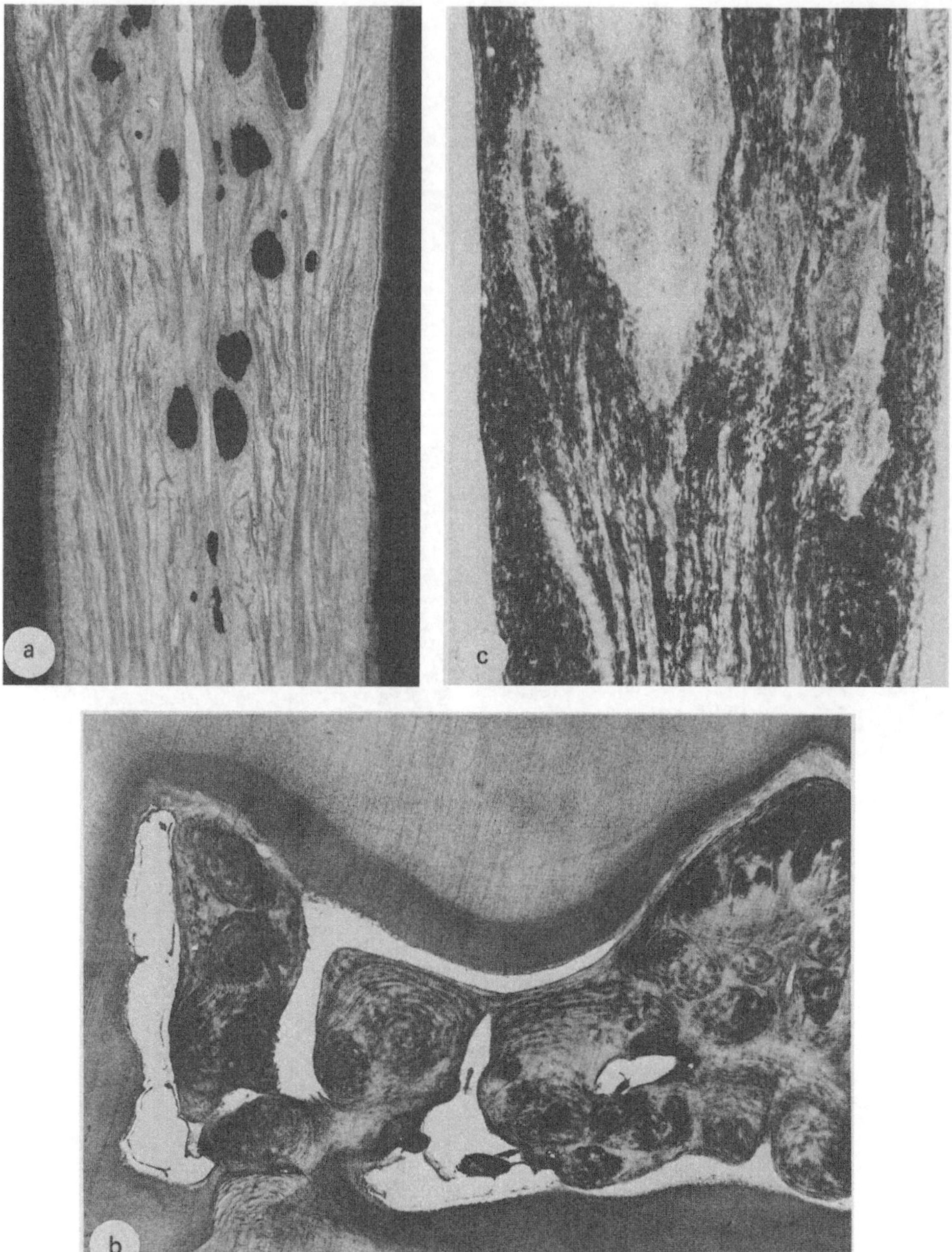

Figure 2. (a) A section of a root pulp from a lower molar tooth obtained from a subject 58 years of age. Note the calcified bodies scattered throughout the root pulp. Iron hematoxylin stain. ×50. (b) Section of an "aged" tooth. The fused calcified masses have nearly filled the pulp chamber. PAS–hematoxylin. ×20. (c) Section of root pulp from an old tooth that was stained with alcian blue (pH 2.5)–PAS combination. The photograph was taken through a red filter to emphasize the alcianophilia of the ground substance. The calcified masses stained red, and appear light in the photomicrograph. ×50.

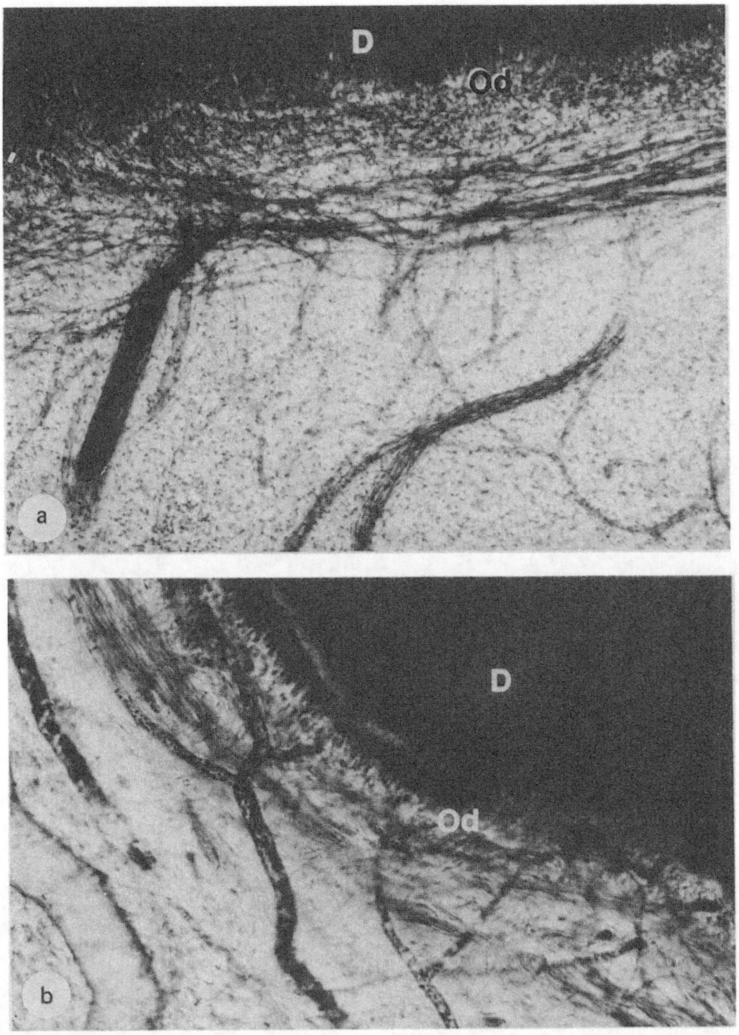

Figure 3. (a) A section of the pulp horn region of a noncarious "young" molar tooth. Note the subodontoblastic plexus of nerves. Nonmyelinated terminals from this plexus pass into the odontoblastic zone. D, dentin; Od, odontoblast zone. Iron hematoxylin stain. ×125. (b) A section of pulp from a young adult. The arterioles are observed entering the odontoblastic layer to divide into terminal capillary networks among the odontoblasts and predentin. D, dentin; Od, odontoblast zone. Iron hematoxylin. ×125.

teeth (Fig. 4b). Other small arteries in the old pulps show arteriosclerotic change characterized by intimal proliferation and a narrowing of the lumina (4c). In the process of diffuse calcification of the root pulp, the adventitia and media of the small blood vessels become calcified, eventually obliterating the vessels (Fig. 4d). At the same time, the connective tissue coverings of the nerves also become calcified (Figs. 5a, 5b). With the

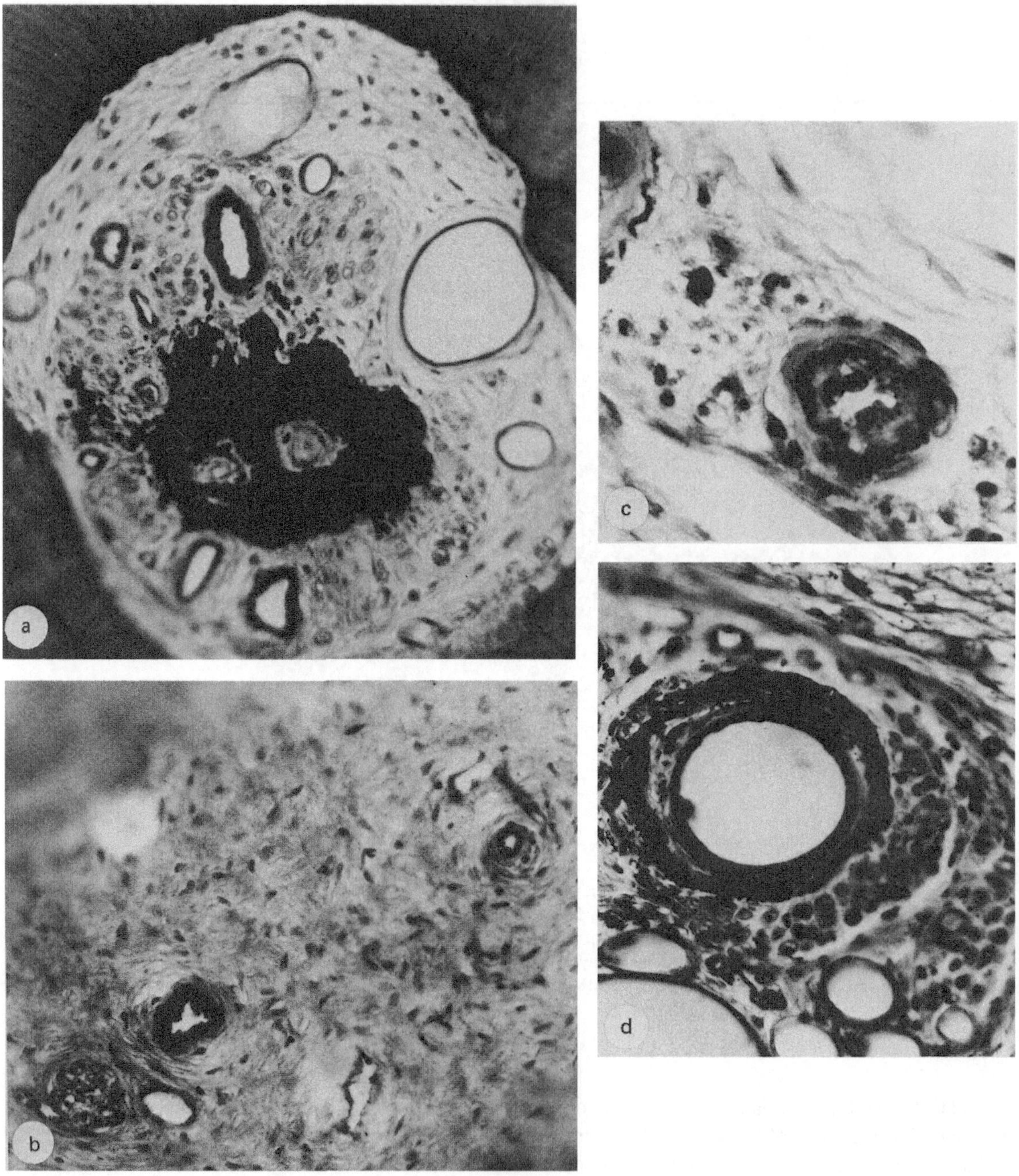
a
b
c
d

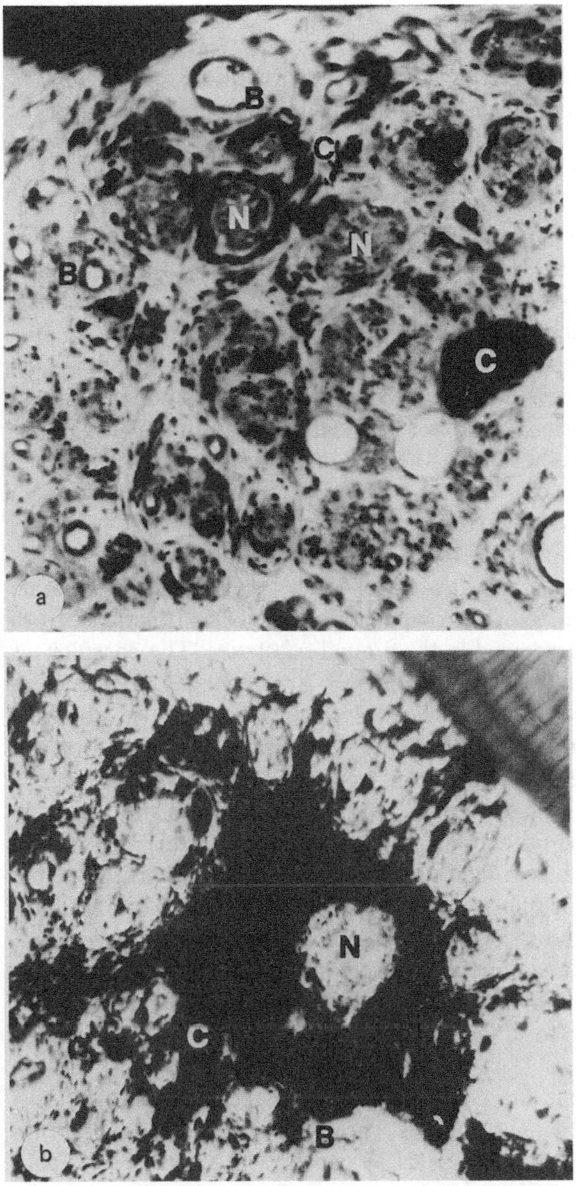

Figure 5. (a) Section of the pulp in the region of the apical third of the root of a premolar from a 50-year-old individual. There are isolated regions of calcification in the perineurium and endoneurium of nerve fasciculi and individual fibers. B, blood vessels; N, nerve fasciculus; C, calcifying mass. Iron hematoxylin. ×150. (b) Section of the pulp in the region of the apical third of the root of a molar tooth from a 60-year-old individual. The calcified mass has obliterated all but one demonstrable nerve fasciculus. B, blood vessels; C, calcifying mass; N, nerve. Iron hematoxylin. ×150.

Figure 4. (a) A section of the root pulp of a first molar from an "aged" individual. The cross section area of the pulp is reduced by the continual deposition of dentin. Notice the blood vessels and nerves concentrated in the core of the pulp and the presence of diffuse calcified masses. The walls of the arteries, and venules stain deep red. A, artery; V, vein. PAS–hematoxylin. ×125. (b) A section of the pulp of a first molar from a young adult. Notice the mesenchymallike tissue that makes up the core of the pulp, and the arterioles, venules, and nerves. A, artery; N, nerve; V, vein. PAS–hematoxylin. ×125. (c) A section of the pulp of a molar tooth from a 70-year-old female. Notice the intimal thickening characterized by proliferation of cells; increase in collagenous and elastic fibers. Iron hematoxylin. ×400. (d) A section of a small artery in the pulp of a molar tooth from an "aged" individual. Notice the extensive calcification of the vessels involving the adventia and media and extending into the subendothelial region. PAS–hematoxylin. ×400.

continual apposition of dentin and diffuse calcification, the circumference and area of the pulp become narrower and smaller so that only a thread of pulp persists. In all the young teeth studied, no calcification either diffuse or discrete is observed in the coronal portion of the pulp (Fig. 6a). In the aged pulp of the crown there is an apparent loss in the number of blood vessels and nerves regardless of whether the pulp exhibits calcification. The sparsity of nerves in the pulp of a tooth from a 60-year-old person is shown in Fig. 6b. The pulp exhibits an apparent increase in collagenous fibers and loci of calcified masses. Cuspal nerves and their terminal branches are absent. The nerve fibers that persist show signs of degeneration such as reticulation, fragmentation, and beading (Fig. 6c). The sparsity of blood vessels in the coronal pulp from a 55-year-old individual is illustrated in Fig. 7a. Only the larger blood vessels persist; yet they, too, exhibit calcification along their walls. In addition, there is a marked loss of terminal capillaries supplying the odontoblasts. On the other hand, there is an extensive vascular network in the young adult teeth (Fig. 7b).

It has been reported that there is an increase in collagen elements of the pulp during the aging process (Bhussry, 1968; Ingle, 1965). Whether this increase in collagen is due to continued deposition of collagen with age or to the precipitation of soluble collagen, or to both, is not clearly defined. The pulpal stroma of young functional teeth consist of fine collagenous fibers interspersed in the ground substance. The conspicuous vascular and neural connective tissue sheaths are the only large collagen fiber bundles present; they follow the pathways of the nerves and blood vessels (Figs. 8a, 8b). The orientation of the vascular and neural sheaths is further emphasized in sections of old pulps that are impregnated with silver nitrate or picrofuchsin. Figure 9a illustrates a silver nitrate impregnated section of an aged tooth. It is apparent that the large collagen fiber bundles are directly associated with the pathways of blood vessels and nerves. When the sections of aged pulps are examined under higher magnification, the pulpal stroma still consist of fine collagenous fibers, and the only thick argyrophilic fiber bundles are the connective tissue sheaths of lost blood vessels and nerves (9b). The prominence of the connective tissue sheaths is further accentuated by the progressive apposition of calcified masses as well as by the loss of additional blood vessels and nerves (Fig. 9c). The fibrosis appears, therefore, to be related to the pathways of the degenerated vessels or nerves.

In summary, it can be said that the coronal pulp of old teeth shows a relative increase in the number of thick collagenous bundles. These fibrous bundles are the connective tissue sheaths of degenerated blood vessels and nerves. The stroma of the pulp still consists of fine collagenous fibers. Changes as the result of aging in collagen in the skin and elsewhere have been summarized by Verzar (1964), Gross (1961), and Milch

→

Figure 6. (a) A thick section of the coronal portion of the pulp from a molar tooth from a "teenage" individual. Pulpal nerve divides into cuspal nerves. There is formation of a subodontoblastic network. PN, pulpal nerve; CN, cuspal nerve; B, blood vessels. Iron hematoxylin. ×50. (b) A thick section of the coronal portion of the pulp from a 65-year-old individual. Note that the pulp contain loci of calcification and increased fibrosis. There is an apparent decrease in the number of demonstrable blood vessels and nerves in the pulp. B, blood vessel; N, nerve; C, calcifying mass. Iron hematoxylin. ×50. (c) A high magnification of a nerve in the pulp illustrated in Fig. 6b. Notice the fragmentation, beading, and reticulation of the nerve fiber characteristic of a degenerating nerve. Iron hematoxylin. ×150.

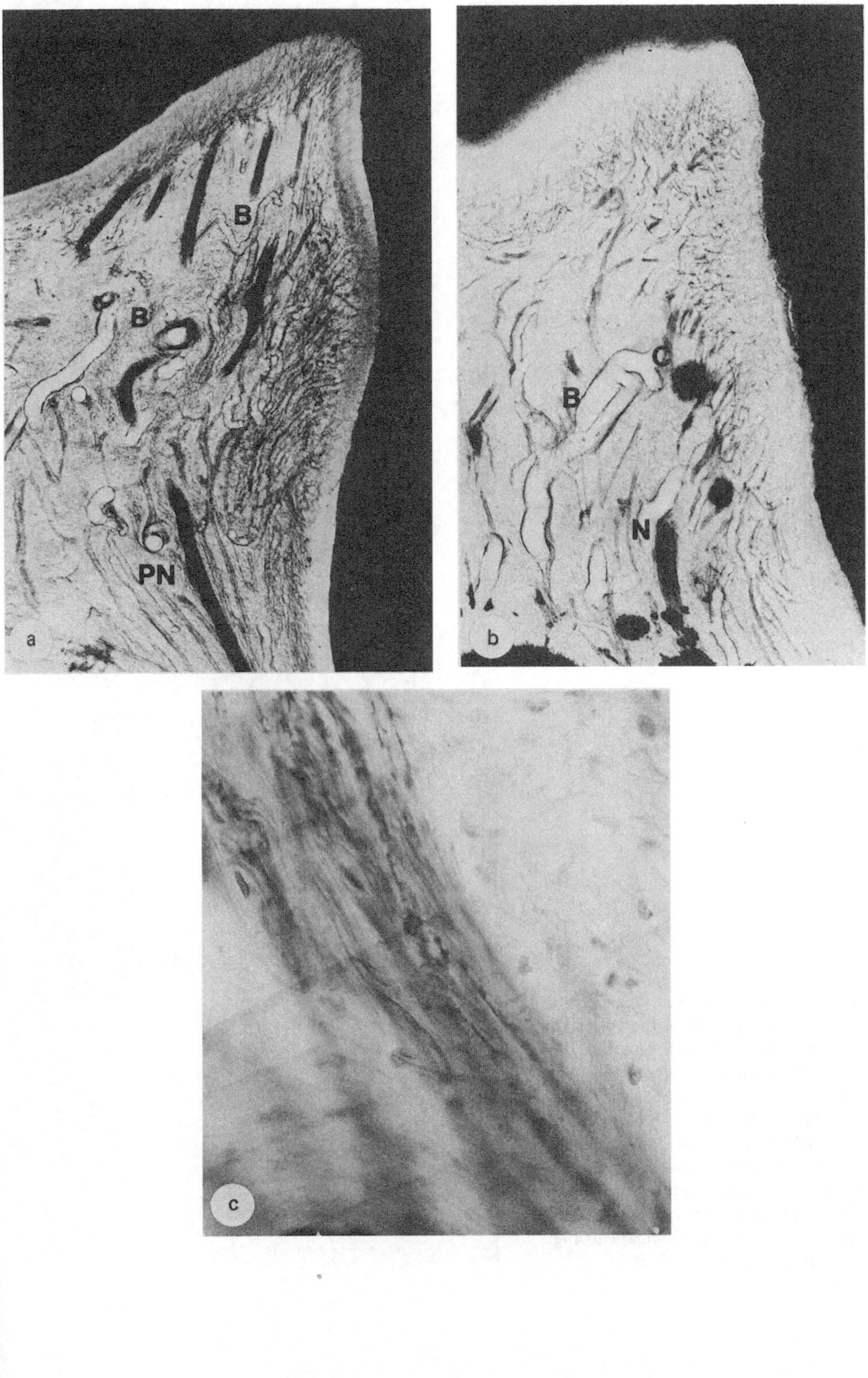
B
B
PN
a
C
B
N
b
c

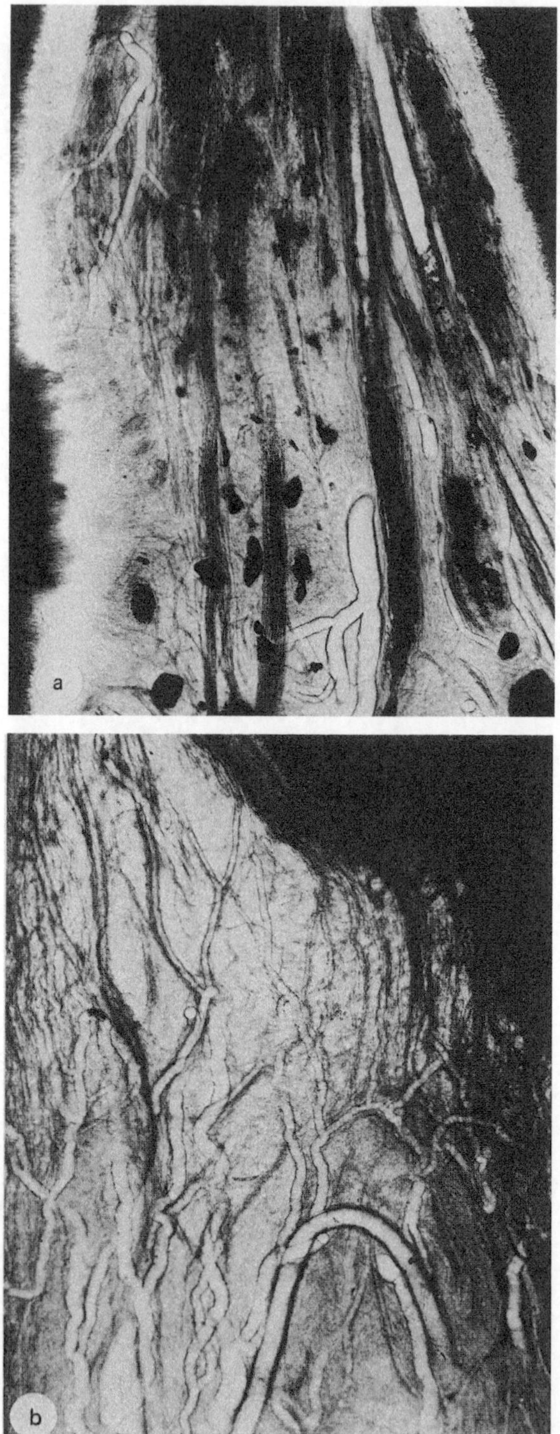

Figure 7. (a) A thick section (150 μm) of the coronal portion of the pulp from an old adult (63 years). Note the diffuse calcification throughout the pulp, the calcified blood vessels, and the loss of demonstrable vessels throughout the pulp. Iron hematoxylin stain. ×50. (b) A thick section (150 μm) of the coronal pulp from a young adult. There is a dense vascular network distributed throughout the pulp. Compare to Fig. 7(a). Iron hematoxylin. ×50.

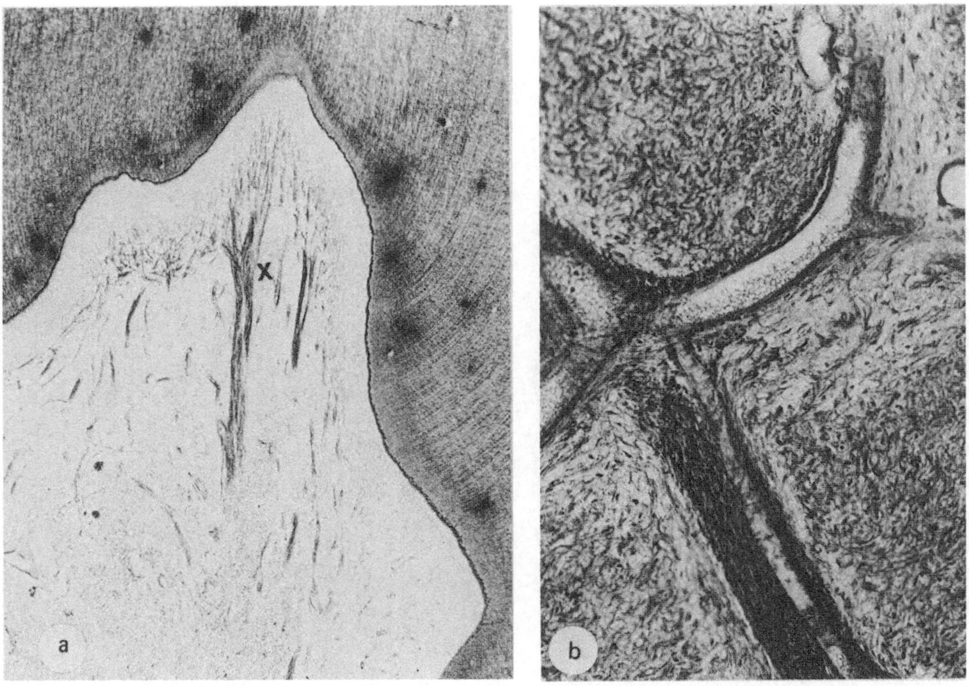

Figure 8. (a) Section of an adult functional tooth from a 30-year-old person. The nerves are very prominent; the stroma consists of fine collagenous fibers. Silver nitrate impregnation. ×35. (b) High magnification of area X shows that the thick collagen fiber bundles are associated with blood vessels and nerves of pulp. Silver nitrate impregnation. ×250.

(1966). They all agree that there is an increase in the thickness of the collagen fibers and changes in the chemical and physical properties of the collagen during the aging process. The pulpal stroma, in contrast to the dermis of the skin, consists mainly of fine collagenous fibers interspersed in the abundant ground substance. At no time during the life of the tooth are thick collagen fibers seen independent of the connective tissue sheaths. The data of the current study indicate that the connective tissue of the pulp is specialized in nature; no comparison can be made between the aging effect of the connective tissue of the pulp and other areas rich in connective tissue such as skin or even gingiva.

Our findings also agree with the conclusions of Shroff (1953) and Stanley and Ranney (1962) that pulpal fibrosis in old teeth is not the result of a continual formation and reorientation of collagen fibers during the aging process. The prominence of fiber bundles in old pulps may be, in part, attributed to the persistence of the connective tissue sheaths in a narrowed pulpal chamber after the vascular and neural structures are decreased or lost.

The cellular composition of the human pulp is modified during the aging process. Frolich (1970) has shown that the number of cells in pulp are reduced by 50% in aged

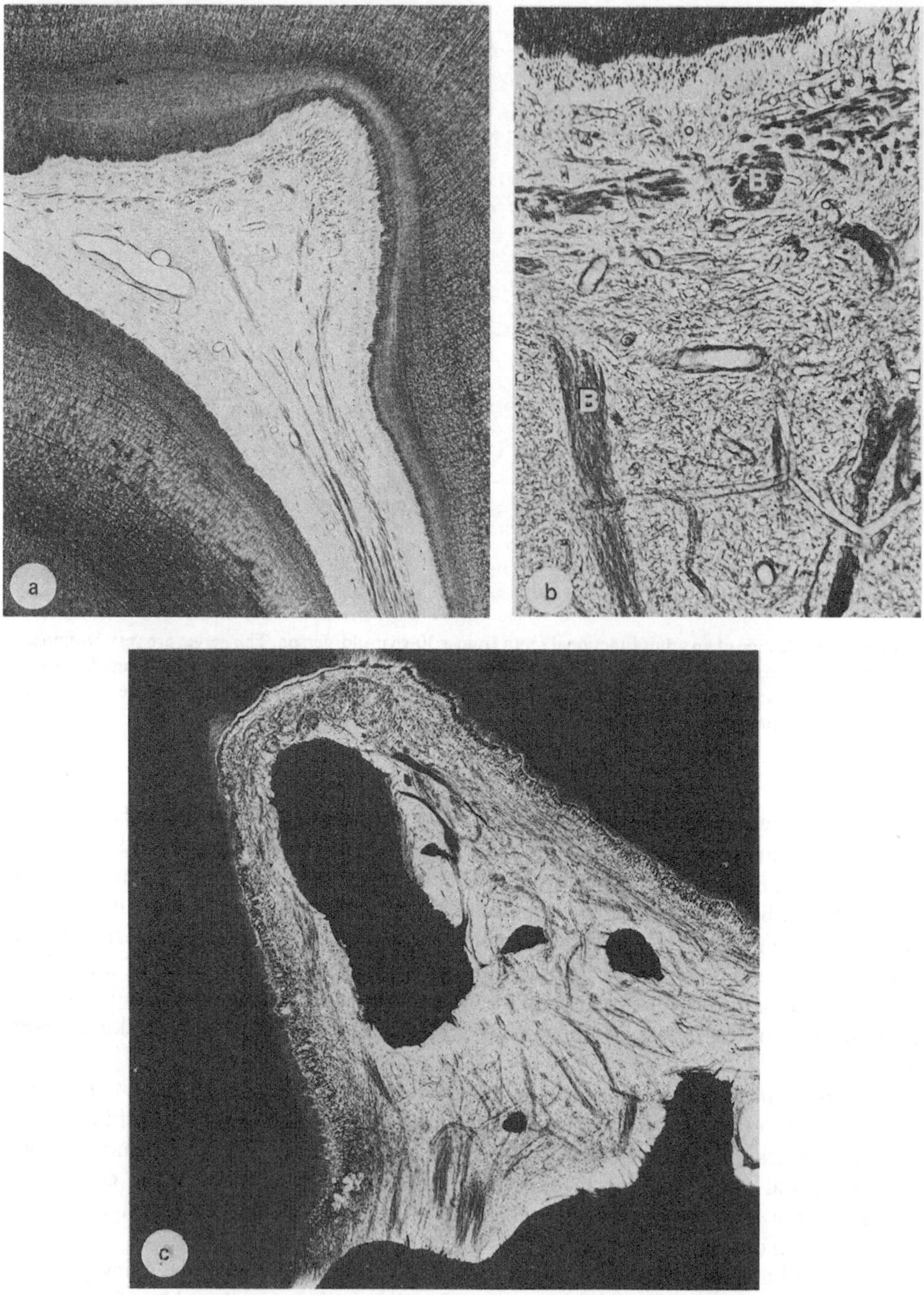
a
B
B
b
c

pulps. In aging rats, for example, there is a loss of 75% of the cells in comparison to cells in "young" pulps (Pinzon *et al.*, 1967). The kinetics of rat molar pulp cells at various ages was investigated by Pinzon *et al.* (1966). They reported that there was a significant decrease in the number of regnerating cells with increasing age.

Fibroblasts show degeneration with increasing age as characterized by smaller size and a decrease in the number of organelles including the rough endoplasmic reticulum, mitochondria, and Golgi complex (Han, 1968). Odontoblasts also exhibit degeneration with age. Ultrastructural studies reveal an increase in vacuole numbers and gradual degenerative changes leading to the absence of cells over some or all of the pulpal surfaces (Bhussry, 1968).

During the aging process, the Von Korf fibers, which are collagenous fibers that emerge from the predentine and pass into the pulp, become accentuated. In nonerupted teeth, fine Von Korff fibers emerge from the predentin to pass through the odontoblastic layer into the zone of Weil (Fig 10a). The argyrophilic fibers become more prominent with eruption and function, and extend into the pupal–odontoblastic border (Fig. 10b). The predentin–pulpal border in the old teeth is very irregular in outline. Concomitantly, there is a fusion of the fine Von Korff fibers. These fibers appear to form short bundles that are limited in their course to the odontoblastic layer (Fig. 10c).

The semiqualitative presence of the protein–acid mucopolysaccharide (G.A.G.) complex is demonstrated by the alcian blue–PAS combination. With this histochemical reaction, the glycosaminoglycans are alcianophilic, whereas the glycoproteins and sialic acid are Schiff-positive. When sections of young functional, and old teeth, respectively, are exposed to this histochemical combination, the predentin and the pulp are alcian blue-positive, whereas the dentin stain pink in the young pulps. In comparing the intensity of alcianophilia between the young and old pulps, little difference appears in the staining response of the pulps. The coronal and root pulpal tissue is alcian blue-positive. The core of the root and crown pulp is alcianophilic in old pulps. The presence of the calcified masses in the root pulp accentuates the Schiff intensity in becoming deep red in color. When there is diffuse calcification in the coronal pulp, the calcified masses stain Schiff-positive in contrast to the surrounding pulpal ground substance that appears deep blue.

Since the pulp is the loose connective tissue in nature, it is rich in proteoglycans. It thus stains intensely blue with alcian blue. The staining response of the pulp of old teeth to the alcian blue and PAS combination do not change drastically from that in the young pulps. The only structures that react differently with this histochemical combination are the vascular and neural connective tissue sheaths. In contrast to the alcianophilia of these collagen bundles in young pulps, the old fibrous bundles are Schiff-

←

Figure 9. (a) Section of "old" molar tooth. Thick fiber bundles follow pathways of nerves and vessels. Silver nitrate impregnation. ×20, reproduced at 90%. (b) High magnification of pulp-odontoblastic region. Thick fiber bundles are prominent and are related to the pathways of blood vessels and nerves. Fibrous stroma still consists of fine collagenous elements. **B**, bundle of collagenous fibers. Silver nitrate impregnation. × 125, reproduced at 90%. (c) Section of an old tooth from an individual over 65 years of age. The fibrosis of pulp is related to persisting connective tissue sheaths of vascular and neural structures. Note the diffuse calcification is associated with the collagenous fibers (area X). Iron hematoxylin–picrofuchsin. ×50, reproduced at 90%.

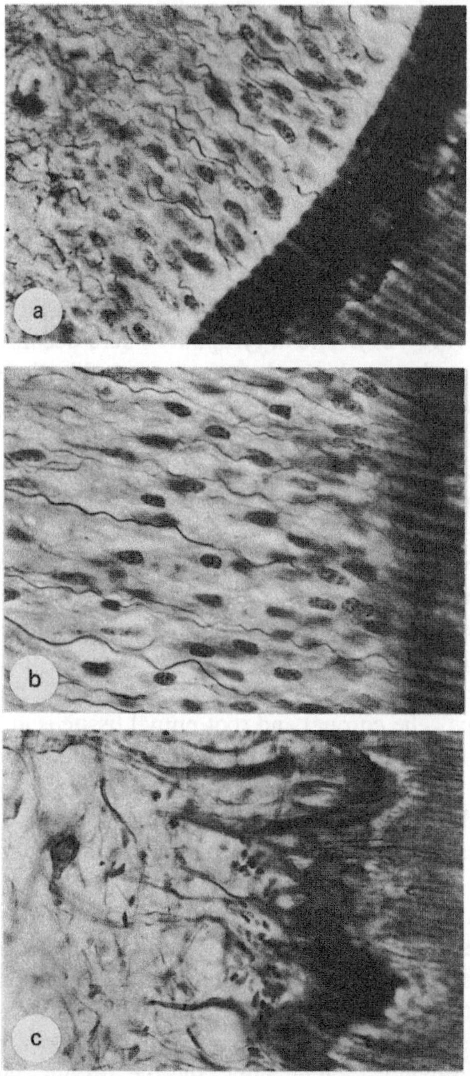

Figure 10. (a) Odontoblastic-predentin region from a nonerupted molar. Note the fine Von Korff fibers (collagen) emerging from predentin and passing into Zone of Weil. Silver nitrate impregnation. ×125. (b) Young adult tooth. The Von Korff's fibers are thicker and pass through Zone of Weil into pulp proper. Silver nitrate impregnation. ×125. (c) An aged tooth. In contrast to long individual fibers that stream through Zone of Weil in the young adult tooth, fibers are thick in the old teeth and appear to fuse into short fiber bundles that terminate in Zone of Weil. Silver nitrate impregnation. ×125.

positive. Bhussrey (1968) and Zerlotti (1964) reported a decrease in mucopolysaccharides in aging pulps, and Mancini and associates (1959), Wentz and associates (1952), and Lorincz (1960) agreed that the amount of amorphous ground substances decreases as the number of collagen fibrils increases in the skin and elsewhere. Since biochemical analysis is not carried out in our study, a definite conclusion cannot be made concerning the amount and nature of the mucopolysaccharide–protein complex in old pulps. However, the semiqualitative histochemical data from the study by Bernick and Nedelman (1975) would indicate that any comparison regarding the loss of ground substance for skin and other connective tissue areas is not valid because the nature and amount of mucopolysaccharides differ in each tissue. In addition, it may be assumed that the decrease in ground substance previously reported for old pulps may partly be due to the decrease in the pulpal area and not due to the nature of the carbohydrate moiety.

2.2. Periodontium

Although the increasing incidence of periodontal disease with advancing age is well documented (Klingsbury and Butcher, 1960; Belting *et al.*, 1953; Sicher and Weinmann, 1944), the problem of intrinsic age changes in the periodontium has received cursory or sporadic attention. Scattered reports deal with age-related apical downgrowth of the epithelial attachment, narrowing of the periodontal ligament, loss of alveolar bone, and isolated changes in the gingiva (King, 1949; Mitchell, 1951; Rushton, 1951; Glickman, 1965).

Age changes affect the following periodontal tissues: (a) vasculature; (b) periodontal ligament; (c) cementum; (d) gingiva and alveolar mucosa; and (e) alveolar bone.

2.2.1. Vasculature

Arteriosclerosis is a frequent finding in aging humans (Grant and Bernick, 1970; Stahl and Fox, 1953). It may be observed in the large muscular vessels within the gingiva, palate, alveolar bone (Figs. 11a, 11b). These vessels are characterized by thickening of the intima which was the result of both cellular and fibrous proliferation.

The small muscular arteries supplying the periodontal ligament also exhibit arteriosclerotic changes. The vascular lesion consists of marked intimal thickening which partially occludes the lumen of the vessel (Fig. 12a). Gingival vessels of aged humans also exhibit severe intimal changes. These arteries are located throughout the palate and lip and exhibit pathology similar to that found in the periodontal ligament (Fig. 12b).

The microvasculature supplying the periodontium features age changes differing from the arteriosclerotic alterations. The walls of both the venules and arterioles within all the tissues of the aged periodontium stain deep red following the exposure to the PAS reaction giving them a "hyalinized" appearance (Fig. 13a). There is no regional predilection to microvasculature changes, for the gingival vessels are affected with the same frequency as are the periodontal ligament vessels. Age alterations are also seen in the capillaries of the area. Their walls are thickened and intensely Schiff positive. These

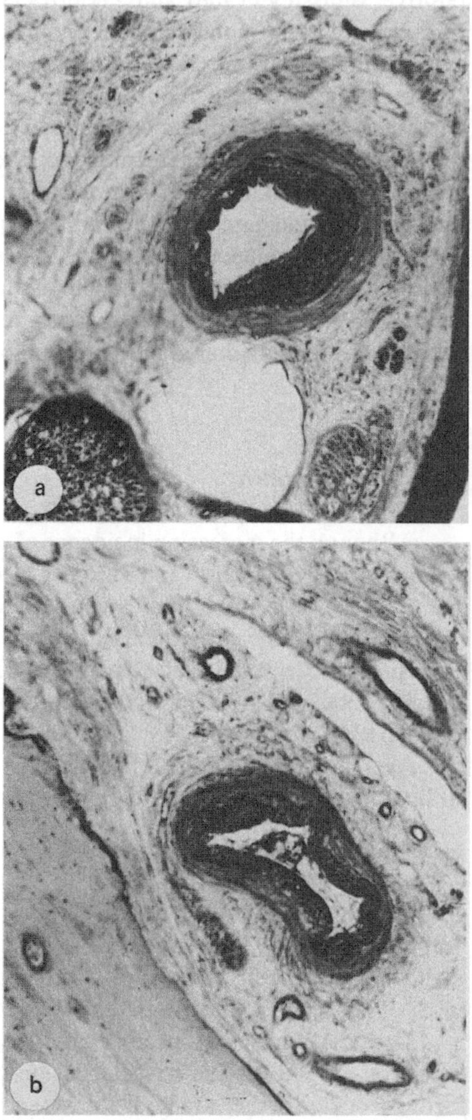

Figure 11. (a) Alveolar artery from a 76-year-old male showing changes characteristic of arteriosclerosis including intimal thickening with cellular and fibrous proliferation. Iron hematoxylin–picrofuchsin. ×125. (b) A large artery located in the palate of a 65-year-old individual. Note the thickening of the intima. PAS–hematoxylin. ×125.

changes are similar to those involving capillaries throughout the body (Ballard, 1979; Bernick *et al.*, 1977). These alterations are not observed in the periodontium of young adults (Fig. 13b).

Calcification of the small arteries and arterioles occur in the periodontal ligament of old individuals. This process progresses from the adventitia inward involving the media and finally the intima (Fig. 13c). This process is in contrast to intimal calcification seen in arteriosclerotic vessels.

In contrast to the age changes to the microvasculature, the lymphatic vessels draining the gingiva and periodontal ligament do not exhibit any morphological alterations in the aged humans. For example, in old inflamed gingivae the lymphatic vessels of all calibers are distended. One may conclude therefore that these vessels are still able to provide a drainage system for the continuous removal of excess fluid proteins and the particulates from both noninjured and injured sites in old gingivae (Figs. 13d, 13e).

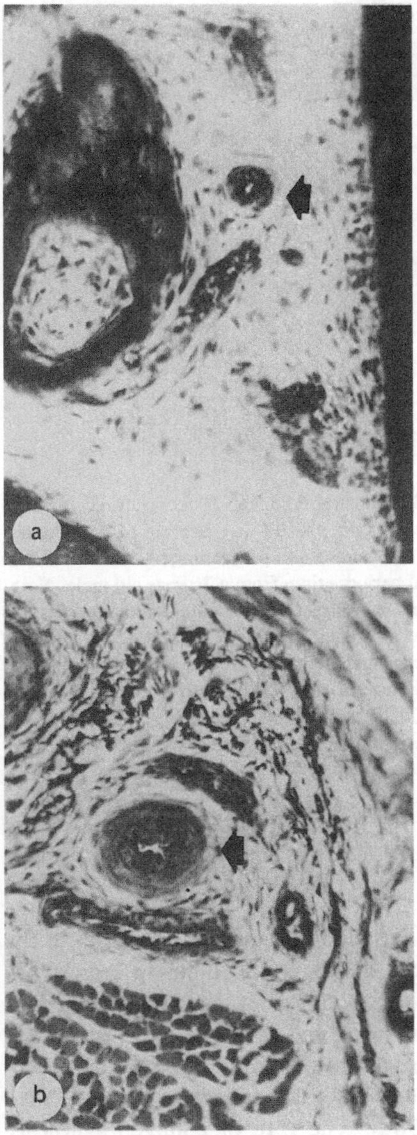

Figure 12. (a) Section of periodontal ligament from a "older" tooth. Note the arteriosclerotic vessel adjacent to the alveolar bone. Insert "b" illustrates a higher magnification (×500) of the sclerotic artery. The narrowing of the lumen is due to intimal proliferation. Iron hematoxylin. ×75. (b) Section of periodontium from an "older" tooth. Note the arteriosclerotic vessel (X) in the gingiva. Insert "b" illustrates a higher magnification (×500) of the sclerotic arteriole (X). The intimal thickening resulted in narrowing of the lumen. Iron hematoxylin. × 75.

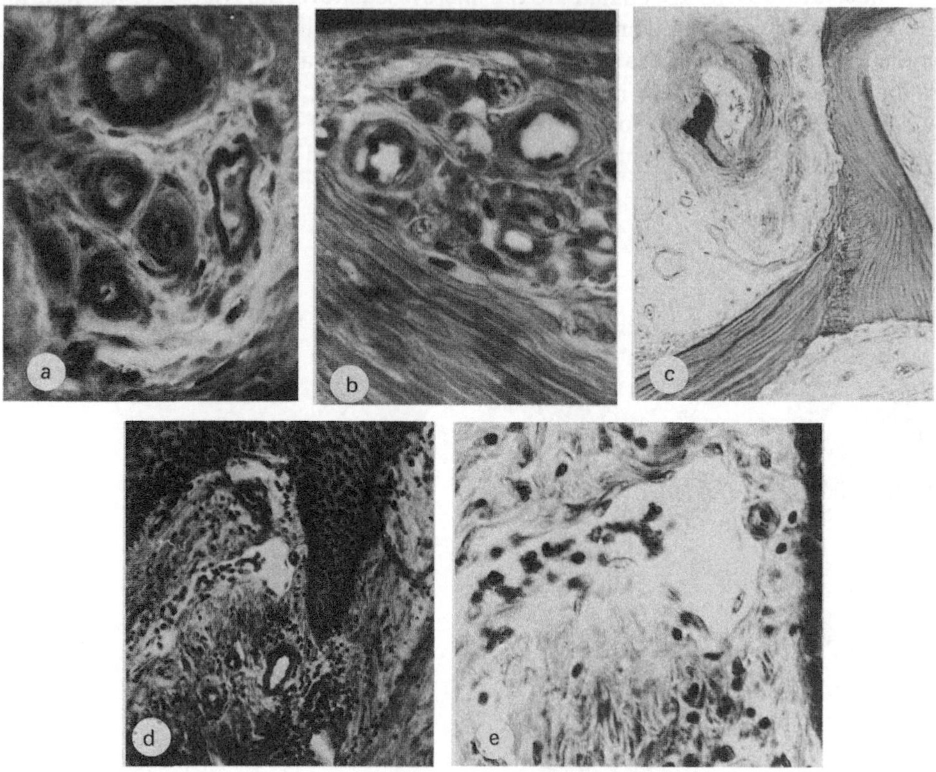

Figure 13. (a) Section of periodontal ligament from an aged individual. The wall of the arteriole appears deep red following exposure to the PAS reaction. PAS–hematoxylin. ×200. (b) Section of periodontal ligament from a 19 year old individual. In comparison to the arteriole seen in preceeding figure, the wall of the vessels stain pink. PAS–hematoxylin. ×200. (c) Section of periodontium from a 76 year old male. Note the process of calcification involving the advertitia and media in contrast to intimal calcification seen in arteriosclerotic vessels. PAS–hematoxylin. ×100. (d) A buccal lingual section of gingiva from a molar tooth of a 76 year old male. An inflammatory infiltrate is seen in the lamina propria. Notice the dilated lymphatic vessel within the papilla. A., arteriole; L., lymphatic vessel. PAS–hematoxylin. ×35. (e) A higher magnification of the lymphatic vessel seen in Figure 13d. The distended lymphatic vessel consists of an endothelial lining and lack an intact basement membrane. Lymphocytes are found within the lumen of the vessel. E., endothelial cell. PAS–hematoxylin. ×250. (Reproduced at 80%.)

2.2.2. Periodontal Ligament

It is well accepted that the thickness of the periodontal ligament decreases in the aged (Coolidge, 1937; Tonna, 1973). Both the cell numbers and labeling indices are found to decrease with the aging of the periodontal ligament as demonstrated in [^{3}H]thymidine studies using large populations of mice and rats (Stahl *et al.*, 1969; Tonna *et al.*, 1972). Studies have shown a decrease in protein-bound hexose and mucoproteins (Fleider, 1962). Other studies describe a steadily decreasing number of DNA-synthesizing cells with age (Jensen and Toto, 1968). The rate of collagen synthesis decreases with age in rats when measured by *in vitro* labeling method (Schneir *et al.*, 1976). In addition, it has been shown that the cells of the periodontal ligament of old mice do not renew as rapidly as those of young animals (Toto and Borg, 1968).

The periodontal ligaments seen in young adults are highly cellular with the fibroblasts oriented along the collagenous fibers. Cementoblasts and osteoblasts are present in abundance. The attachment fibers are deeply embedded in the alveolar bone proper (Fig. 14a). In contrast, the fiber bundles in the "aged" periodontal ligaments are broad and wavy. Few fibroblasts, osteoblasts, and cementoblasts are present (Fig. 14b). The

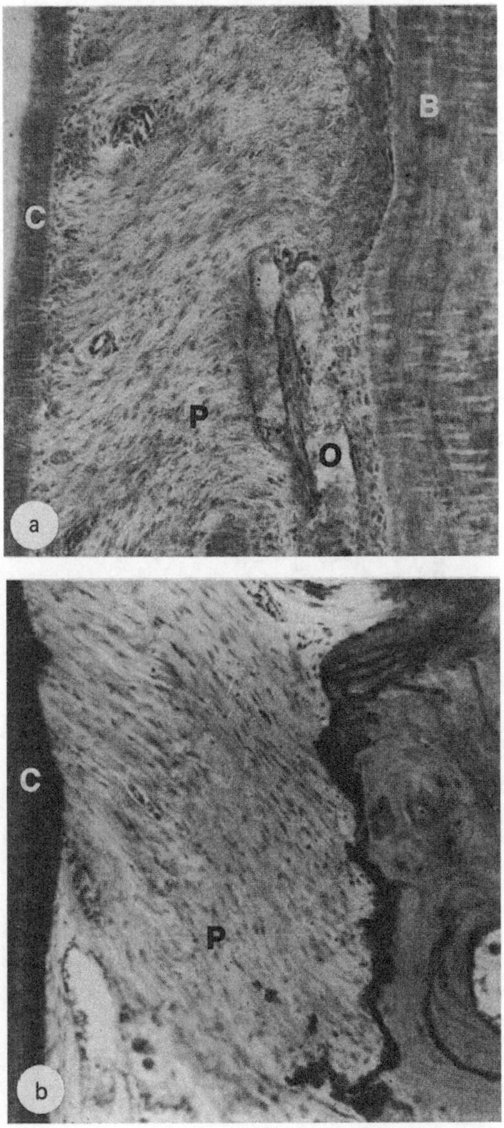

Figure 14. (a) Section of periodontal ligament, bone, and cementum of a "young" tooth. Well-defined Sharpey's fibers are seen emerging from the lamellar bone to become distinct attachment fibers of the periodontal ligament. Note the cellularity of the ligament. B, bone; C, cementum; O, osteoblasts; P, periodontal ligament. PAS–hematoxylin. ×125. (b) Section of periodontal ligament, bone, and cementum of an "aged" tooth. Note the thin darkly stained margin at the alveolar bone proper. There is an apparent decrease in demonstrable osteocytes, osteoblasts, cementoblasts, and other cells of the ligament. Sharpey's fibers are poorly visible, and the attachment fibers stain intensely Schiff positive. B, bone; C, cementum; P, periodontal ligament. PAS–hematoxylin. ×125.

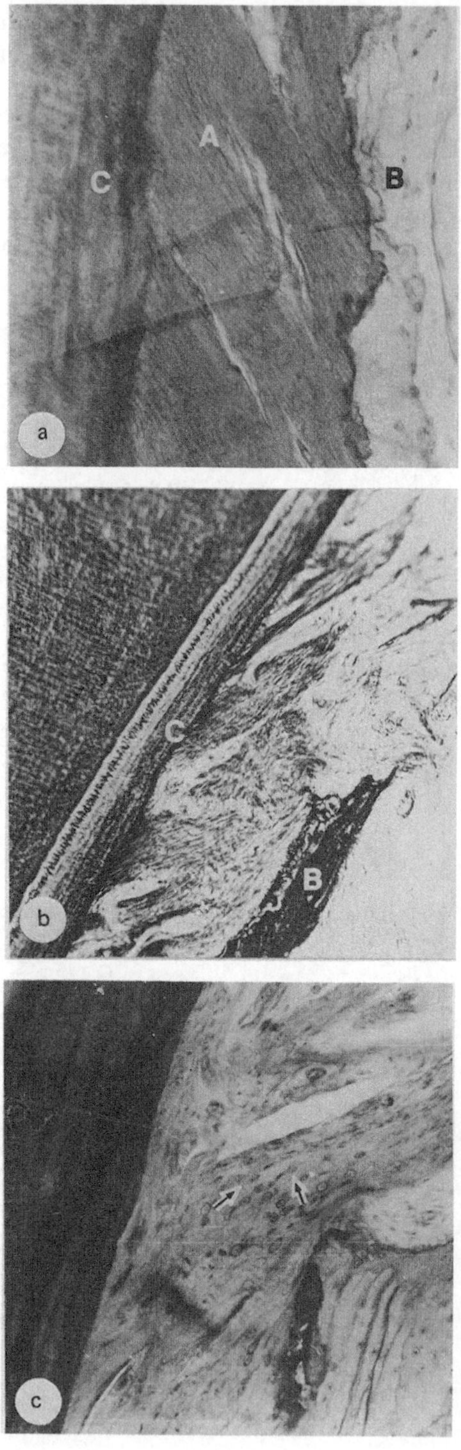

Figure 15. (a) Increased fuchsinophilia of ligament fibers is found in old specimens exposed to alcian blue–PAS combination. This may indicate an increased concentration of neutral carbohydrates in the ground substance of the ligament space. C, cementum; B, alveolar bone; A, PAS positive ligament fibers. ×125. (b) A mild argyrophilia is exhibited by periodontal fibers in the 76-year-old man. Principal fibers are well organized, but individual fibers are less distinct. Sharpey's fibers are not demonstrated in bone. C, cementum; B, alveolar bone. Silver nitrate impregnation. ×125. (c) Cartilage-like tissue is seen in the periodontal ligament of the 72-year-old man in areas where past injury could be discerned. The chondroid tissue is bordered by hyalinized areas in the periodontal ligament. Note the lacunae in the area. PAS–hematoxylin. ×125.

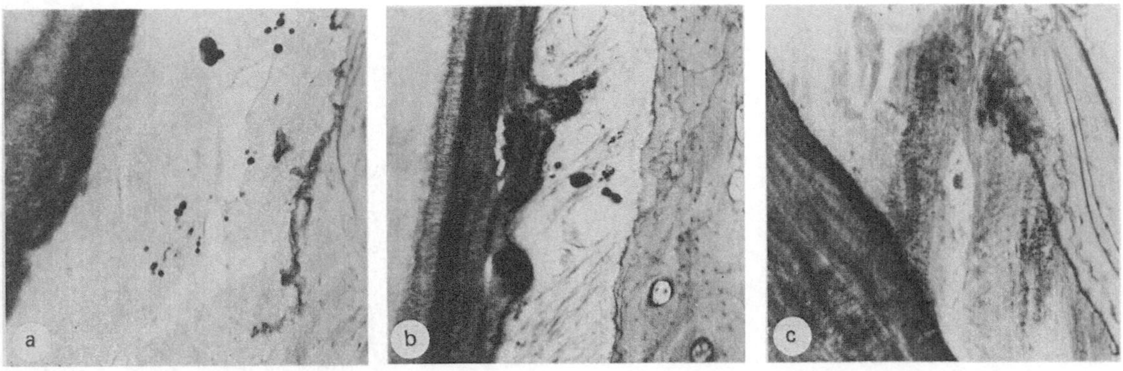

Figure 16. (a) Calcified bodies are frequently found in the periodontal ligaments of old individuals. Globular calcifications of different sizes are on and between fiber bundles. X, globular calcified bodies. PAS–hematoxylin. ×125, reproduced at 80%. (b) Spurring of bone and cementum was seen. Small calcospherites are present near bone and appeared to coalesce to form larger, rounded or ovoid bodies in a molar tooth from a 72-year-old man. PAS–hematoxylin. ×125, reproduced at 80%. (c) The calcifications sometimes are present along an entire fiber bundle, extending between bone and cementum to ankylose the tooth. PAS–hematoxylin. ×125, reproduced at 80%.

aging attachment fibers stain intensely with the PAS reaction in contrast to the light pink coloring of "young" collagen fibers. When sections of "aged" ligaments are stained with the alcian blue–PAS combination, the ground substance stains red in contrast to the alcianophilia exhibited in the ligaments obtained from younger adults (Fig. 15a). In addition, the attachment fibers in aging ligaments appear less distinct following silver nitrate impregnation than the black appearance of the fibers in young individuals (Fig. 15b).

It is not uncommon to find "hyalinized" areas in the aging periodontal ligament. The hyalin areas are strongly Schiff positive and are chondroid in nature. Cells within lacunae are scattered throughout the "hyalinized" matrix (Fig. 15c). The appearance of these cartilagelike areas could be associated with traumatic injury of the ligament.

Calcified bodies within the ligament are common in the aging ligaments. Rounded calcospheroidlike bodies are demonstrable on and between fiber bundles. Two types of globular calcified bodies are seen, small round bodies and large ovoid or irregular structures (Figs. 16a, 16b). Tiny calcospheritelike bodies seem to emerge from bony and cemental surfaces and extend along fiber bundles, sometimes coalescing into larger, rounded masses that eventually completely calcify the fiber bundle (Fig. 16c).

Epithelial nests in the periodontal ligament show altered forms of aggregation in aging. Epithelial nests aggregates tend to contain more cells with both proliferative and degenerative morphology (Reeve and Wentz, 1962). The aggregates are frequently enclosed by a thickened basement membrane (Fig. 17a). The epithelial aggregates are found irregularly located in the ligament space, near the cementum, midway in the space, and near the bone. It is not unusual to find nests undergoing degeneration, or becoming completely calcified (Figs. 17c, 17b).

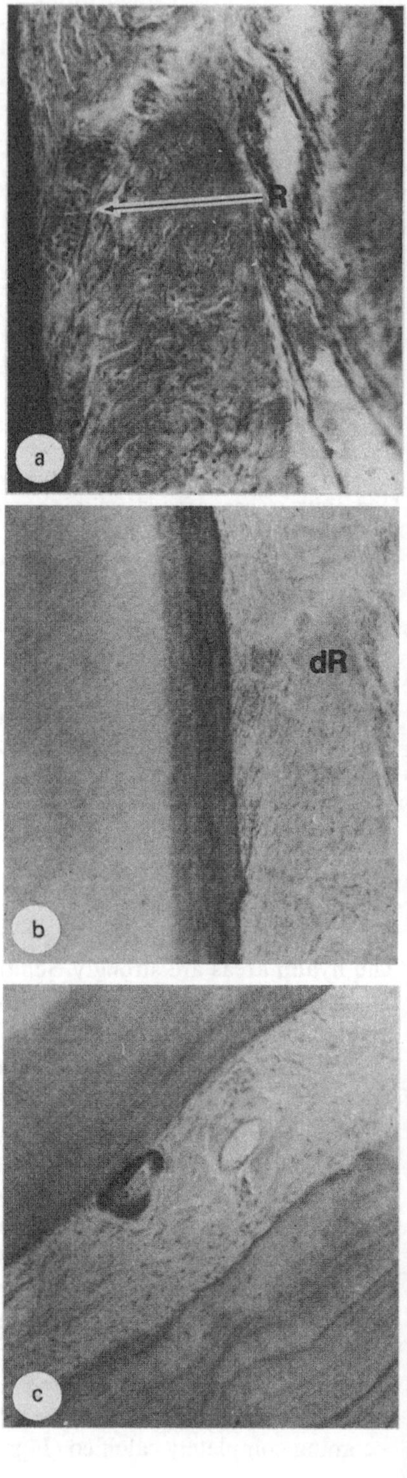

Figure 17. (a) Epithelial rest groupings are irregularly located in the periodontal ligament from aged individuals. This enlarged congregate shows many cells, some degenerating, surrounded by a thickened PAS positive basement membrane. PAS–hematoxylin. ×100. (b) Periodontal ligament from a 76-year-old male. Degenerating epithelial rests are seen throughout the periodontal ligament. A loss of basement membrane around the rest aggregates accompanied by cell degeneration as individual cells become indistinct. PAS–hematoxylin. (c) Periodontal ligament from a 73-year-old female. Note that an epithelial nest aggregate adjacent to the cementum is becoming calcified, forming a cemental spur. PAS–hematoxylin. ×100.

2.2.3. Cementum

Cementum deposition appears to be continuous throughout life. A straight line relationship has been shown between age and cementum thickness (Klingsberg and Butcher, 1960; Zander and Hurzler, 1958). Cemental deposition is less near the cemento–enamel junction and greater in the apical area. The laminar deposition results in a broad zone of cementum covering the root (Fig. 18a). Furthermore, in the aging

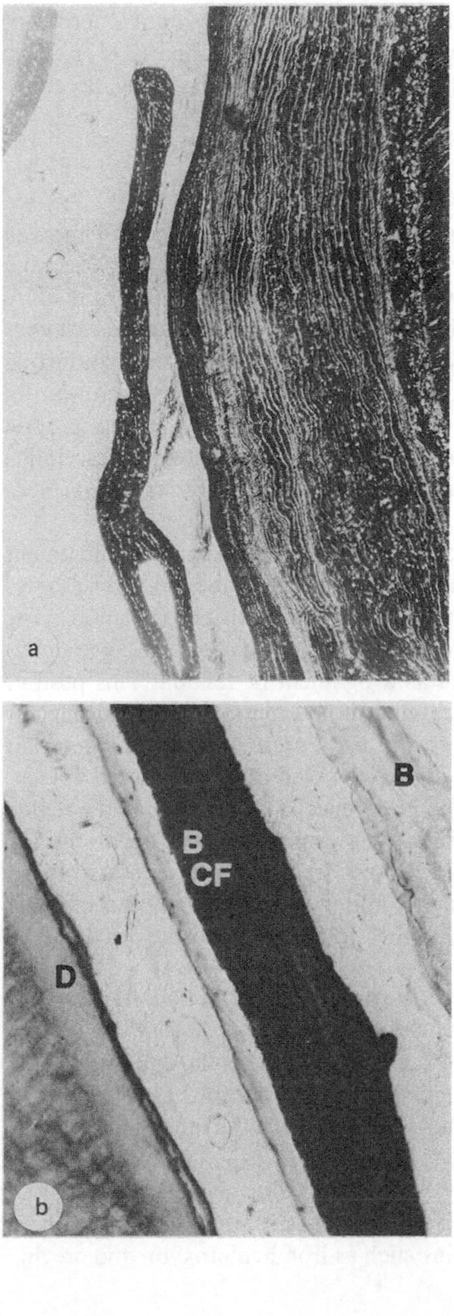

Figure 18. (a) Section of the root of a premolar of a 92-year-old woman. The continuous rhythmic deposition of cementum is so extensive that the original form of the root was altered. PAS–hematoxylin. ×75. (b) Midroot area of a premolar tooth from a 76-year-old male. A large fragment of cementum has now separated from the tooth and forms a part of the septal bone. Bone has been deposited on the ligament side of the cementum, and the principal fibers extend from this new bone to the tooth. B, bone; CF, cemental fragment; D, dentin; C, detached cemental fragment. PAS–hematoxylin. ×100.

humans, attachment of cementum to dentin may be weakened; the frequent cemental tears seen in specimens from aged humans may be related to changes in the organic matrix of the cementum, to reduce vascular supply, or to the thickened and less extensible ligament fibers embedded in the cementum (Fig. 18b). Spurring of cementum is sometimes the result of the fusion of calcospheroid bodies near cementum or of the calcification of epithelial nest aggregates. It is not unusual to find ankylosis between the cementum and alveolar bone.

Both light and electron microscopic studies have revealed that cellular cementum undergoes degeneration with age. Gottlieb (1942) demonstrated by light microscopy that the nuclei within lacunae in aging cementum are pyknotic. Tonna (1977), using electron microscopy, has shown that there are similarities between the aging of cementocytes and osteocytes.

2.2.4. Gingiva

One of the changes reported in aged gingiva is the decrease in the thickness of keratinization (Courtney, 1972). Nedelman *et al.* (1970) compared the gingivae from aged individuals that were dentulous; some wore dentures while others were denture free. They noted that the strateum corneum became progressively thicker in non-denture-bearing mucosa while in those individuals that wore dentures, the keratinizing layer was thinner progressively related to the length of time the denture was worn in the mouth. Age-related changes to the components of the underlying lamina propria are few in number (Wentz *et al.,* 1952; Stahl and Fox, 1953). The following description will deal with age changes in human specimens that we have previously examined (Nedelman and Bernick, 1978; Grant and Bernick, 1972).

The connective tissue of the lamina propria of all the dentulous and edentulous specimens show marked changes. When sections are exposed to the PAS reaction of glycoproteins the connective tissue elements stain intensely PAS positive (Fig. 19a). The collagenous elements are thick and arranged in dense parallel bundles similar in architecture to regular dense connective tissue such as a ligament or tendon. This positive histochemical response by the connective tissue gives it at times a hyalinized appearance. In addition, there are regions adjacent to the bony crest or spurs of bone that appear cartilagelike or chondroid. The fibroblasts are rounded up and enclosed in lacunar spaces. The intercellular material appears homogeneous in nature, giving the region a chondroid appearance. The appearance of the lamina propria in the 14-year-old boy differs from that of the "old" gingivae in that the fibers are thinner in diameter, irregular in distribution, and have a greater amount of ground substance between the fiber bundles (Fig. 19b). In addition, it stains light pink following exposure to the PAS reaction.

The changes found in the lamina propria of the aging gingiva are similar to those changes described in aging connective tissue elsewhere in the body (Clausen, 1966; Verzar, 1964; Gross, 1961; and Milch, 1966). These changes to the collagenous elements as well as the ground substance could alter the physical properties of these elements so that exchange and diffusion of nutrients from the circulation into the interstitium might be hindered.

One of the unusual findings in the alterations to the collagenous elements of the lamina propria of aged gingiva is the presence of an abnormal number of thick fibers that stain positive with various elastic tissue stains such as iron hemotoxylin and orecin.

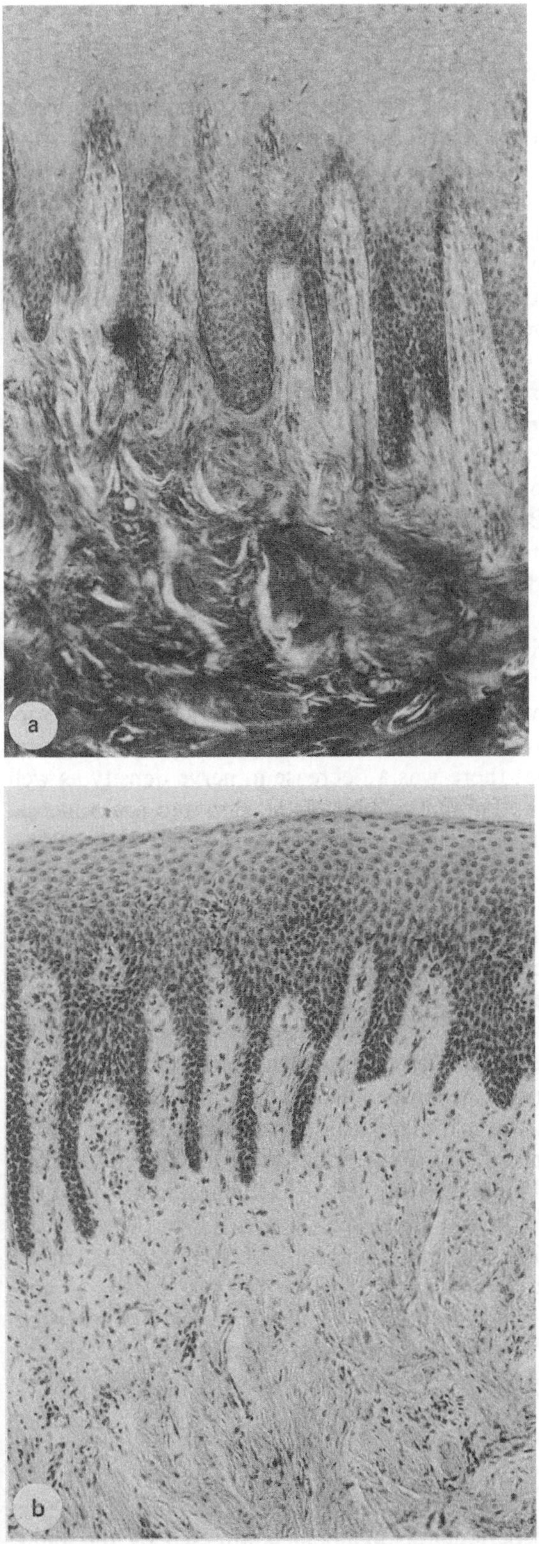

Figure 19. (a) Section of gingiva from a dentulous individual 65 years of age. The collagen bundles in the zona reticularis are thick and the region stains intensely PAS positive. PAS–hematoxylin. ×125. (b) Section of gingiva from a dentulous individual 14 years of age. Note that the fibers are thinner, more irregular in distribution, and have a greater amount of ground substance. The area stains light pink. PAS–hematoxylin. ×125.

The finding of these aberrant number of fibers staining positive with specific elastic tissue stains was not anticipated as it is believed that the gingiva contain only a sparse number of elastic tissue fibers that are associated with the blood vessels. The positive elasticlike fibers are thick and usually radiate into the connective tissue papillae as long bundles (Fig. 20a). Other thick fibers course in all directions through the zona reticularis. When alternate sections from all blocks of tissues were first pretreated with acid hydrolysis or collagenase before staining with elastic tissue stains, the resultant stained sections showed a marked loss of positive staining fibers throughout the lamina propria of both the attached and free gingivae (Fig. 20b). A few thin elastic fibers are still present in the adventitia of the blood vessels. Silver nitrate impregnation of alternate sections reveal that the elasticlike fibers are argyrophilic, a characteristic of collagenous fibers (Fig. 20c). Elastic fibers, on the other hand, do not exhibit argyrophilia when treated with silver nitrate. The only other connective tissue site in which there is an increase in the number and thickness of elasticlike fibers is in the dermis of aging skin that has been exposed to actinic rays (Banfield and Brindey, 1963). The nature of these elastinlike fibers is still controversial. It is felt, based on our findings, that these elastinlike fibers in the gingivae are probably degenerating collagen fibers that are able to stain with elastic tissue stains.

Although changes to the vascularity of the gingivae from older humans have been reported earlier (Grant and Bernick, 1969), the alterations to the nerves have been largely overlooked. The prevailing conviction was that there were no substantial structural changes in nerves and their terminal endings once their development was completed. Winkelman (1965) reported that there was a decrease in nerve density as well as widely separated nerve endings in the lamina propria of the skin and oral mucosa. In our study, the large nerves exhibit thick neural connective tissue sheaths (Fig. 21a). In addition, the smaller-sized nerves appear fibrotic in nature (Figs. 21b, 21c). Fibrosis of all nerves is more apparent with the loss of dentition. Degenerative characteristics such as reticulation and fragmentation of nerves are prominent in the lamina propria of older gingivae whether the jaws are dentulous or edentulous (Fig. 21d). There are also structural changes in the nerve terminals so that only a few nerve fragments, or none, are seen in the connective tissue papillae of the older gingivae (Fig. 22b) in contrast to what is seen in young gingivae (Fig. 22a). Figure 23a illustrates a section of gingivae from an adult individual stained for nonspecific esterase. The demonstration of positive granules in the connective tissue papillae is similar to cholinesterase localization demonstrated by Avery and Rapp (1959). Notice that there is marked decrease in the staining of cholinesterase in aged gingivae (Fig. 23b). The neural loss seen in the older gingivae is similar to what occurs in the aging skin. Cauna (1965) studied the alteration in the receptor organs in aging skin and came to the conclusion that Meissner's corpuscles are progressively reduced in number. Nerve endings within the corpuscle underwent changes from the neural fibrillar network to fine reticulated irregular nonmyelinated nerves. The decrease in nerves in "older" gingivae in both dentulous and edentulous jaws in older individuals may lead to a loss of sensitivity of gingivae to external stimuli.

2.2.5. Alveolar Bone

Alveolar bone associated with the periodontal ligament is affected by the aging process. The cribriform plates are narrow with a strongly basophilic staining with hem-

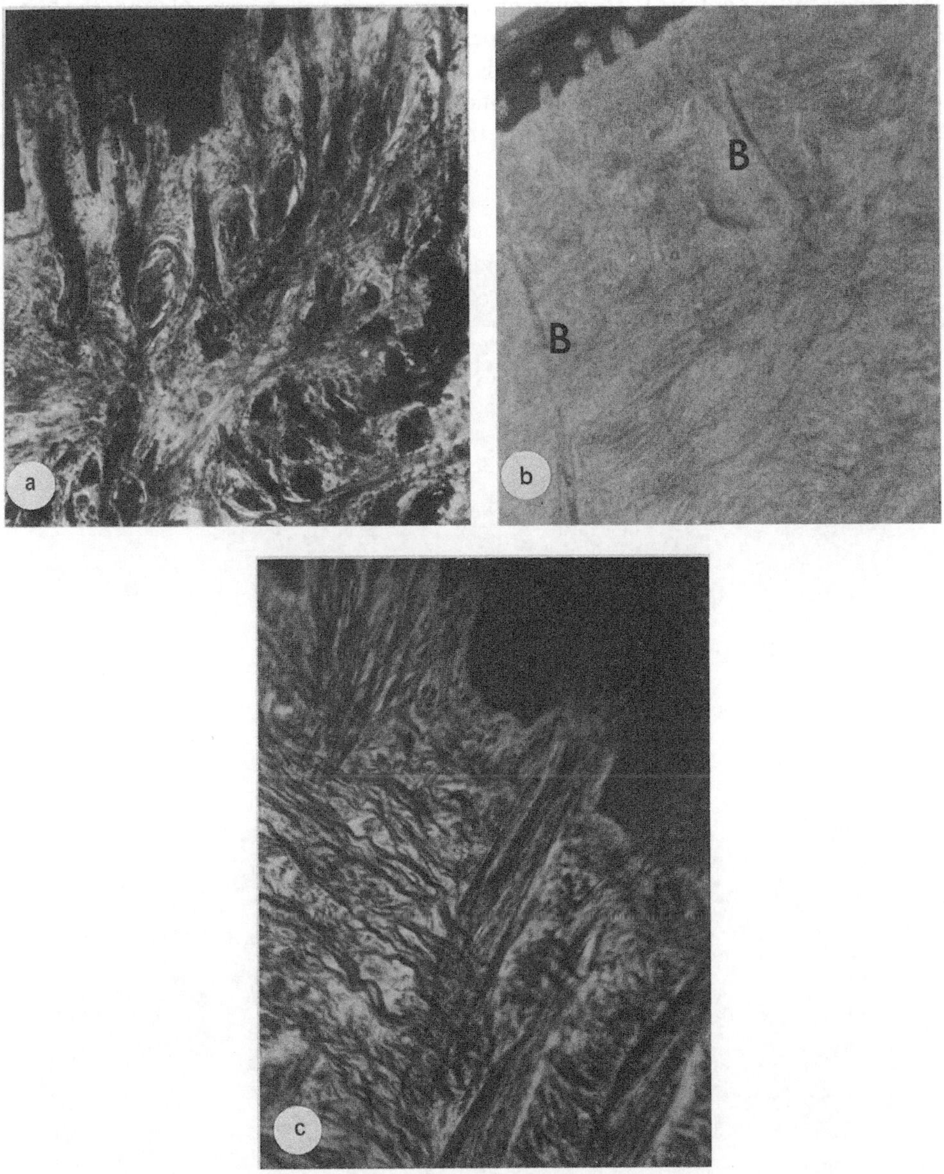

Figure 20. (a) Section of the lamina propria of gingiva of an upper molar tooth from a 72-year-old male. Notice the fibers which stain positive for elastin radiating into the connective tissue papillae, in zona reticularis, and in aggregates. Iron hematoxylin. ×75. (b) Alternate section of the gingiva illustrated in the preceding figure that was pretreated with 1.0 *N* HCl before being exposed to iron hematoxylin staining. In contrast to Fig. 20a, the section shows a loss of black staining fibers throughout the lamina propria. Only elastic fibers associated with blood vessels are present. B, blood vessels. Iron hematoxylin. ×50. (c) Section of gingiva from a 72-year-old male. The section was exposed to silver nitrate impregnation for the demonstration of collagenous elements. Notice the argyrophilic fibers that radiate into the connective tissue papillae as well as in the zona reticularis are similar to the pseudoelastic fibers seen in Fig. 20a. ×100.

otoxylin (Fig. 14b). There is a reduction in the width of the bone with a decrease in bone cellularity in contrast to the number of cells seen in young bone. Neither the lamellar arrangement of the alveolar bone proper nor the embedded ends of the attachment fibers are demonstrable tincitorially. It has been shown clinically that there is a significant alveolar bone resorption but insignificant crest height reduction with age (Dean *et al.*, 1973). Tonna (1973) has shown in the aging mouse that alveolar bone crest loss

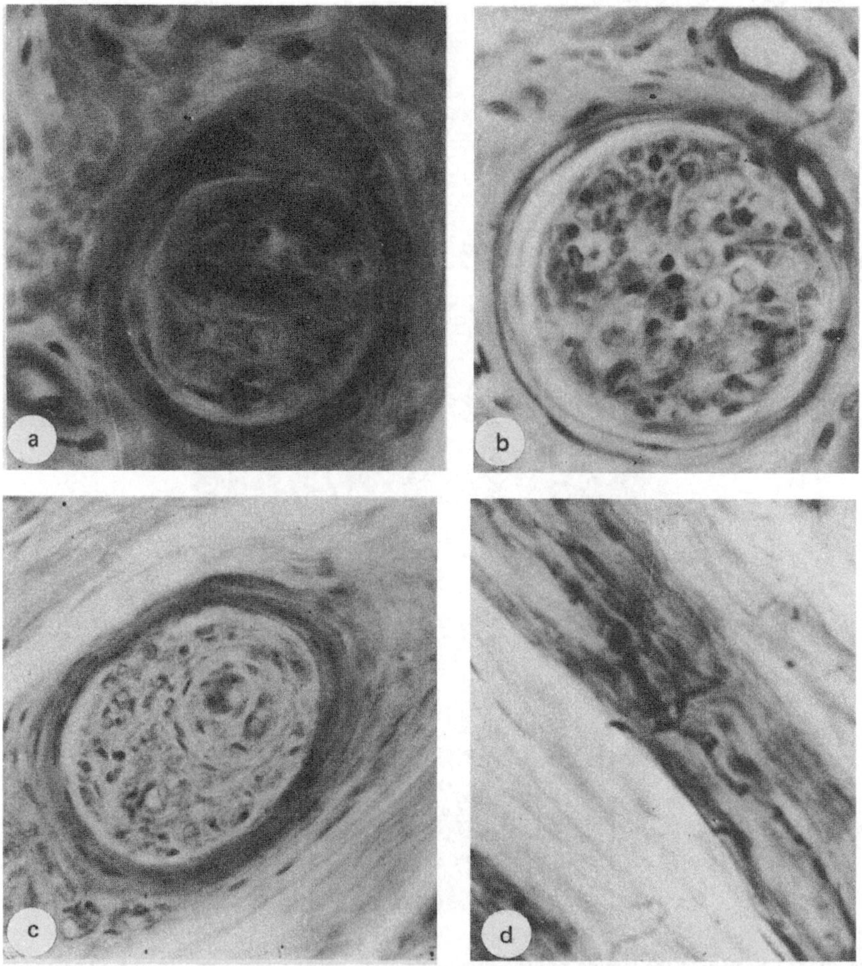

Figure 21. (a) Section of gingiva from edentulous jaws of 82-year-old man. The neural connective tissue sheath is thick and stains deep red with the PAS reaction. PAS–hematoxylin. ×100. (b) Section of gingiva from an 80-year-old male. Note the thick connective tissue sheath, the thickened myelinated nerve fiber, and the PAS positive arteriole. ×100. (c) Section of gingiva from an 80-year-old male. Notice the thickened neural connective tissue sheath and the fibrotic appearance of the nerve. PAS–hematoxylin. ×100. (d) Longitudinal section of nerve found deep in the lamina propria from the gingiva illustrated in Fig. 21a. Notice the wide connective tissue sheath, and a reticulated and fragmented nerve fiber within the sheath. PAS–hematoxylin. ×200.

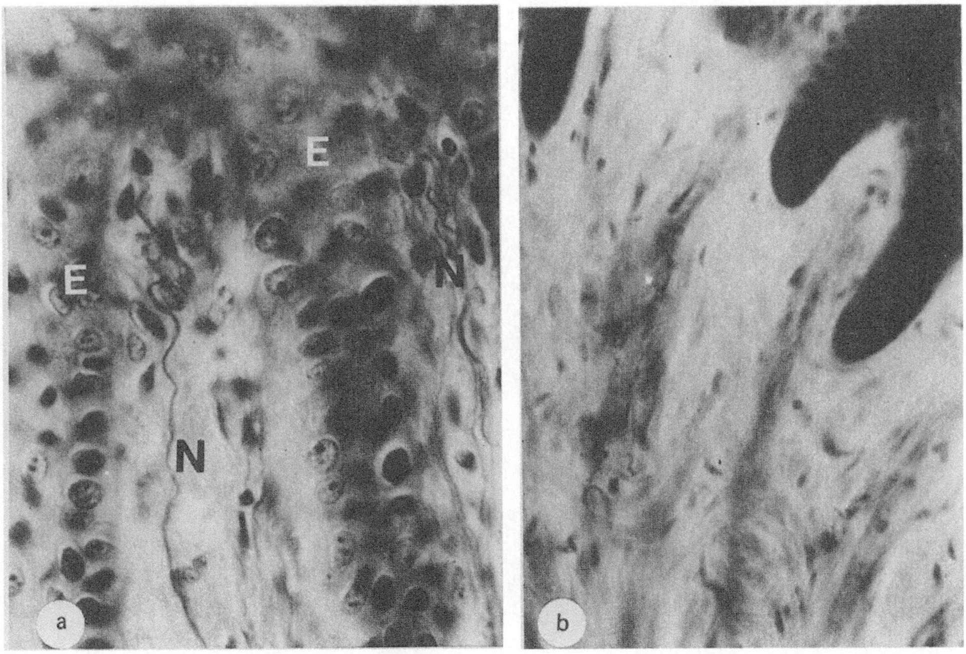

Figure 22. (a) Area of a connective tissue papilla in gingiva from a young individual. Note the nonmyelinated nerve fibers in the papilla. E, epithelial cells; N, nerve. Iron hematoxylin. ×200. (b) Area connective tissue papilla in gingiva from dentulous jaw of 76-year-old man. Note that only nerve fragments are found in papilla. Iron hematoxylin. ×200.

is not the result of periodontal disease, as these animals are normally free of periodontal disease.

3. MAXILLA AND MANDIBLE

3.1. General Consideration

There has been an extensive number of light and electron microscopic reports during the last two or three decades concerning the structure of the cellular and matrical components of young and adult bone in animals and man. However, during this period there has been a paucity of reports that dealt with the morphological changes in aging bone. Tonna (1981), in his excellent review on electron microscopy of skeleton aging, summarizes by saying, "although electron microscopic studies of skeletal tissues are numerous especially, investigations of aging bone, cartilage, and associated soft connective tissue are appallingly exigious, and most of the information comes from several laboratories" (Tonna, 1981). In his series of studies over two decades on the cellular components of old mice bones, he and co-workers have shown degeneration of the osteocytes, loss of osteoblastic action, and decrease in osteoclastic activity. The following description is of changes in aging human maxillae and mandibles studied in our laboratory using varying histochemical reactions.

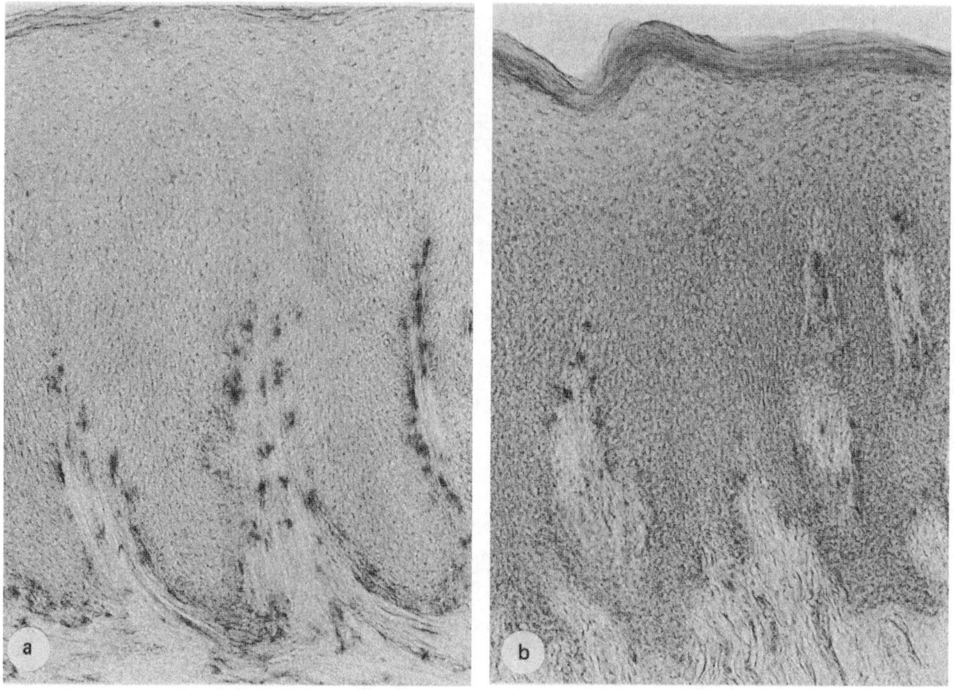

Figure 23. (a) A section of gingiva from a 19-year-old male. Note the presence of nonspecific esterase granules in the connective tissue papillae. The location of the enzyme corresponds to the nerve fibers and ending as demonstrated with nerve stain. Gomoris Napthol A.S. acetate method. ×125. (b) A section of gingiva from a 58-year-old male. Note the overall decrease of the enzyme loci throughout the connective tissue papillae. Gomoris Napthol A.S. acetate method. ×125.

3.2. Osteoporosis

Osteoporosis is a common feature in both the mandibles and maxillae of aged individuals over 60 years of age (Atkinson and Woodhead, 1968). The osteoporosis seen in the mandible or maxilla is similar in intensity whether or not the subject is dentulous or edentulous (Figs. 24a, 24b). There is no difference in the degree of osteoporosis between the mandible or maxilla (Fig. 24c). The bones obtained from females over 60

Figure 24. (a) Section of an edentulous mandible from a female 70 years of age. Note the osteoporosis of the spongiosa, and thinning of the cortical dense bone. Iron hematoxylin. ×35. (b) Section of a dentulous mandible from a male 72 years of age. Note the osteoporosis of the body of the mandible. Iron hematoxylin. ×35. (c) Section of an edentulous maxilla from a male 75 years of age. Note the extreme thinness of the cortical plate, and trabeculae of the spongiosa. G, gigiva; CP, cortical plate; MS, maxillary sinus; S, spongiosa. PAS–hematoxylin. ×35.

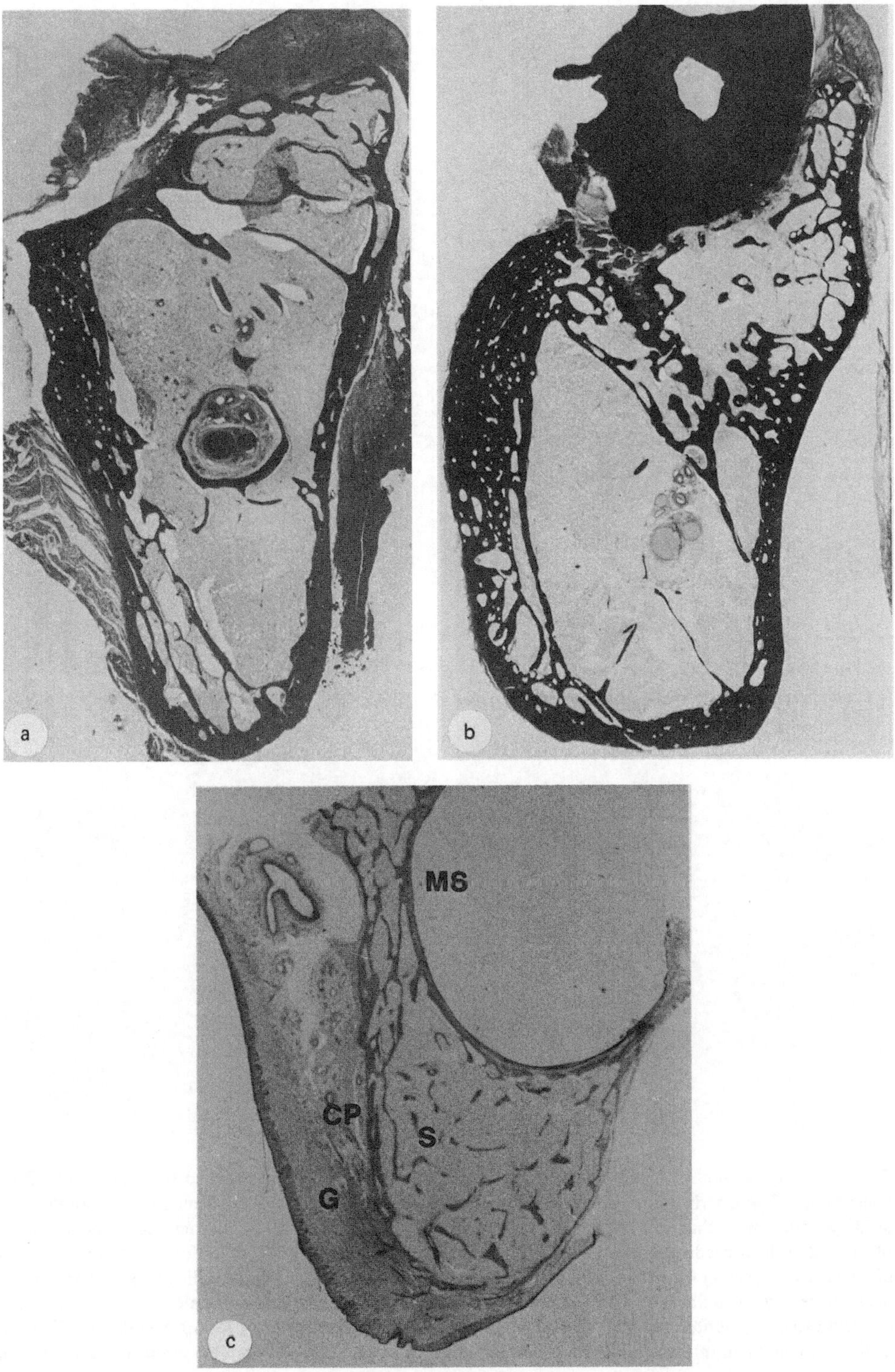
a
b
MS
CP
S
G
c

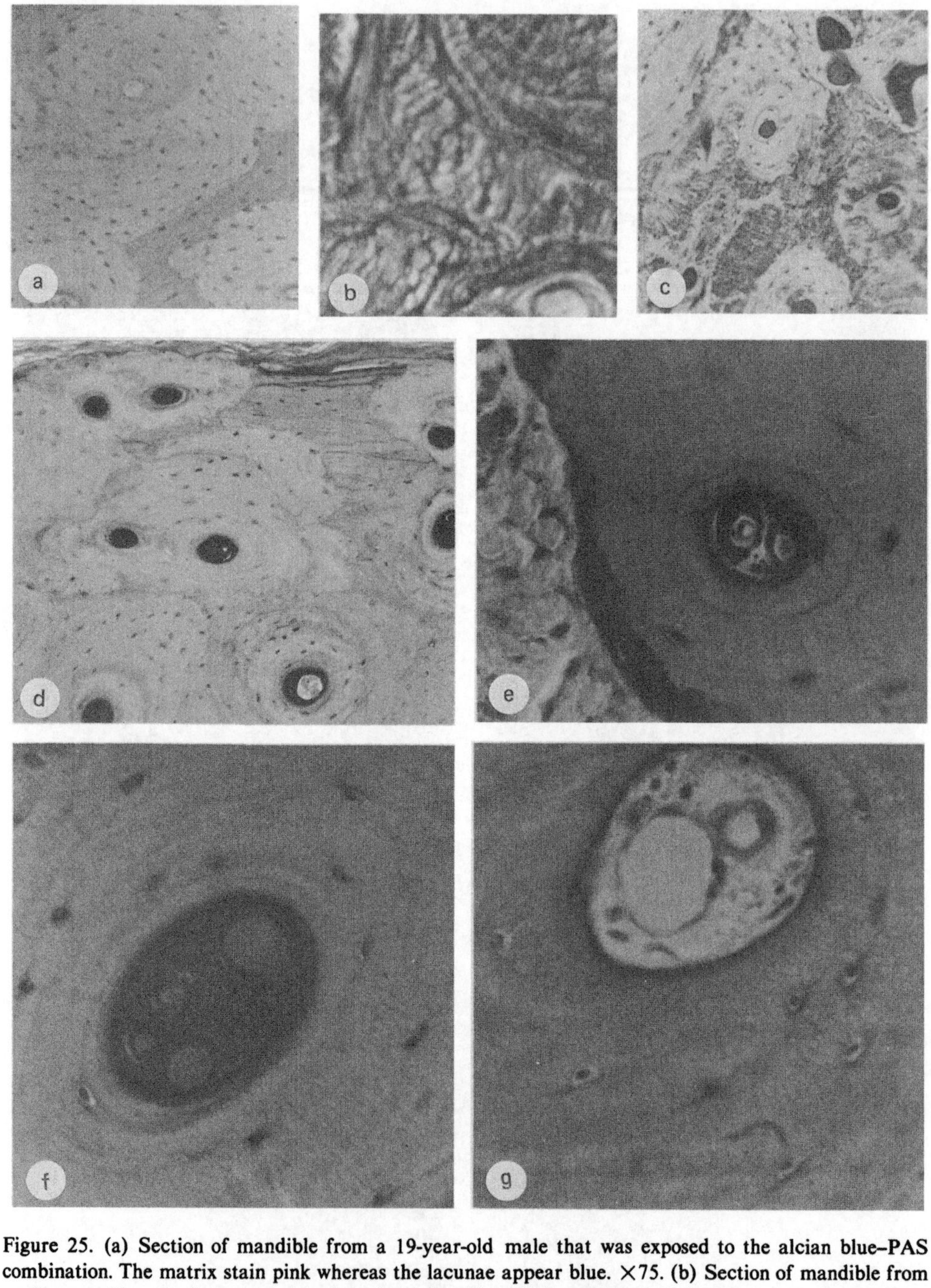

Figure 25. (a) Section of mandible from a 19-year-old male that was exposed to the alcian blue–PAS combination. The matrix stain pink whereas the lacunae appear blue. ×75. (b) Section of mandible from an 81-year-old male that was exposed to the alcian blue–PAS combination. In contrast to the specimen illustrated in the preceding figure, the matrix exhibits areas of alcianophilia. ×75. (c) An alternate section of the specimen seen in Fig. 25b that was stained with silver nitrate. The alcianophilic areas consist of thick disorganized collagen fibers. ×125. (d) Section of mandible from a 55-year-old male. Note that all of the nutrient canals are obliterated by PAS positive material. PAS–hematoxylin. ×75. (e) A high magnification of the nutrient canal in a section of an aged mandible. Note the thick wall capillary, and the calcification

years of age exhibit more severe bone loss than the bones from male individuals (Figs. 24a, 24b).

An interesting finding in the aged mandibles is the absence of an external circumferential lamellae. In contrast, these lamellae are present in the young lamellae. The loss of these external lamellae is similar to what is seen in the long bones throughout the body.

3.3. Histochemical Changes to Bone Matrix

There is a paucity of knowledge concerning the maintenance of the glycosaminoglycans in aging bone matrix. In bone, about one third of the hexosamine in made up of chondroitin-4 sulfate, and small amounts of hyaluronic acid. Little has been reported about the synthesis and degradation of glycosaminoglycans consisting of chondroitin-4 sulfate and a small amount of hyaluronics in aging bone. The other two thirds represent various glycoproteins such as sialoproteins (Herreny, 1964). Herring (1964) and Vajlens (1971) reported that the glycosaminoglycan concentration per weight of collagen is about 0.8% to 0.6% in bones of older individuals. Analysis of cancellous bone from vertebrae showed a decrease in total hexosamine content with advancing age (Casuccio *et al.*, 1962). In rat bone epiphysis ^{35}S sulfate uptake diminishes with advancing age, implying a decrease in sulfate glycosaminoglycans (Dziewiatkowski, 1954; Tonna and Cronkite, 1960).

When sections of "young" bone are exposed to the alcian blue–PAS combination, the bone matrices appear pink in color, whereas the stroma of the nutrient canals and lacunae are alcianophilic (Fig. 25a). In contrast, there are areas of alcianophilia in the matrices, and many of the nutrient canals are partially or completely obliterated by Schiff-positive material (Fig. 25c). When sections of aged mandible are impregnated with silver nitrate, the alcianophilic regions consist of thick disorganized collagen fibers (Fig. 26c).

Protein synthesis by aging bone cells has been investigated in the mouse using [^{3}H]histidine, [^{3}H]glycine, and [^{3}H]proline by Tonna and co-workers (Tonna, 1961, 1966; Tonna and Pavelec, 1971; Tonna *et al.*, 1983). A decrease in incorporation of these precursors into protein was observed in the older animals. In addition these workers have shown a decrease in RNA staining in aging osteoblasts. In the human studies by the present author, using the Ninhydrin–Schiff reaction for the presences of alpha amino acids, there is a marked decrease in Ninhydrin–Schiff staining in the matrices of the aged mandibles and maxillae in contrast to the intense positive staining in young bone.

←

around the blood vessels. PAS–hematoxylin. ×150. (f) A high magnification of the nutrient canal in a section of mandible from a 76-year-old male. Notice that the nutrient canal and lacunae are filled with PAS positive material. PAS–hematoxylin. ×150. (g) A section of mandible from a 19-year-old male. In comparison to Fig. 25f, the lacunae are open with the nuclei of the osteocytes still present. PAS–hematoxylin. ×150, reproduced at 85%.

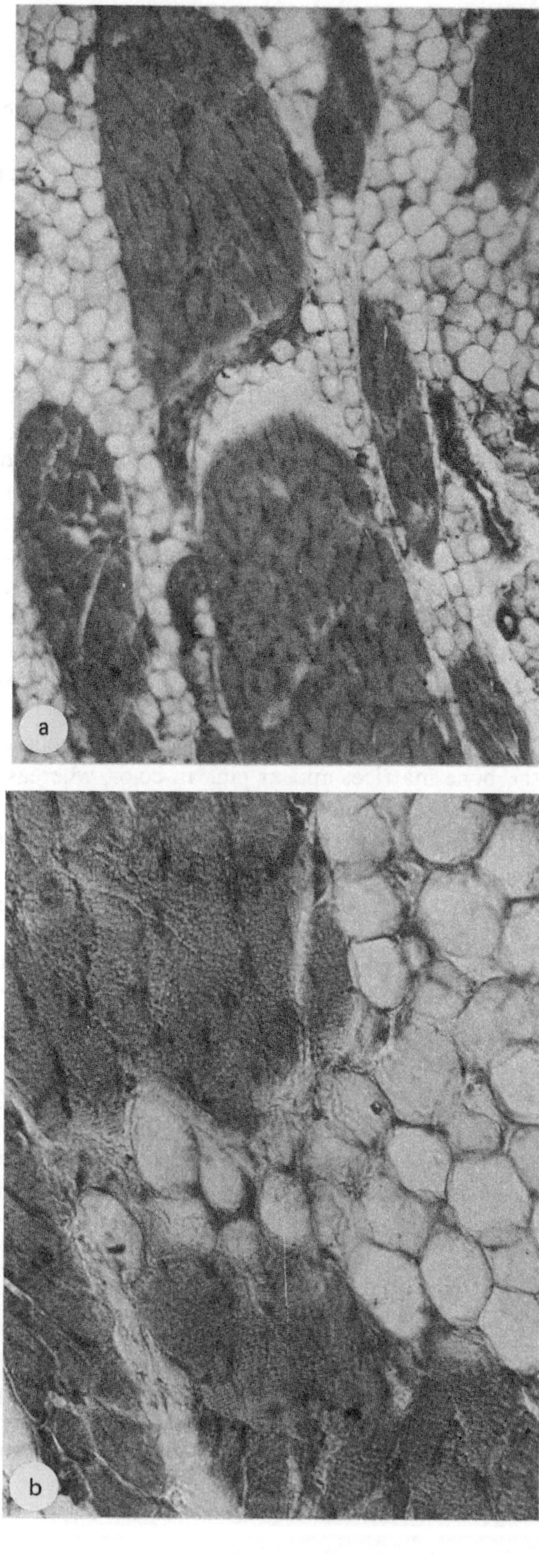

Figure 26. (a) Section of intrinsic muscles in the tongue of a 63-year-old male. Note the infiltration of fat cells resulting in a loss of the continuity of muscle fiber. PAS–hematoxylin. ×125. (b) A high magnification of a muscle fiber and adjacent fat cells. Note the fat cells intimately associated to the muscle fiber. PAS–hematoxylin. ×250.

3.4. Changes in Vascularity

There has been no previous study concerning the vascularity supplying bone. In our study on aging human mandibles, changes are observed in the nutrient canals of the osteons. In "young" osteons, the nutrient canals consist of a loose connective tissue stroma with spindle-shaped cells lining the adjacent bony surfaces. The walls of the arterioles and venules stain pink in color following the PAS reaction or alcian blue–PAS combination. The blood capillary usually found near the bone surfaces appears as a small thin red staining vessel due to presence of an intact basement membrane that abuts against the endothelial layer. In the "aged" osteons, the walls of the arterioles and venules stain deep red with the PAS reaction. The capillaries are easily demonstrable because of their thickened Schiff-positive wall. Following the vascular changes, there is a progressive deposition of Schiff-positive material in the connective tissue stroma leading to complete obliteration of the nutrient canal elements (Figs. 25d, 25f).

Varying degrees of diffuse calcification of the canals is seen in many of the "aged" osteons. It is not unusual to find osteal nutrient canals completely calcified, which may or may not obliterate the microvasculature (Fig. 25e).

Concomittant with changes to the nutrient canals, there are changes in the osteal lacunae. The lacunae consist of Schiff-positive deposits in which pyknotic nuclei are embedded (Fig. 25f). The lacunae in the peripheral third of the osteons are empty of cells or deposits. These structural changes are not evident in the osteons of young mandibles (Fig. 25g).

4. TONGUE

4.1. Mucosal Changes

Detailed studies describing age changes to the human tongue are few in number. The few studies available have shown that there is an atrophy of the surface epithelium with alterations to the underlying lamina propria (McCarthy and Shklar, 1961). Shklar (1966) also reported that the polysaccharides of human oral mucosa increased with age. Pedreira (1951) has shown that there is a decrease in the keratinization of the oral mucosa including the tongue in individuals over the age of 60 years. Clinically it has been suggested that there was no loss of sensitivity to salty tastes, but a decreased sensitivity to sweet, sour, and bitter tastes in older humans (Hermel, 1970). It has also been demonstrated that there is a loss of the various papillae from the dorsum of the aging tongue leading in many instances to a smooth appearance of the tongue (Massler, 1971). Decrease in the number of taste buds has been described by Arey *et al.* (1935) and Smith and Sethi (1975). Other studies have noted that, in females over the age of 40 years, there is a beginning loss of taste buds, while, in the male, it began at the age of 60 years (Allera, 1939; El Bardi and Bourne, 1951; Mochizuki, 1939; Conger and Wells, 1969).

In our study of 80 human tongues (deMignard *et al.*, 1981) obtained from individuals varying in age from one to 81 years, there is a flattening of the surface epithelium, atrophy of the filiform and fungiform papillae, and an increase in intraepithelial glycogen. At the same time, there is flattening of the circumvallate papillae with a marked loss of the taste buds. These findings are not seen in the tongues of individuals under the age of 40 years. The connective tissue components of the lamina propria of the old tongues consist of thick collagen bundles that stain deep red with the PAS reaction. There are areas in the lamina propria which appear glassy or hyalinized.

4.2. Muscle Changes

Muscular and neural changes are evident in the tongues obtained from individuals over the age of 40 years. Fat cell infiltration of the submucosal area is present in specimens studied from the older age group (Fig. 26a). The infiltration of the fat cells becomes so extensive that it disrupts the continuity of the muscle fiber bundle (Fig. 26b). In this situation it is not unusual to find fat cells in direct opposition to the fragmented edges of the muscle fiber.

Age changes to intrinsic and extrinsic muscles of the "aged" tongue are similar to those findings described by Gutman (1977) and Shafiq *et al.* (1981) for skeletal muscles throughout the body. Gutmann has summarized the physiological, biochemical, and ultrastructural changes. In the old tongues that we have examined, there is atrophy of the intrinsic muscle fibers, loss of striations, and thickened basement membranes of the muscle fibers. Figures 27a and 27b illustrate the difference between a young and old muscle fiber bundle in the tongues studied. Notice that there is a loss of striations in some of the muscle fibers in the old tongue in comparison to the uniform staining of the striations in the young tongues. In addition, when sections of tongues from the two age groups are exposed to the Ninhydrin Schiff reaction for the demonstration of the alpha amino acids, there is a decrease in the reaction intensity in the older age group. This finding is in agreement with the work of Breuer and Florini (1965), Kanungo *et al.* (1970), and Britton *et al.* (1972). All have demonstrated a decrease in proteosynthesis in aged muscle. In addition, in our study, there is a deposition of Schiff-positive material within the connective tissue sheaths of the muscle bundles.

4.3. Neural Changes

Changes in the innervation of the senescent muscle fibers are easily apparent. The connective tissue neural sheaths of all motor nerves of varying calibers are thickened and stained intensely with PAS reaction (Fig. 28a). This finding is in agreement with Johnson *et al.* (1978) who demonstrated such changes in the nerves of the muscles of extremities. The myelinated nerve fibers within the large nerve are characterized by a thick capsule consisting of either or both the endoneurium and basement membrane which stain positive with the PAS reaction (Fig. 28a). In addition, fragmentation and reticulation of the myelinated nerves are common findings in the "aged" motor nerves (Fig. 28b). These characteristics are similar to what has been reported by Arnold and Harriman (1970) in peripheral nerves.

Some effects of aging on the morphology of the motor end plates have been emphasized by Harriman *et al.* (1970). They reported that, in aging human muscles, many of the terminal sprays of axons showed "regressive" changes and terminal and subterminal

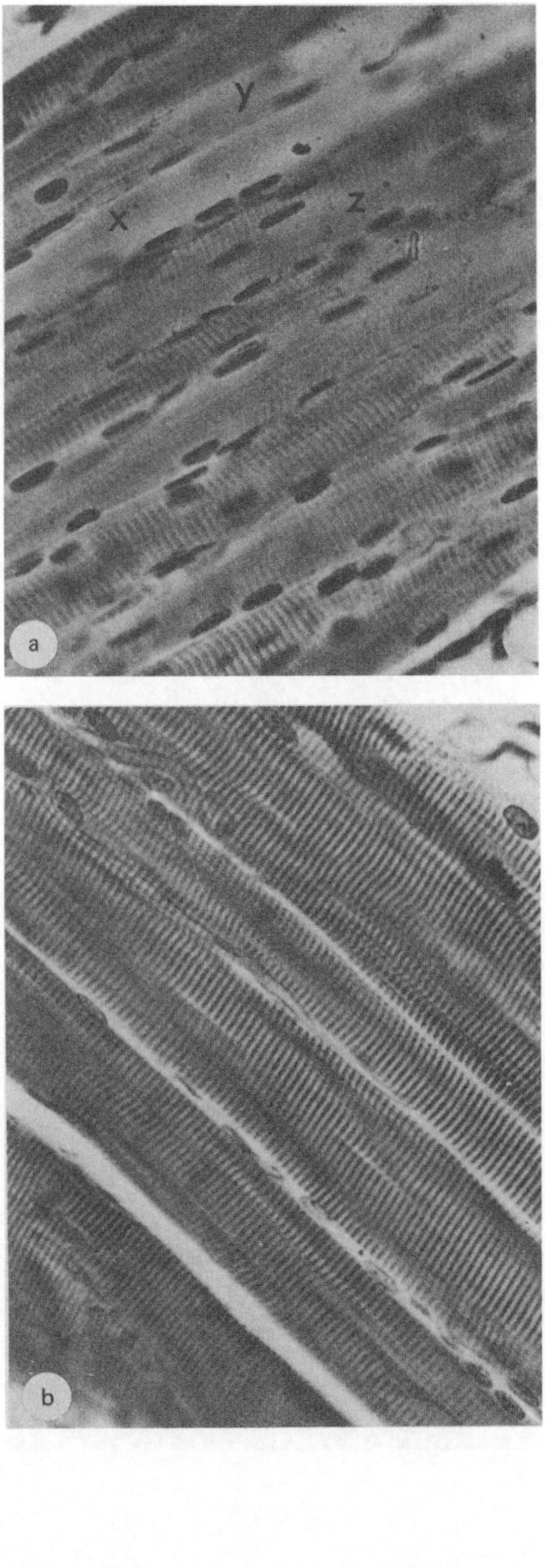

Figure 27. (a) Section of intrinsic muscles of the tongue of a 75-year-old male. Note that fibers at x, y, and z show loss of striations. Silver nitrate impregnation. ×200. (b) Section of intrinsic muscles of the tongue of a 28-year-old male. Note the uniform staining of the striations. Compare to the preceding figure. Silver nitrate impregnation. ×200.

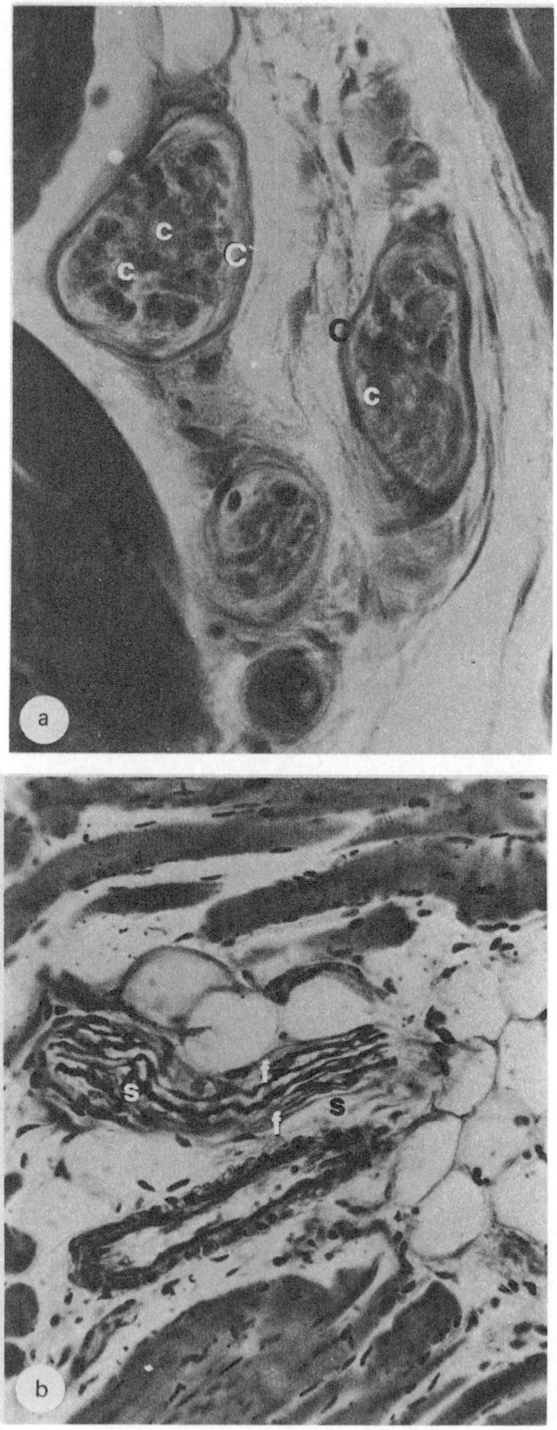

Figure 28. (a) A motor nerve in the muscle layer of the tongue of a 75-year-old male. The neural connective tissue sheaths are thickened, and the individual myelinated nerve covering consists of a thick sheath that stains PAS positive. S, connective tissue sheath; C, covering of myelinated nerve. PAS–hematoxylin. ×150. (b) A segment of a motor nerve between muscle fibers of a 63-year-old male. Notice the fragmentation of nerve fibers (f), areas of swelling (s), and the thick connective tissue sheath. Silver nitrate impregnation. ×100.

swelling. Electrophysiological studies have claimed a progressive fall in the number of motor units (Brown, 1972; McComas *et al.*, 1973). In our studies of the intrinsic muscles of the "aged tongues" there is an apparent loss of the terminal sprays as well as degeneration of the proximal segment of the motor nerve (Fig. 20a).

Neuromuscular spindles have been examined in aging muscles. Spindle pathology, as result of age, is reported by Swash and Fox (1972), who found a linear increase in mean capsular thickness and a decrease in the mean number of intrafusal fibers. The findings in our study show the same changes as illustrated in Fig. 29b. In addition to the thickening, the arteriole usually associated with the spindle is altered. Its walls are glassy in appearance and stained with the PAS reaction. These alterations are not demonstrable in the muscle spindles in young muscle fibers.

4.4. Vascular Changes

The microvasculature associated with the muscles, as well as the lamina propria of the muscosa, exhibit age changes. Although morphometric studies were not undertaken, there is an apparent decrease in the number of capillaries within the endodermal spaces as well as the lamina propria. The walls of the capillaries present are thick and stain deep red following the PAS reaction. In addition, the walls of both the arterioles and venules are Schiff-positive, giving the vessels a glassy appearance. As mentioned earlier, the large and small arteries are arteriosclerotic in nature, characterized by thickened intima.

5. SALIVARY GLANDS

Although there are studies reported in the literature on the effect of age on the structural components of the salivary glands, human studies are limited in number. This paucity is probably due to technical difficulty in obtaining adequately fixed material for either light or electron microscopy. Hamperl (1931), while studying the submandibular glands of aged humans, reported the occurrence of serous acinar cells which exhibted large cell bodies and pyknotic or hypertrophied nuclei. He called these enlarged cells "onocytes." Church (1955) reported that, in the salivary glands of aged rats, he also found an increased number of onocytes. This was also seen by Andrew (1949b). Both Church and Andrew noted that duct cells also exhibited small pyknotic nuclei. Histochemical data reported by Bogart (1967) revealed that there was a decrease in the concentrations of succinic dehydrogenase, nonspecific esterase, cholinesterase, and alkaline phosphatase. He further noted that there is a decrease in acid mucopolysaccharides in serous acini, increase in lipofuchsin granules, and acid phosphatase (1970). Kuyatt and Baum (1981) demonstrated by biochemical means that there is an age-related decrease in the contents of protein, sialic acid, and neutral sugar, but not in DNA in aged male rats.

Kim (1981) and King *et al.* (1980) noted that the cellular level of amylase activity and the incorporation of [^{3}H]leucine into acid-soluble protein declines with age in the rats' parotid glands. Sharaway *et al.* (1982) reported that in old female rats, morpho-

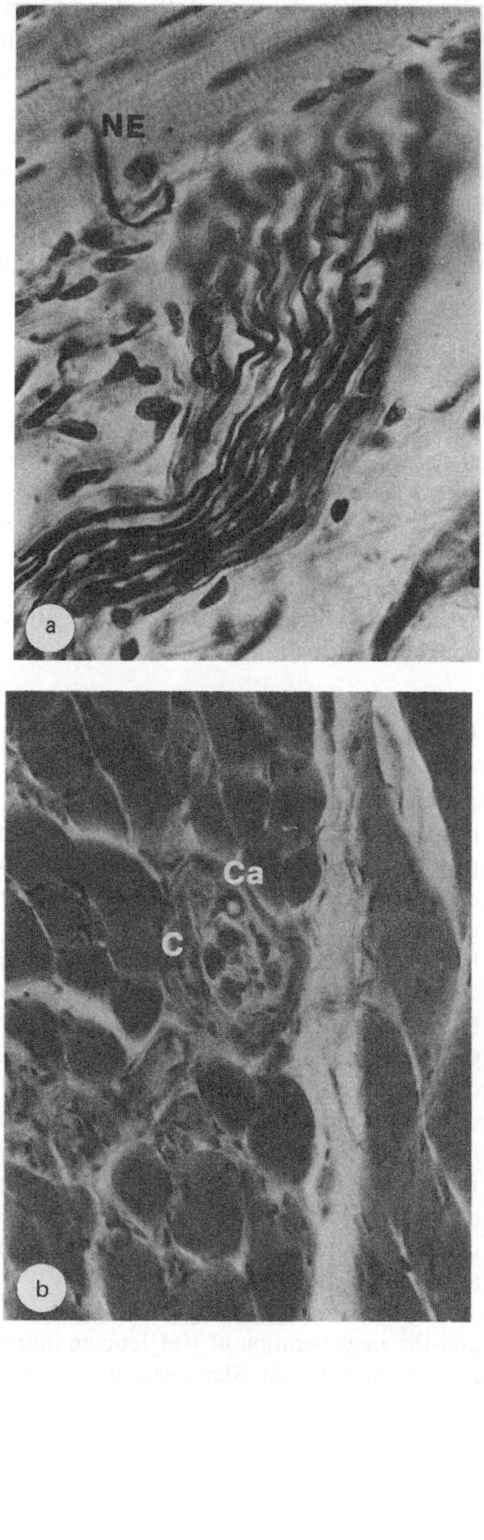

Figure 29. (a) Section of muscle with adjacent nerve fibers from an aged individual. Note the blunt end of a nerve in an area of a motor end plate. N, nerve. Silver nitrate impregnation. ×150. (b) A muscle spindle in an intrinsic muscle fiber of the tongue of a 63-year-old male. Note the thick capsule, the fibrotic appearance of the spindle, and the adjacent thick-walled capillary. C, capsule; Ca, capillary. PAS–hematoxylin. ×75.

metric studies revealed that there was a reduction in volume of serous acine as well as a diminished number of zymogen granules within the serous acinar cells. They also found an increase of connective tissue throughout the gland. A high portion of the parotid glands of aged human subjects show fatty changes and afibrosis (Bauer, 1950).

6. SUMMARY

In this presentation we have undertaken to give a brief description of age-related changes in various organs of the oral cavity. We have tried to limit our survey to human material. However, due to the difficulty of obtaining adequately fixed specimens for certain structures and organs, animal data, reported in the literature, have been included in the presentation.

The age alterations seen in aged dental pulp, periodontal ligament, and the lamina propria of the gingiva and tongue are not surprising. The cellular, fibrillar, and ground substance changes of these structures are similar to those described in other connective tissues of the body (Clausen, 1966; Verzar, 1964; Gross, 1961; and Milch, 1966). All agree that changes take place in the physical and chemical properties of collagen. There is an increase in the tensile strength of collagen fibrils, a decrease in extensibility, and an increase in thermal contraction. There is also an increased resistance of proteolytic enzymes and a decrease in protein bound hexase and mucoproteins. A decrease in the chondroitin sulfate to hyaluronic acid ratio has been noted in skin as well as in the gingiva and tongue with advancing age (Milch, 1966).

The diffuse calcification found in the pulp and periodontal ligament is non unusual, for Milch (1966) postulated that the apparently increased cross-linkage of collagen with advancing age enhances its tendency for mineralization. As described previously, the thick collagen bundles seen in pulp and periodontal ligament were the loci of calcification in an environment that is potentially a calcifying medium. The blood vessels found in all the oral tissues in aged humans exhibited the usual alterations observed throughout the body. Arteriosclerosis of the muscular arteries with varying diameters is a common finding in all the tissues. Changes in the microvasculature, including the thick wall capillaries seen in the various organs described above, may impede the transfer of material from the capillary to the interstitium and finally to the target cells.

Neural changes seen in all the tissues including the pulp, peridontal ligament, gingiva, and muscles of the tongue were consistent with the degeneration of peripheral sensory and motor nerves reported for skin and skeletal nerves (Cauna and Mannan, 1961; Winkelmann, 1965; Spencer and Ochoe, 1981).

An understanding of the morphologic alterations that occur with age in the various organs comprising the oral cavity is important, for such knowledge will aid in the understanding of the functional age changes that may lead to decreased activity in the aged as well as predispose aged individuals to disease.

REFERENCES

Andrew, W., 1949a, Age changes in the parotid glands of Wistar Institute rats with special references to the occurrence of onocytes in senility, *Am. J. Anat.* **85**:157–197.

Andrew, W., 1949b, Age changes in the salivary glands of Wistar Institute rats with particular references to the submandibular gland, *J. Gerontol.* **4**:95–103.

Andrew, W., 1952, A comparison of age changes in the salivary glands of man and of the rat, *J. Gerontol.* **7**:178–190.

Arey, L., Tremain, M. J., and Monzingo, F. L., 1935, The numerical and topographical relation of taste buds to human circumvallate papillae throughout the life-span, *Anat. Rec.* **64**:9–25.

Arnold, N., and Harriman, P. G. F., 1970, The incidence of abnormality in control human peripheral nerve studies by single axon dissection, *J. Neurol., Neurosurg., Psychiatry* **33**:55.

Atkinson, P.J., 1969, Structural aspects of ageing bone, *Gerontologia* **15**:171–173.

Atkinson, P. J., and Woodhead, C., 1968, Changes in human mandibular structure with age, *Arch. Oral Biol.* **13**:145–163.

Avery, J. K., and Rapp, R., 1959, Pain conduction in human dental tissues, *Dent Clin. N. Am.* 481–501.

Ballard, K. W., Bernick, S., and Sobin, S. S., 1979, Changes in the human microcirculation with age, *Microvasc. Res.* **17**:511.

Banfield, W. G., and Brindley, D. C., 1963, Preliminary observations on senile elastosis using the electron microscope, *J. Invest. Derm.* **41**:9–17.

Bauer, W. H., 1950, Old age in human parotid glands with special references to peculiar cells in uncommon salivary gland tumors, *J. Dent. Res.* **29**:686.

Belting, C. M., Schour, I., Weinmann, J. P., and Shepro, M. S., 1953, Age changes in the periodontal tissues of the rat molar, *J. Dent. Res.* **32**:332–353.

Bernick, S., 1967a Age changes in the blood supply to human teeth, *J. Dent. Res.* **46**:544–550.

Bernick, S., 1967b, Effect of aging on the nerve supply to human teeth, *J. Dent. Res.* **46**:694–699.

Bernick, S., Levy, B. M., and Patek, P. R., 1969, Studies on the biology of the periodontium of marmosets VI, arteriosclerotic changes in the blood vessels of the periodontium, *J. Periodont.* **40**:355–358.

Bernick, S., and Nedelman, C., 1975, Effect of aging on the human pulp, *J. Endo.* **1**:88–94.

Bernick. S., Sobin, S. S., and Lindal, B. A., 1978, Changes in the microcirculation with age, *Microvasc. Res.* **16**:215–223.

Bernick, S., and Sobin, S., 1982, The lymphatic and blood vessels of the aged human gingiva, *Gerodontology* **1**:65–71.

Bhussry, B. R., 1968, Modification of the dental pulp during development and aging, in: Biology of the dental pulp (S. B. Finn, ed.), pp. 146–165, University of Alabama Press, University Alabama.

Bogart, B. I., 1967, The effect of aging on the histochemistry of the rat submandibular gland, *J. Gerontol.* **22**:272–275.

Bogart, B. I. 1970, The effect of aging on the rat submandibular gland. An ultrastructural cytochemical and biochemical study, *J. Morphol.* **130**:337–352.

Breur, C. D., and Florini, J. R., 1965, Amino acid incorporation into protein by cell-free system from skeletal muscle. IV. Effect of animal age, androgens, and anabolic agents on activity of muscle ribosomes, *Biochemistry* (N.Y.) 1544–1549.

Britton, V. J., Sherman, F. G., and Florini, J. R., 1972, Effect of age on RNA synthesis by nuclei and soluble RNA polymerase from liver and muscle of C57B1/6J mice, *J. Geront.* **27**:188–192.

Brown, W. F., 1972, A method for estimating the number of motor units and the changes in motor unit count with aging, *J. Neurol. Sci.* **16**:417–432.

Casuccio, C., Bertolin, N., and Falzi, M., 1962, Dell' osteoporosi, senile, *Clin. Ortop.* **14**:1–57.

Cauna, N., 1965, The effects of aging on the receptor organs of the human dermis, in: *Advances in Biology of Skin (Aging),* Vol 6, (W. Montagna, ed.), p. 63, Pergamon Press, Oxford.

Church, L. E., 1955, Age changes in the nucleics of salivary glands of Wistar Institute rats, *Oral Surg., Oral Med., Oral Path.* **8**:301–314.

Clausen, B., 1966, Aging of connective tissue, in: *Hormones and Connective Tissue* (Asboe-Hansen, ed.), Williams and Wilkins Co., New York.

Conger, T., and Wells, M. H., 1969, Radiation and aging effect on taste structure and function, *Radiat. Res.* **37**:31.

Coolidge, E. D., 1937, The thickness of the human periodontal membrane, *J. Am. Dent. Assoc.* **24**:1260–1270.

Courtney, R., 1972, Age changes in the periodontium and oral mucous membrane, *J. Mich. Dent. Assoc.* **54**:335–342.

Dean Boyle, W., Jr., Via, W. F., Jr., and McFall, W. T., 1973, Radiographic analysis of alveolor crest height, and age. *J. Periodontol.* **44**:236–243.

DeMignard, V. A., Dreizen, S., and Bernick, S., 1981, Aging changes in human tongue components, *J. Dent. Res.* **60**:545.

Dziewiatkowski, D. D., 1954, Effect of age on some aspects of sulfate metabolism in the rat, *J. Exp. Med.* **99**:283–298.

El Bardi, A., and Bourne, G., 1951, Theory of tastes and odors, *Science* **11**:660.

Everett, F. G., and Bruckner, R. J., 1970, Cartilage in the periodontal ligament, *J. Periodont.* **41**:165.

Flieder, D. G., 1962, Cytochemistry of human oral mucosa, *J. Dent. Res.* **41**:112.

Frolich, E., 1970, Alter Veranderungen der Pulp und des Paradontium, *Dent. Zahnaergtl. Z.* **25**:175–183.

Glickman, I., 1965, *Clinical Periodontology*, 3rd ed., Saunder's, Philadelphia.

Gottlieb, B., 1942, Biology of the cementum, *J. Periodontol.* **13**:13–17.

Grant, D. A., and Bernick, S., 1970, Arteriosclerosis in periodontal vessels of aging humans, *J. Periodontol.* **41**:170–176.

Grant, D., and Bernick, S., 1972, The periodontium of ageing humans, *J. Periodontol.* **43**:660–667.

Gross, J., 1961, Aging of connective tissue; The extracellular components, in: *Structural Aspects of Ageing* (G. H. Bourne, ed.), pp. 177–195, Hafner, New York.

Gutmann, E., 1977, Muscle, in: *Handbook of the Biology of Aging* (C. E. Finch and L. Hayflick, eds.) pp. 445–469, Von Nostrand Reinhold, New York.

Han, S. S., 1968, The fine structure of cells and intercellular substance of the dental pulp, in: *Biology of the Dental Pulp Organ* (S. B. Finn, ed.), pp. 103–139, University of Alabama Press, University Alabama.

Hamperl, H., 1931, Beitrage zur normalen und pathologischen Histologie menschlicher Speicheldrusen, *Z. Mikro-Anat. Forsch.* **27**:1–55.

Hansen, G. C., 1973, An epidemiologic investigation of the effect of biologic aging on the breakdown of periodontal tissue, *J. Periodontol.* **44**:269–277.

Hermel, J., Schonwether, S., and Samueloff, S., 1970, Taste sensation and age in man, *J. Oral. Med.* **25**:39–42.

Herring, G. M., 1964, Chemistry of bone matrix, *Clin. Orthop.* **36**:169–181.

Hill, T. J., 1949, *A Textbook of Oral Pathology*, pp. 211–217, Lea and Febiger, Philadelphia.

Ingle, J. I., 1965, Etiology of pulp inflammation necroses or dystrophy, in: *An Atlas of Pulp and Periapical Biology* (A. L. Ogilvie and J. I. Ingle, eds.) p. 289, Lea and Febiger, Philadelphia.

Jenkins, G. N., 1966, Permeability and age changes in the dental tissues in the physiology of the mouth, F. A. Davis Co., Philadelphia.

Jensen, J. L., and Toto, P. D., 1968, Radioactive labeling index of the periodontal ligament in aging rats, *J. Dent. Res.* **47**:149.

Johnson, P. C., Yoshimora, M. P., and Gaines, J. A., 1978, Thickening of the permanent basement membrane in diabetes and aging. Cited in *Pathology of Peripheral Nerves* (G. Asbury and P. C. Johnson, eds.), Saunders and Co., Philadelphia.

Kanungo, M. S., Koui, Q., and Reddy, K. A., 1970, Concomitant studies on RNA and protein synthesis in tissue of rats of various ages, *Exp. Geront.* **5**:261–269.

Kim, S. K., 1981, Age-related changes in the cellular level of amylase and protein synthesis in the rat parotid gland, *J. Dent. Res.* **60**:738–747.

Kim. S. K., Weinhold, P. A., Han, S. S., and Wagner, D. J., Age-related decline in protein synthesis in the rat parotid gland, *Exp. Gerontol.* **15**:77–85.

King, J. D., 1949, Histological observations on periodontal disease in the golden hamster, *J. Path. Bact.* **61**:413–425.

Klingsburg, J., and Butcher, E. O., 1960, Comparative histology of age changes in oral tissues of rat, hamsters, and monkey, *J. Dent. Res.* **39**:158–169.

Kuyatt, B. L., and Baum, B. J., 1981, Characteristics of submandibular glands from young and aged rats, *J. Dent. Res.* **60**:936–941.

Levy, B. M., Dreizen, S., and Bernick, S., 1972, Effect of aging on the marmoset periodontium, *J. Oral Path.* **1**:61–65.

Loring, A. L., 1960, Physiology of the aging skin, *Ill. Med. J.* **117**:59.

Mancini, R. E., Vilar, O., Stein, E., and Fiorini, H., 1959, Histochemical study of postnatal development of skin connective tissue, *Rev. Soc. Argent. Biol.* **35**:196.

Martinez, I. R., and Pekarthy, J. M., 1974, Ultrastructure of encapsulated nerve endings in rat gingiva, *Am. J. Anat.* **140**:137.

Massler, M., 1971. Oral aspects of aging. *Postgraduate Medicine* **49**:179–183.

McCarthy, P. L., and Shklar, G., 1961, Disease of oral mucosa, in: Normal Anatomy, Histology, and Histochemistry of the Oral Mucosa, Chap. 1-10, McGraw-Hill, New York.

McComas, A. J., Upton, A. R. M., and Sica, R. E. P., 1973, Motorneural disease and aging, *Lancet* **2**:1477.

Milch, R. A. S., 1966, Aging of connective tissue, in: *Perspectives in Experimental Gerontology* (N. W. Shock, ed.), pp. 109–124, C. C. Thomas, Springfield.

Mitchell, R. W., 1967, Chronic solar dermatosis, a light and electron microscopic study of the dermis, *J. Invest. Derm.* **48**:203–220.

Mochizuki, Y., 1939, Papilla folita of Japanese, *Folia Anat. Jpn.* **18**:37.

Moskow, B. S., 1971, Origin, histogenesis, and fate of calcified bodies in the periodontal ligament, *J. Periodont.* **42**:131.

Nedelman, C. I., and Bernick, S., 1978, A significance of age changes in human alveolar mucosa and bone, *J. Prost. Dent.* **39**:495–501.

Nedelman, C. I., and Bernick, S., 1981, Changes in nerve supply to aging human gingiva, *Prosth. Dent.* **45**:364–370.

Nedelman, C. I., Gamer, S., and Bernick, S., 1970, The alveolar ridge mucosa in denture and non-denture wearers, *J. Prosth. Dent.* **23**:265–273.

Pavelec, M., Tonna, E. A., and Fand, I., 1967, The localization and distribution of protein-bound sulfhydryl and disulfide groups in skeletal tissues of mice during growth and aging, *J. Geront.* **27**:185–190.

Pedreiva, R. A., 1951, A study of keratinizations of the oral mucosa of aged males, *J. Dent. Res.* **6**:88–91.

Pinzon, R. D., Toto, P. D., and O'Malley, J. J., 1966, Kinetics of rat molar pulp cells at various ages, *J. Dent. Res.* **45**:934–938.

Pinzon, R. D., Kozlov, M., and Burch, W. P., 1967, Histology of rat molar pulp at different ages, *J. Dent. Res.* **46**:202–208.

Reeve, C. M., and Wentz, F. M., 1962, The prevalence, morphology, and distribution of epithelial rests in the human periodontal ligament, *Oral Surg.* **15**:785.

Rushton, M., 1951, Epithelial downgrowth on molar teeth of golden hamsters, *British Dent. J.* **90**:87–93.

Schneir, M., Furute, D., and Berger, K., 1976, Collagens of oral soft tissue, *J. Periodont. Res.* **11**:235.

Seltzer, S., and Bender, I. B., 1965, Retrogressive and age changes of the dental pulp, in: *The Dental Pulp: Biologic Considerations in Dental Procedures,* pp. 212–233, J. B. Lippincott, Philadelphia.

Shafiq, S. A., Lewis, S., Leung, B., and Schutta, H. S., 1981, Fine structure of aging skeletal muscle, in: *Aging and Cell Structure,* Vol. 1, (J. E. Johnson, Jr., ed.), pp. 333–345, Plenum Press, New York.

Sharawy, M., Holovack, N., Larke, V., and Pennington, C., 1982, Age related changes in female rat parotid gland, *Anat. Rec.* **202**:174A.

Shklar, G., 1966, The effects of aging upon oral mucosa, *J. Invest. Dermatol.* **47**:115–120.

Shroff, F. R., 1953, Physiologic pathology of changes in the dental pulp. Senile pulp atrophy, *Oral Surg.* **6**:1455.

Sicher, H., and Weinmann, J. P., 1944, Bone growth and physiologic tooth movement, *Am. J. Orth. Oral Surg.* **30**:109.

Smith, B. H., and Sethi, P. K., 1975, Aging and the nervous system. *Geriatrics* **30**:109.

Spencer, P. S., and Ochoa, J., 1981, The mammalian peripheral nervous system in old age, in: *Aging and Cell Structure,* Vol. 1 (J. E. Johnson, Jr., ed.), pp. 35–103, Plenum Press, New York.

Stahl, S. S., and Fox. L. M., 1953, Histological changes of the oral mucosa associated with certain oral disease, *Oral Surg.* **6**:339.

Stahl, S. S., Tonna, E. G., and Weiss, R., 1969, The effects of aging on the proliferative activity of rat periodontal structures, *J. Periodontol.* **24**:447–450.

Stahl, S. S., Tonna, E. G., and Weiss, R., 1970, Autoradiographic evaluation of gingival response to injury. V. Surgical trauma in low protein fed mature rats, *J. Dent. Res.* **49**:537–545.

Stanley, H. R., and Ranney, R. R., 1962, Age changes in the human pulp. The quality of collagen, *Oral Surg.* **15**:1396.

Swash, M., and Fox. K. P., 1967, The effect of age on human skeletal muscle. Studies of the morphology and innervation of muscle sprindles, *J. Neurol. Sci.* **16**:417–432.
Tonna, E. A., 1960, Periosteal osteoclasts, skeletal development and aging, *Nature* **185**:405–406.
Tonna, E. A., 1961, The cellular complement of the skeletal system studies autoradiographically with triated thymidine (H^3TDR) during growth and aging, *J. Biophys. Biochem. Cytol.* **9**:813–824.
Tonna, E. A., 1964, An autoradiographic evaluation of the aging cellular phase of mouse skeleton using tritiated glycine, *J. Geront.* **19**:198.
Tonna, E. A., 1965, Skeletal cell aging and its effects on the osteogenetic potential, *Clin. Orthop.* **40**:57–81.
Tonna, E. A., 1966, A study of osteocyte formation and distribution in aging mice complemented with H^3 proline autoradiography, *J. Geront.* **21**:124–130.
Tonna, E. A., 1972, Paradontal inflammation and aging in the laboratory mouse, *J. Periodont.* **43**:403–410.
Tonna, E. A., 1973a, Electron microscopic study of skeletal cell aging. II The osteocyte, *Exp. Geront.* **8**:9–16.
Tonna, E. A., 1973b, Histological age changes associated with mouse paradental tissues, *J. Geront.* **28**:1–12.
Tonna. E. A., 1977, Aging of skeletal-dental systems and supporting tissues, in: *Handbook of the Biology of Aging* (C. E, Finch and L. Hayflick, ed.), pp. 470–495, Van Nostrand Reinhold Co., New York.
Tonna, E. A., 1981, Electron microscopy of skeletal aging, in: *Aging and Cell Structure,* Vol. 1 (J. E. Johnson, Jr., ed.), pp. 251–304, Plenum Press, New York.
Tonna, E. A., and Cronkite, E. P., 1959, Histochemical and autoradiographic studies on the effect of aging on the mucopolysaccharides of the periosteum, *J. Biophys. Biochem. Cytol.* **6**:171–177.
Tonna, E. A., and Cronkite, E. P., 1960, Autoradiographic studies of changes in S^{35} sulfate uptake by femoral diaphyses during aging, *J. Geront.* **15**:377–382.
Tonna, E. A., and Pavelec, M. 1971, An autoradiographic study of H^3-proline utilization, by aging mouse skeletal tissues. Bone cell compartments, *J. Geront.* **26**:310.
Tonna, E. A., and Stahl, S. S., 1974, Comparative assessment of the cell proliferative activities of injured paradental tissues in aging mice, *J. Dent. Res.* **53**:609–622.
Tonna, E. A., Weiss, R., and Stah, S. S., 1972, The cell proliferative activity of paradental tissues in aging mice, *Arch. Oral Biol.* **17**:969–982.
Tonna, E. A., Cronkite, S. P., and Pavelec, M., 1983, An autoradiographic study of the localization and distribution of tritiate histidine in bone, *J. Histochem. Cytochem.* **10**:601.
Toto, P. D., and Borg, M., 1968, Effect of age changes on the premitotic index in the periodontium of mice, *J. Dent. Res.* **47**:40.
Vajlens, L., 1971, Glycosaminoglycans of human bone tissues I. Pattern of compact bone in relation to age, *Calc. Tiss. Res.* **7**:175–190.
Verzar, F., 1964, Ageing of the collagen fibers, in: *International Review of Connective Tissue Research,* Vol. 2 (D. D. Hall, ed.), pp. 224–282, Academic Press, New York.
Wentz, F. W., Maier, A. W., and Orban, B., 1952, Age changes and sex differences in the clinically "normal" gingiva, *J. Periodont.* **23**:13.
Winkelmann, R. K., 1965, Nerve changes in aging skin, in: *Advances in Biology of Skin (Aging),* Vol. 6, p. 51, Pergamon Press, Oxford.
Zander, H. A., and Hurzler, B., 1958, Continuous cementum apposition, *J. Dent. Res,* **37**:1035.
Zerlotti, E., 1964, Histochemical study of the connective tissue of the dental pulp, *Arch. Oral Biol.* **9**:149.

6

Morphological Changes in the Aging Mammalian Pancreas

BERNARD G. SLAVIN, WENDELIN J. PAULE, and SOL BERNICK

1. INTRODUCTION

A comprehensive description of morphological and cytological changes in the aging mammalian pancreas has not been reported. What works have been published consist of light microscopic descriptions of a variety of histopathologic lesions in both the aging exocrine and endocrine pancreas. While this chapter deals specifically with the area of age-related cytomorphologic changes in the mammalian pancreas, emphasis will be placed on a description of the endocrine pancreas. This is not difficult to understand in light of the relationship of the endocrine pancreas to diabetes mellitus, a disease mainly of the aged population.

The impetus for writing this chapter arose from one of us (SB) who has devoted many years to the study of aging tissues and cell structure. Most of the light microscopic data herein were derived from numerous virgin Sprague–Dawley male rats at various ages (2–29 months) in the laboratory of the latter investigator. The ultrastructural data presented in this report arose from a study of male Sprague–Dawley rats (both virgin and retired breeders 3–16 months of age) and from aging virgin female C57BL/6J mice.

BERNARD G. SLAVIN, WENDELIN J. PAULE, and SOL BERNICK • Department of Anatomy and Cell Biology, University of Southern California School of Medicine, Los Angeles, California 90033

2. AGE-RELATED MORPHOLOGICAL CHANGES IN THE EXOCRINE PANCREAS

2.1. Acinar and Ductal Changes

The most common age-related morphological change in the exocrine pancreas is lobular (acinar) atrophy reported by Andrew (1944), Berg (1967), Houcke *et al.* (1968), Kendrey and Roe (1969), and Boorman and Hollander (1973) in man and in a variety of rat strains. Acinar atrophy is manifested by shrinkage of individual acini and concomitant interstitial fibrotic changes. Corroboration of acinar atrophy of the pancreas has been reported in 40% of outbred Sprague–Dawley retired male breeder rats (Cohen *et al.*, 1978). A lower incidence of age-related acinar atrophy (4.9–6%) has been reported in the pancreas of virgin Fischer 344 rats (Coleman *et al.*, 1977; Cohen *et al.*, 1978). In two commercially produced stocks of Sprague–Dawley rats, acinar atrophy as well as acinar fibrosis has been reported with progressive aging (Anver and Cohen, 1979; Anver *et al.*, 1982). The former investigators demonstrated a reduction in acinar size, acinar cell vacuolization, and absence of intracellular secretory (zymogen) granules. Kim *et al.* (1981) found that the number of secretory granules in 18-

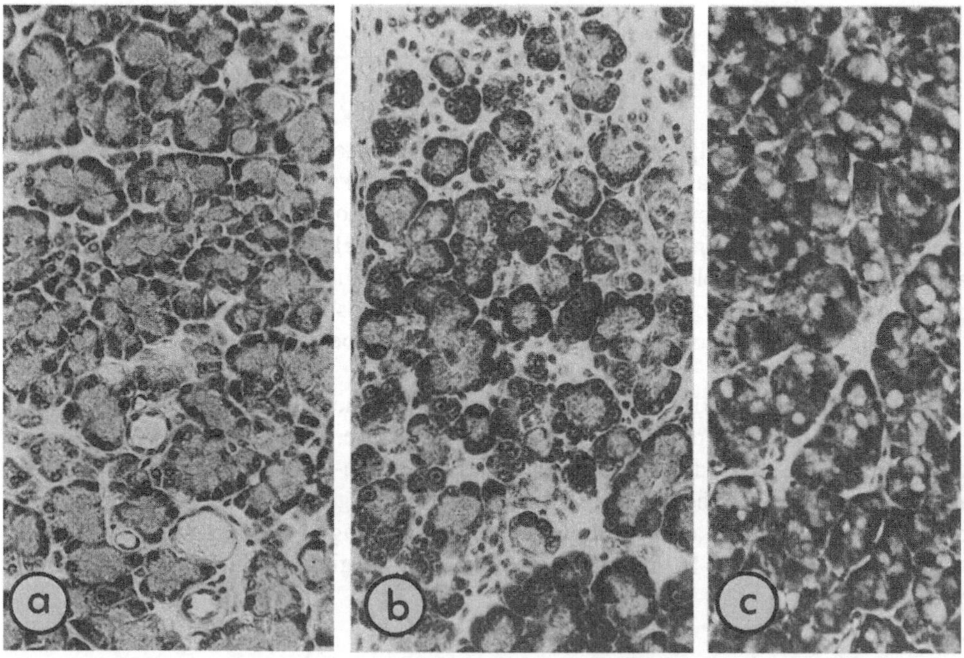

Figure 1. Rat exocrine pancreas. (a) Section from a 2-month-old rat. Note the normal close relationship between adjacent acini. (b) In a 19-month-old rat, note acinar degeneration and fibrosis especially in the upper half of field. (c) At 29 months of age the rat exocrine pancreas consists of regions of vacuolated acinar cells. Hematoxylin and Eosin. ×250.

and 30-month-old rats was greater than that in rats at two months. This finding was correlated with a higher level of amylase within acinar cells of the older rats. In our laboratory, acinar fibrosis and degeneration and scattered acinar cell vacuolization were noted in most Sprague–Dawley rats usually over 28 months of age (Figs. 1a–1c), although a small degree of vacuolization and degeneration was noted in acinar cells of C57BL/6J mice at 13.5 months. A decrease in cytoplasmic basophilia was reported in acinar cells of aging rats (Kendry and Roe, 1969). In addition, Kendrey and Roe (1969), Boorman and Hollander (1973), and Anver and Cohen (1979) observed the transformation of acinar cells into dilated cystlike forms lined by squamous or cuboidal cells in aging Wistar, Sprague–Dawley, and August hooded rats. In the report of Kendrey and Roe (1969), single acini or groups of acini were converted to dilated glandular structures lined by cuboidal epithelium (Fig. 2a). In advanced pathological stages whole lobules were found transformed into microcystic configurations (Fig. 2b). We were unable to confirm the latter findings in aged rats or mice except for occasional, small, isolated areas of the exocrine pancreas where a few dilated glandular structures lined by cuboidal cells were found mingled with acini (Fig. 3).

Age-related alterations in the morphology of exocrine pancreatic ducts have been reported in the August hooded rat (Kendrey and Roe, 1969). These investigators demonstrated ductal dilitation with the altered ducts being tortuous and often filled with PAS-positive inspissated material. In aging man and rats, Andrew (1944) described the metaplasia of exocrine ductal epithelium characterized by the transformation of columnar cells to stratified squamous cells. In addition, the latter investigator noted locular formation or cavitation of ducts in aging animals. We were unable to confirm these findings in our Sprague–Dawley rats.

2.2. Stroma-Vascular Changes

Anver *et al.* (1982) reported a high incidence (37% and 55%) of interstitial hemosiderin deposition in two stocks of barrier-reared male Sprague–Dawley rats. Using the PAS staining method we were able to demonstrate abundant lipofuscin-containing cells usually located perivascularly (Fig. 4).

Chronic inflammatory lesions have been described in the exocrine pancreatic connective tissue of aging rats (Kendrey and Roe, 1969). This was manifested in some cases by periductal accumulation of lymphocytes, plasma cells, and eosinophils. In addition, fibrosis appeared to accompany the inflammatory lesions especially in the vicinity of excretory ducts. Zyss (1964) and Kyogoku and Okamoto (1968) reported a high incidence of intra- and extralobular fibrosis in the exocrine pancreas of aging man. Infiltration of atrophic acinar lobules by lymphocytes or adipocytes was reported in aging rats by Kendrey and Roe (1969), Boorman and Hollander (1973), Cohen *et al.* (1979), and in man by Andrew (1944) and Houcke *et al.* (1964). Our studies on aging Sprague–Dawley rats and C57BL/6J mice have confirmed some of these findings. Progressive infiltration of the pancreas by adipose tissue was found in rats and mice beginning at 12 months. In the oldest rats, adipose tissue appeared to replace some of the exocrine parenchyma. This was generally a random type of infiltration usually surrounding

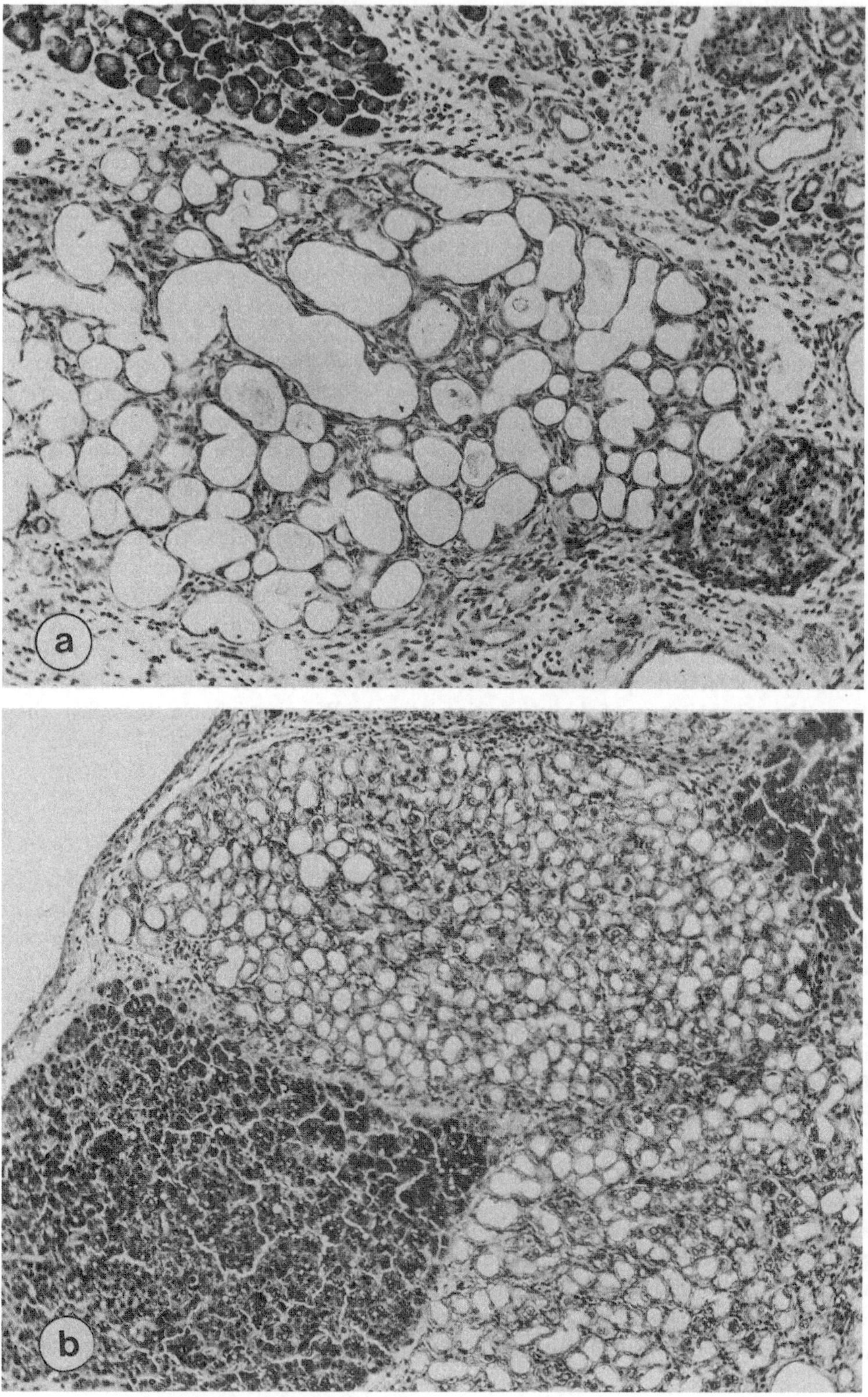
a
b

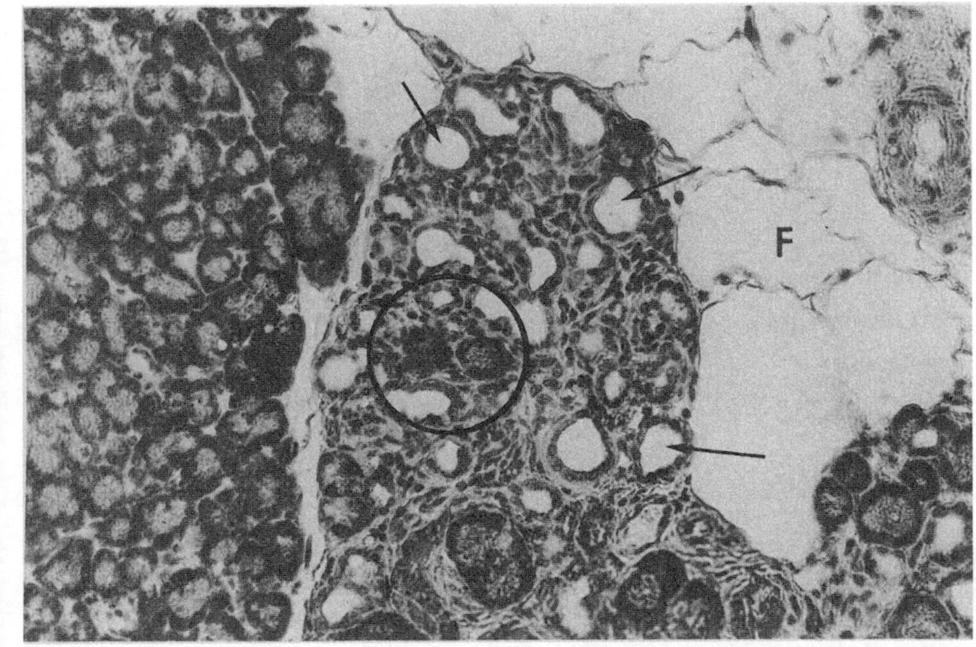

Figure 3. Portion of exocrine pancreas of a 19-month-old rat. Note formation of dilated glandlike structures (arrows). Area at left side of micrograph consists of the normal acinar pattern. Remaining in the area of dilated glands are a few normal acini (encircled). Also, note the fatty infiltration of the pancreas, F. Hematoxylin and Eosin. ×250.

larger ducts and blood vessels (Fig. 5a). Lymphocytic infiltration of the exocrine pancreas was observed in the interstitium of rats usually after 20 months. In addition, perivascular inflammatory changes were noted in the pancreas of these rats characterized by the presence of abundant lymphocytes mostly located on one side of the blood vessel (Figs. 5b and 5c).

Vascular lesions in the exocrine pancreas of aging rats involving both small and medium-sized arteries were observed by Kendrey and Roe (1969). However, these vascular changes occurred far less frequently than parenchymal or stromal lesions. When present, vascular lesions were characterized by fibrotic thickening of the vascular wall, intimal necrosis, and invasion by mononuclear cells and occasional eosinophils. These age-related pancreatic changes in rats appear similar to those occurring with polyarteritis nodosa in man.

←

Figure 2. (a) Microcystic transformation of pancreatic lobules. The acinar cells have lost their normal structure and become cuboidal or flattened. Seventeen-month-old CTS Wistar female rat. PAS. ×100. (b) Marked distention of acini in a pancreatic lobule. The acini are lined with flattened cells and contain inspissated material. Note inflammatory changes of the interstitium. Fourteen-month-old female August hooded rat. Hematoxylin and Eosin ×130. [From Kendrey and Roe (1969) by permission from Laboratory Animals, LTD. London, England.]

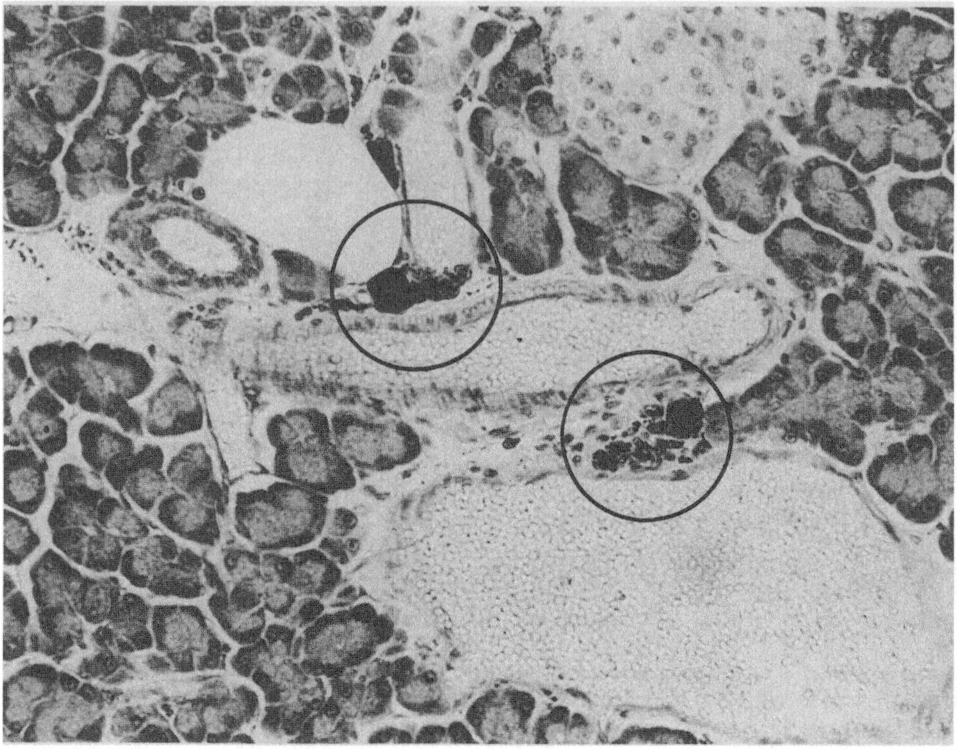

Figure 4. Section of pancreas from a 26-month-old rat stained with PAS and Hematoxylin which demonstrates lipofuscin-laden cells (encircled) along the vasculature. ×300.

2.3. Ultrastructural Changes in the Exocrine Pancreas

No ultrastructural studies of the aging exocrine pancreas have been reported. We have had only a limited opportunity to examine the fine structure of the exocrine pancreas during the aging process. Our preliminary findings show that in the 12-month-old rat the fine structure of acinar cells appears to differ little from that at two or three months (Figs. 6 and 7). Ultrastructural features of these acinar cells consisted of a basally located dense aggregation of flat cisternae of granular endoplasmic reticulum and scattered spheroid, oval, or elongated mitochondria. Zymogen granules abounded at the apical pole of the cells. Other organellar features typical of mammalian acinar cells were noted as described in current textbooks of histology. The only difference noted was a slight but definite increase in the number of lysosomes in the older rat (Fig. 7). Similar findings were observed in the aging mouse. In addition, in sections examined from several 23-month-old mice we noted an abundance of ceroid (lipofuscin) material within the cytoplasm of acinar cells (Fig. 8). The finding of ceroid pigment within acinar cells is not surprising since it has been reported in a variety of tissues in aging animals (Strehler *et al.*, 1959; Samorajski and Ordy, 1967). Ceroid pigment is thought to be present in cells as a lesion produced by the peroxidation of polyunsaturated lipid

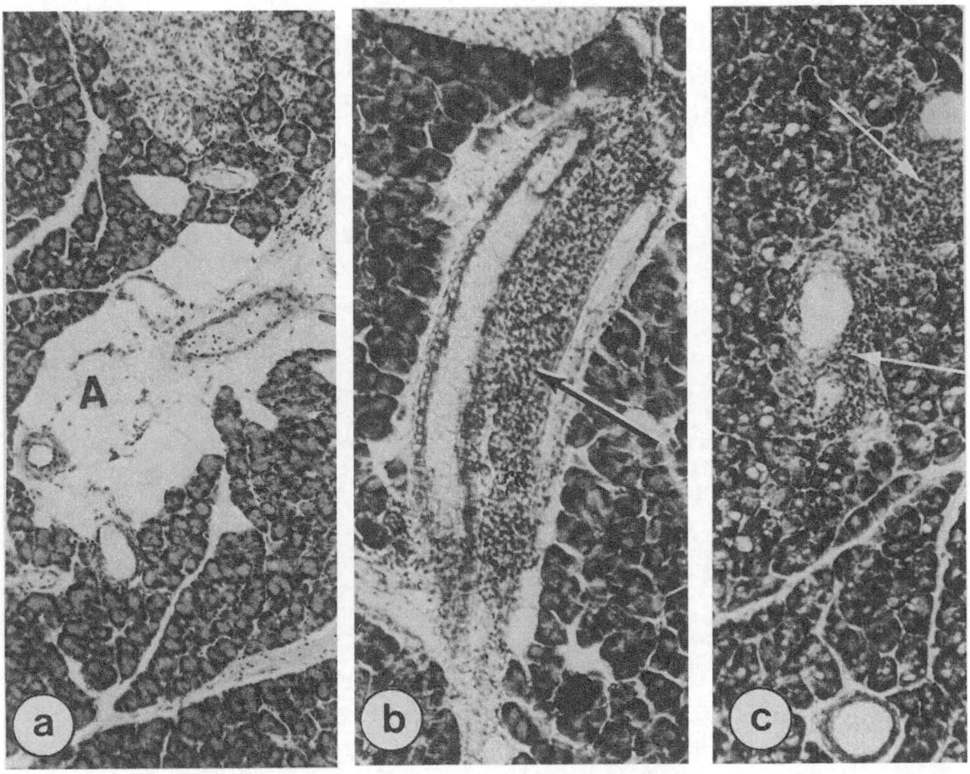

Figure 5. (a) Rat, 26 months. This section shows infiltration of the pancreas by adipose tissue, Aldehyde fuchsin, Hematoxylin, and Eosin ×89. (b),(c) 27-month-old rat pancreas showing extensive perivascular lymphocytic infiltration (arrows). PAS and Hematoxylin. ×127.

in the membranes of cellular organelles such as mitochondria, ergastoplasm, and lysosomes (Chio *et al.*, 1969; Tappel, 1975; Sandi, 1977).

Fine structural examination of the ductal system (centro-acinar cells, intralobular and interlobular ducts) in the exocrine pancreas was essentially similar in young and aged rats and mice except for that described in Section 3.2.

2.4. Age-Related Functional Correlates

In an attempt to correlate physiological changes in the aging pancreas it was shown in older humans that there was a decrease in volume and enzymatic content of pancreatic juice (Meyer and Necheles, 1940; Necheles *et al.*, 1942). Ritter (1964) and Fikry (1968) reported a depressed secretion from the stimulated pancreas of aged individuals. Snook (1975) found that the total pancreatic enzyme content (chymotrypsinogen, trypsinogen, amylase, and lipase) of aged rats rose significantly from 1.5 months to 12 months of age, then declined dramatically from 12 months to 28 months. Additional evidence of decreased enzymatic secretion in the aged human pancreas was reported by Bartos and Groh (1969), who showed that repeated stimulation of the pan-

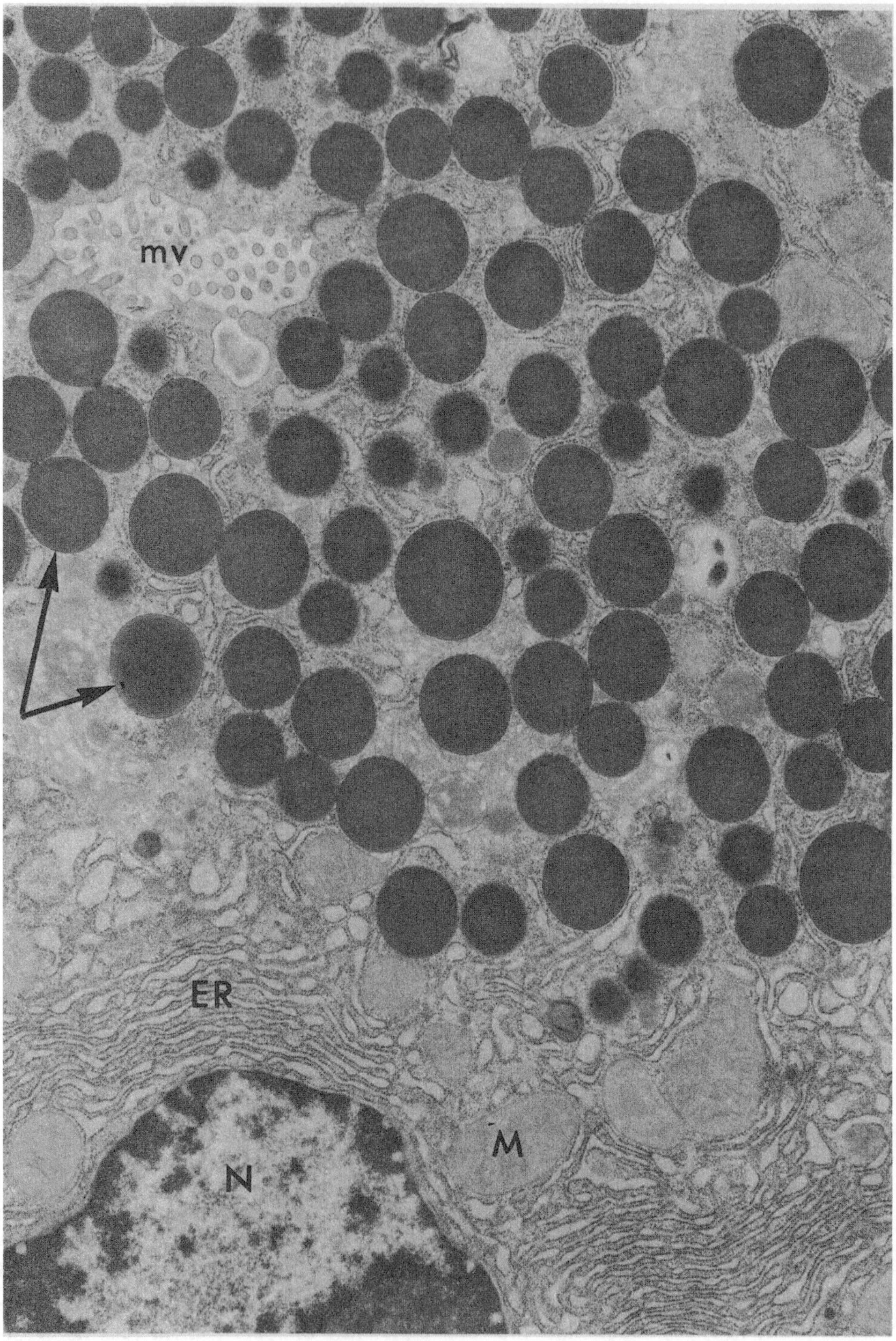

Figure 6. Electron micrograph of a typical acinar cell in a 3-month-old rat pancreas. Note the basally located nucleus, N, and endoplasmic reticulum, ER. Also note the abundance of zymogen granules (arrows) towards the apex of the cell, and the lumen of the acinus containing profiles of microvilli, mv. Mitochondrion, M. ×17,500, reproduced at 95%.

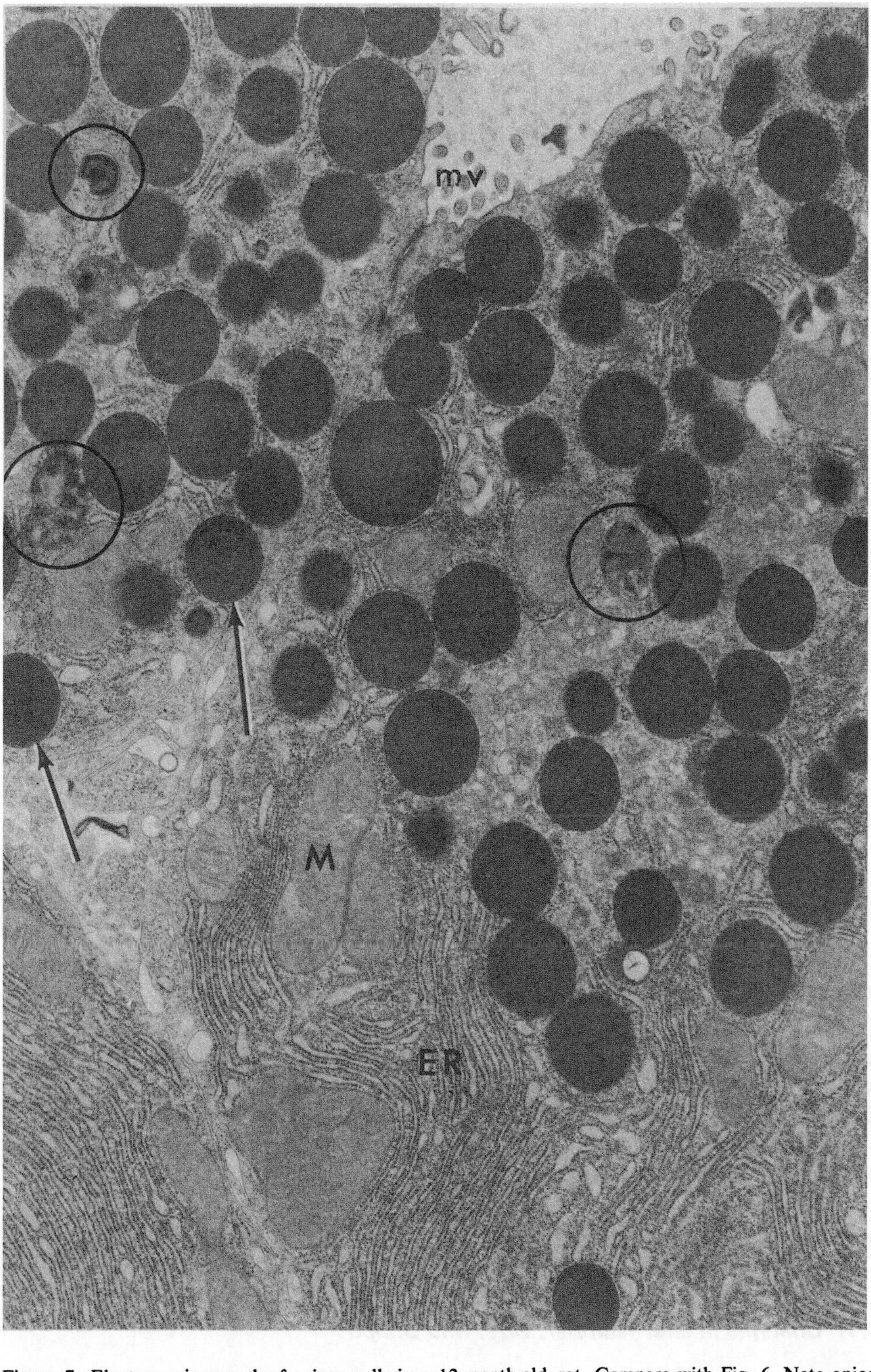

Figure 7. Electron micrograph of acinar cells in a 12-month-old rat. Compare with Fig. 6. Note apical zymogen granules (arrows), mitochondria, M, endoplasmic reticulum, ER, and microvilli, mv. Numerous lysosomal bodies are encircled. ×17,500, reproduced at 95%.

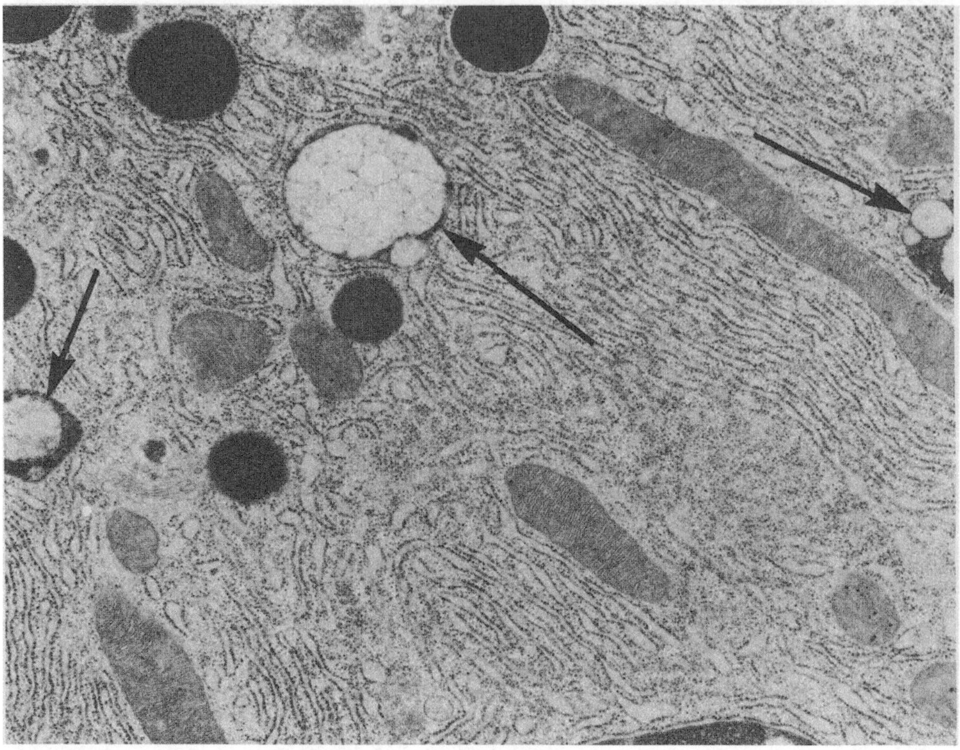

Figure 8. Arrows indicate ceroid or lipofuscin pigment in an acinar pancreatic cell of a mouse 13.5 months old. ×20,000.

creas by pancreozymin and secretin resulted in the reduction of amylase secretion although the secretion of bicarbonate was the same as in younger controls.

A decline in protein synthesis in the pancreas of aged rats has been reported by Kim *et al.* (1981). These investigators also found that the cellular level of amylase was higher in older rats compared to younger ones, and correlated this finding with the demonstration of more secretory granules in ancinar cells of older rats. Based on the finding of fewer zymogen granules in the younger animals, they suggested that this reflects a higher rate of secretory activity in these animals. Of interest was the finding that age-related exocrine pancreatic lesions had no apparent effect on the general health of Sprague–Dawley rats, e.g., signs of pancreatic insufficiency (steatorrhea, etc.) were not present (Anver *et al.*, 1982).

3. AGE-RELATED MORPHOLOGICAL CHANGES IN THE ENDOCRINE PANCREAS

3.1. Changes in Islet Number, Size, and Shape

In aging rats (12 months or older) it has been reported that there is a marked increase in the number and size of pancreatic islets compared to younger controls (Hellman, 1959b; Wexler and Fischer, 1963; Hajdu and Rona, 1967; Remacle *et al.*, 1977; Reaven *et al.*, 1979; Slavin, 1981). Similar results have been demonstrated in the mouse (Bunnag, 1966) and man (Houcke *et al.*, 1964).

In our laboratory, we have had the opportunity to study the endocrine pancreas both in male virgins and retired breeder rats ranging in age from 2 to 29 months. In adult virgin rats at 4 and 7 months of age, pancreatic islets were spherical to oval in shape, variable in diameter, and randomly distributed (Fig. 9a). At 10 and 13 months of age a marked increase in the size and number of islets was observed (Fig. 9b). Progressively from 13 to 18 months there was a further increase in islet hypertrophy and number (Fig. 9c). This latter increase appeared to be due in part to an increase in the number of very small islets (Figs. 9b and 9c). Islet hypertrophy and an increase in islet number have also been observed in young adult animals and man under a variety of abnormal conditions, some of which include obesity, partial pancreatectomy, sulfonylurea treatment, prediabetes, diabetes, hormonal injection (ACTH, adrenal steroids), force-feeding high carbohydrates, dietary restriction, Cushing's syndrome, and pregnancy (Wexler and Fischer, 1963; Logothetopoulos, 1972; Hellerstrom, 1977).

The shape of islets also changes with age. In the younger rat the islets are generally spherical or slightly oval whereas in older rats bizarre, irregular-shaped islets are seen

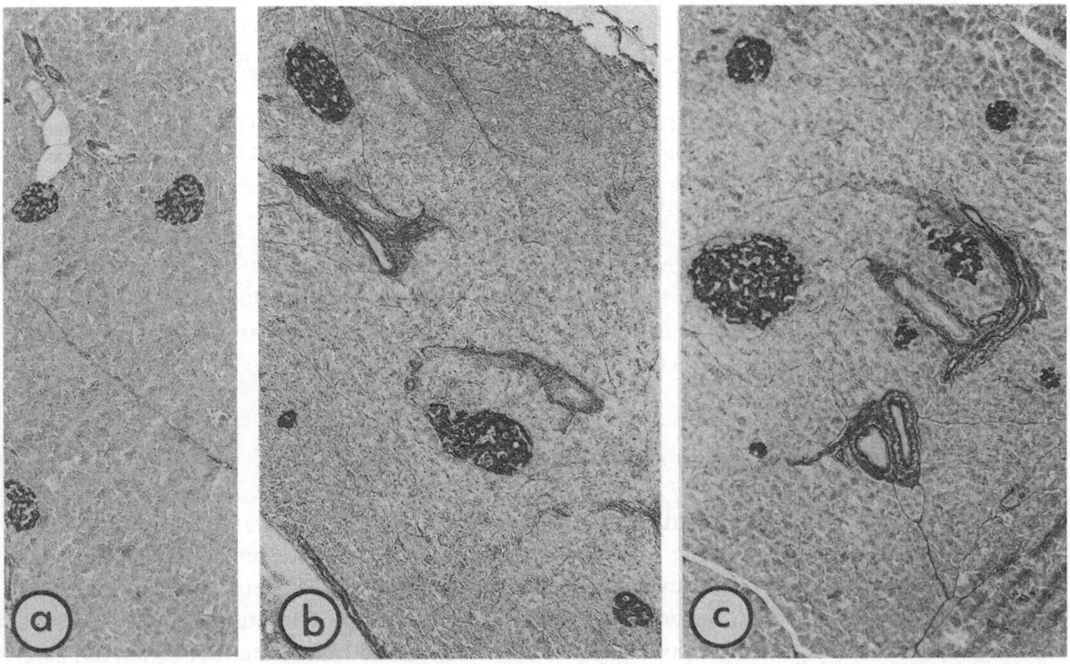

Figure 9. These three figures represent sections of the pancreas stained with aldehyde fuchsin for the demonstration of B cells. (a) In the 4-month-old rat, note the uniform size of the three islets which is typical for animals at this age. (b) In the 13-month-old rat, note the hypertrophy of islets as well as the appearance of small aggregates of B cells (left lower center). (c) In the 18-month-old rat, note further hypertrophy of islets. In addition, islet hyperplasia can be seen in the form of small aggregates of B cells scattered through the acinar tissue. ×200.

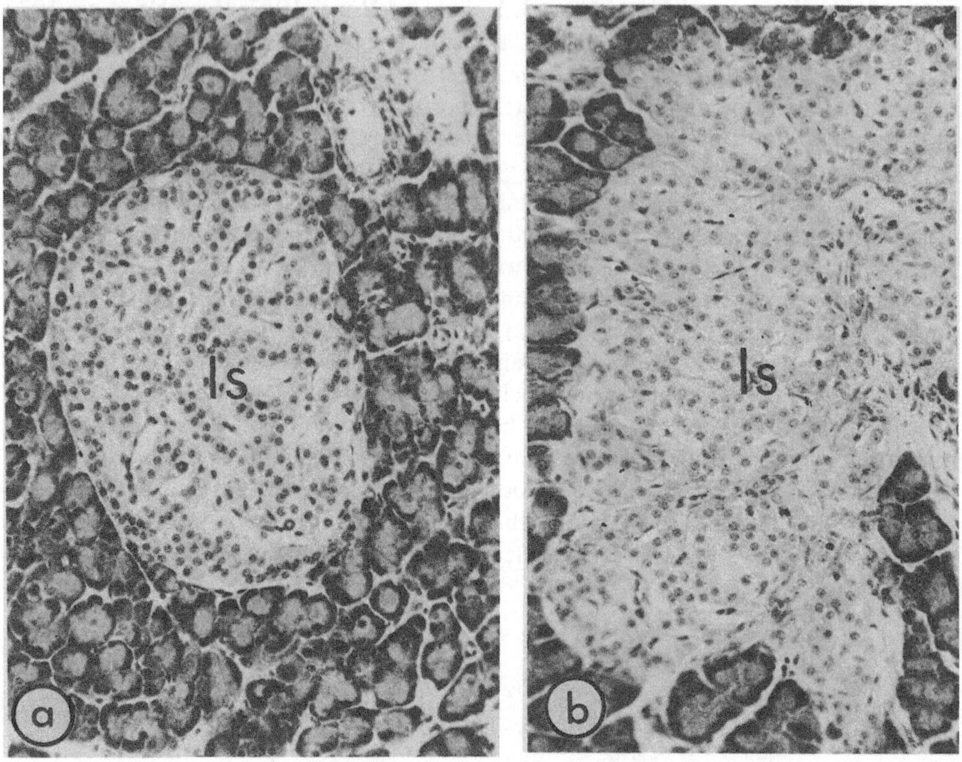

Figure 10. (a) Pancreatic islet, Is, in a 3-month-old rat. Note the slightly oval shape of the islet and the clear demarcation of islet from surrounding exocrine pancreas. (b) Islet from a 29-month-old rat which has undergone extensive hypertrophy. Note that the boundary between exocrine tissue and islet is much less clear. PAS and Hematoxylin. ×200.

with increasing frequency (Figs. 10a, 10b, 11a and 11b). It appears that islets have expanded to such a degree as to cause islet invasion of adjacent acinar tissue with possible erosion of acinar cells (Figs. 11a–11d).

3.2. Islet Neogenesis in Aging Rats

It is well accepted that during prenatal development, pancreatic islets arise from ductal cells (Conklin, 1962; Like and Orci, 1971; Pictet and Rutter, 1972; Hellerstrom, 1977; Rutter, 1980). However, the source of new and larger islets with aging has not been fully established. Several theories have been proposed to account for the increased number and size of islets postnatally. These include the transformation of ductal, periductal, acinar, and centro-acinar cells into islet cells as well as the replication of islet cells by mitosis of preexisting islet cells (Bunnag, 1966; Hajdu and Rona, 1967; Slavin *et al.*, 1981).

In the present report, we have indirect morphologic evidence suggesting that new islet cells can arise from direct transformation of ductal, acinar, or centro-acinar cells. However, no direct proof is offered. In tracing the age-related development of islet cells from exocrine ducts it was found initially that intralobular ducts of the exocrine pancreas in virgin animals less than 10 months old consisted of low cuboidal epithelial cells with scanty pale-staining cytoplasm (Fig. 12a). A thin connective tissue interstitium separated the ducts from the adjacent acini. In animals 13 months of age and older a number of intralobular ductal cells were seen to undergo transformation into islet cells. The earliest recognizable change appeared to be an increase in the number of ductal cells (Fig. 12b). Apparently, the increase in number of ductal cells was due to mitosis (Fig. 12b). Selective staining of the pancreas of 12-month-old rats with aldehyde fuchsin demonstrated the presence of occasional ductal cells stained for intracellular granules typical of those within the B cells of endocrine islets (Fig. 12c).

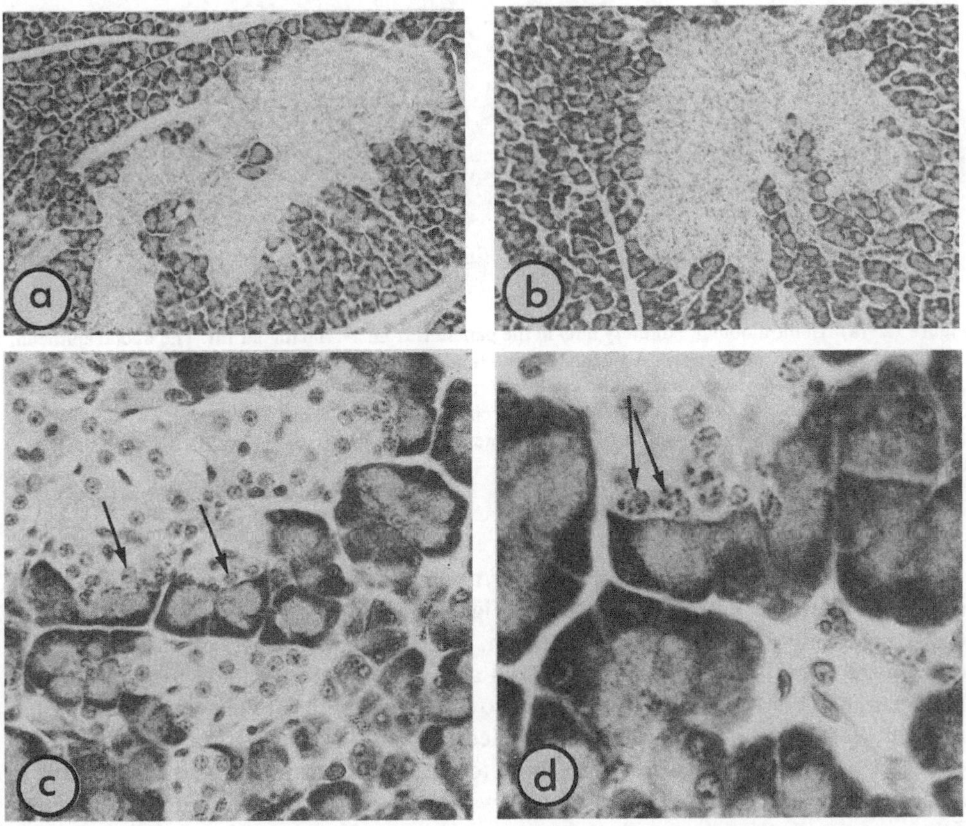

Figure 11. All sections on this plate were stained with PAS-hematoxylin. Rats were 24 months old. (a,b) Note the large, irregularly shaped islets containing projections into the surrounding acinar tissue. Also, note, especially in (a) the entrapment of acini within the islet. ×190. (c) Higher magnification of an area at the edge of the islet similar to that seen in the previous two figures. Islet cells (arrow) appear to have invaded the adjacent acini resulting in the apparent erosion of acinar cells. ×380. (d) Higher magnification of an area similar to that in previous figure. Note the encroachment of islet cells (arrows) onto the acinar cells. ×760.

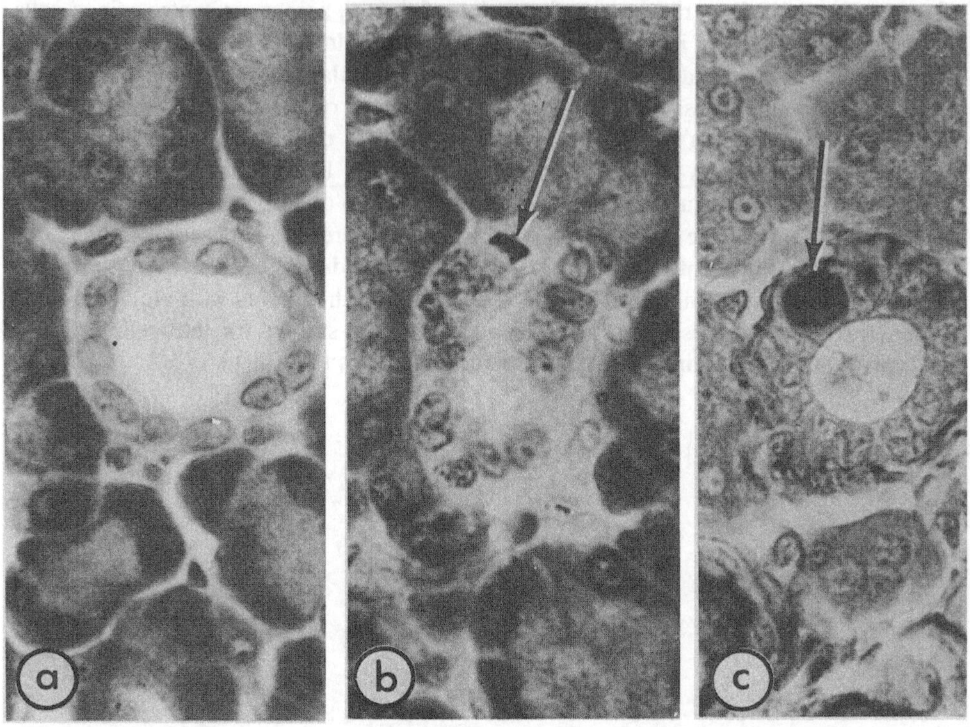

Figure 12. (a) An intralobular excretory duct in the pancreas from a 4-month-old rat. The ductal epithelial cells are of the low cuboidal type. PAS-hematoxylin stain. ×760. (b) Section of an intralobular duct demonstrating a ductal cell undergoing mitotic division (arrow). Thirteen-month-old rat. PAS-hematoxylin. ×760. (c) Pancreatic section of an intralobular duct from a 13-month-old rat. The section was stained with aldehyde fuchsin for the demonstration of B cells. Note the intensely stained cell within the duct indicating the presence of B cell granules. ×760.

Supporting evidence for the transformation of ductal cells into islet cells was obtained by ultrastructural observation. Figures 13 and 14 indicate the various processes occurring during this transformation. At 13 months of age, various intralobular ducts were found to consist of both ductal and islet cells surrounded by a continuous basal lamina (Figs. 13 and 14). It was also apparent that islet cells proliferated from the duct to form an individual islet nourished by blood capillaries (Fig. 14).

Newly formed islet cells appeared also to arise by direct transformation either from acinar or centro-acinar cells. This process was noted only in rats 13 months and older. The acinar or centro-acinar cells that appeared to transform into islets were never adjacent to intralobular ducts, or preexisting islets as determined by serial section analysis. The first morphologic indication of acinar-islet transformation was the appearance of a single B cell within an acinus at the light microscopic level (Fig. 15, inset). At the fine structural level several islet cells within or immediately adjacent to the acinus were observed (Figs. 15 and 16).

In the literature a still unsettled controversy exists as to whether or not acinar or centro-acinar cells can transform into pancreatic endocrine cells (Liu ano Potter, 1962; Pictet *et al.,* 1967; Logothetopoulos, 1972; Hellerstrom, 1977). Embryologically, it is well accepted that both exocrine and endocrine cells are derived from undifferentiated progenitor cells associated with primitive pancreatic tubules (ducts). Further prenatal and neonatal maturation of islet cells is believed to be due to mitotic division of preexisting islet cells (Logothetopoulos, 1972; Pictet and Rutter, 1972) or to the proliferation of ductal and periductal cells (Bunnag, 1966). These investigators presented no evidence for the transformation of acinar cells into islet cells. However, one cannot rule out the possibility that new islet cells can arise from acinar or centro-acinar cells based on our findings of intimate morphological association between these cells.

Mitosis within preexisting islets during progressive aging was noted only rarely in our study. This scanty evidence of islet cell mitosis may be the result of a rapid mitotic cycle or circadian rhythm. but it is possible that islets expanded due to increased numbers of islet cells having undergone replication. Irregular-shaped, enlarged islets noted in this report and by Hajdu and Rona (1967) could have formed by mitosis of preexisting islet cells or perhaps by a combination of other sources (ducts, acini). Conklin

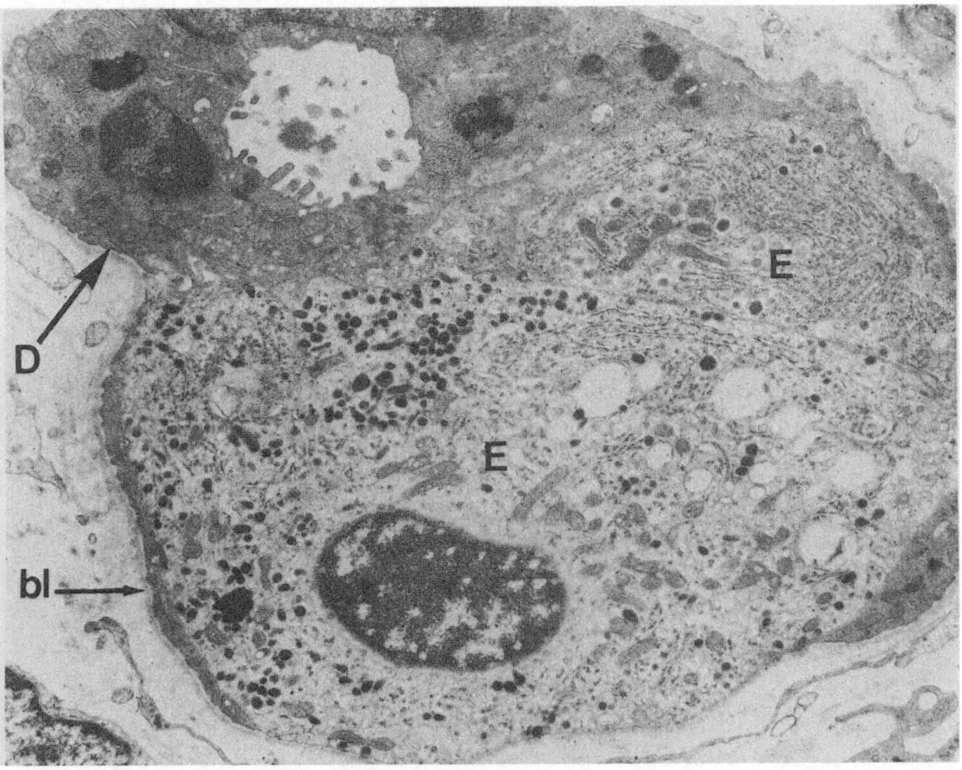

Figure 13. Electron micrograph showing two endocrine islet cells, E, associated with an intralobular duct, D. Note that one of the islet cells (containing a nucleus) appears somewhat immature while the adjacent cell contains abundant rough endoplasmic reticulum and a few mature B cell granules. Note also thin extensions of ductal cells, partially surrounding the newly formed endocrine cells. A basal lamina, bl, can be seen to surround ductal and islet cells. Rat, 13 months old. ×7900.

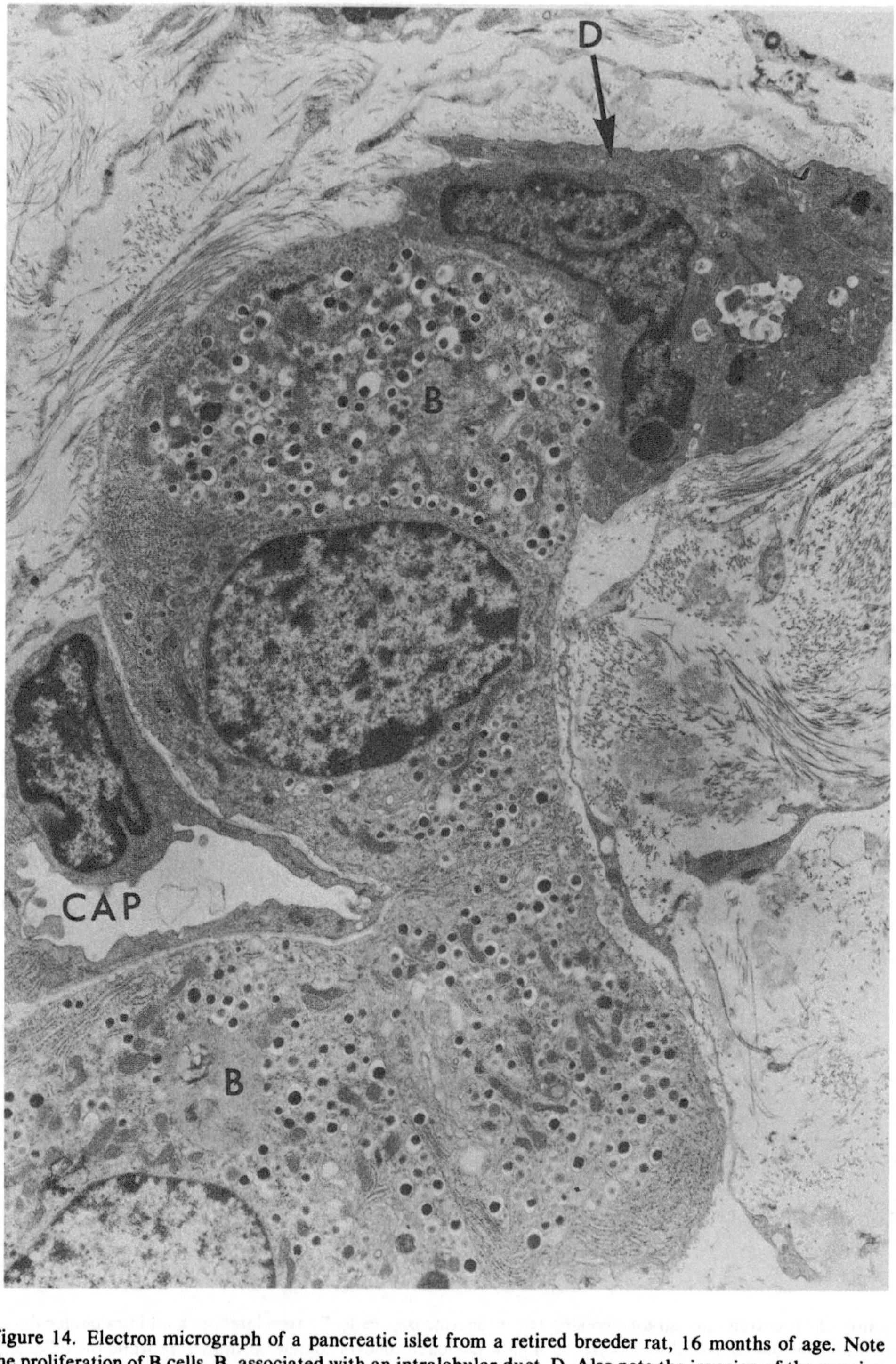

Figure 14. Electron micrograph of a pancreatic islet from a retired breeder rat, 16 months of age. Note the proliferation of B cells, B, associated with an intralobular duct, D. Also note the invasion of the growing islet by a capillary, CAP. Although not clearly shown in this figure, there is continuity of the basal lamina from duct to islet cell. ×7200, reproduced at 95%.

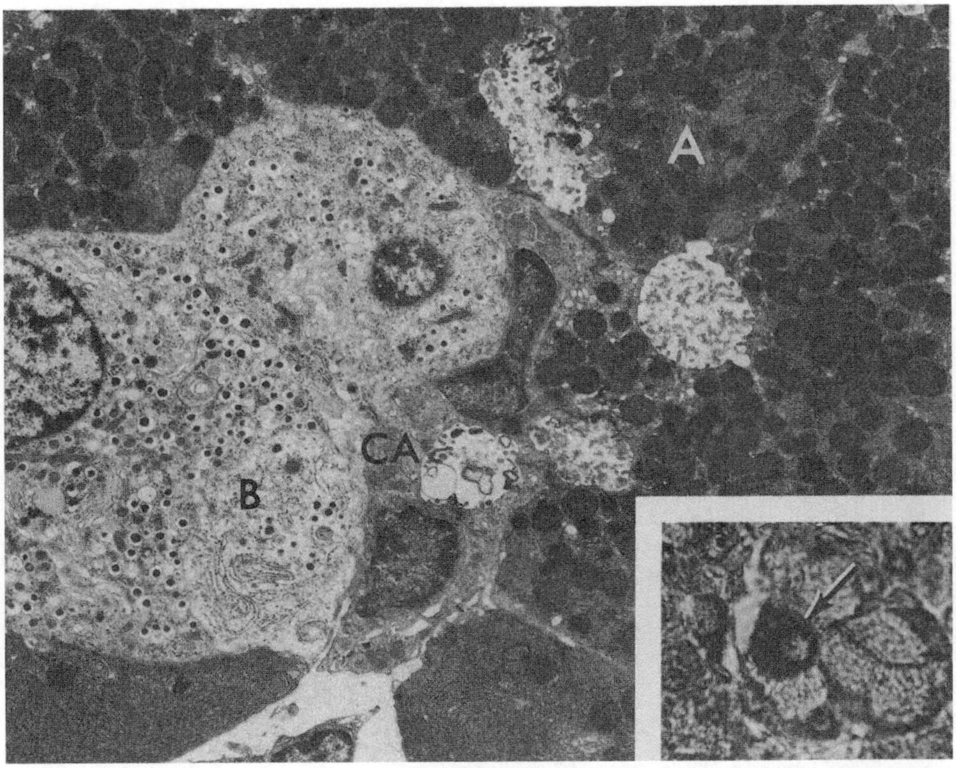

Figure 15. Electron micrograph showing the close relationship between acinar cells, A, and beta cells, B, in the pancreas of a 13-month-old rat. CA, centro-acinar cells. ×5250. Light micrograph (inset) shows an intensely stained cell (arrow) intimately associated with an acinus in the pancreas of a 16-month-old rat. Aldehyde fuchsin. ×875.

(1962), in human fetal pancreata, demonstrated mitotic activity in preexisting islet cells to account for islet growth secondary to the formation of islet cells from primitive exocrine ducts. Pancreatic islets of diabetic mutant mice underwent hypertrophy by mitosis of preexisting B cells when mice were put on a restricted diet followed by ad lib feeding (Like and Chick, 1969). No mitotic activity was noted in exocrine ducts. Thus, these investigators stressed the decreased importance of ductal cells as precursors for new islet cells. Similar findings have been reported in other animals with experimental diabetes (Like and Jones, 1967; Like and Miki, 1967) or in animals injected with adrenal corticoids (Like and Chick, 1974). In addition, Logothetopoulos and Bell (1966) described a four- to six-fold increase in the incorporation of tritiated thymidine in islets of mice injected with antibody to insulin.

Another proposed method to account for islet hypertrophy with aging involves the coalescence of small islets into a large single islet. A similar mechanism of islet formation was described by Hard (1944) in prenatal rats. This process of islet coalescence might explain the presence of acinar tissue within the hypertrophied islets described by Hajdu and Rona (1967) and the present investigators.

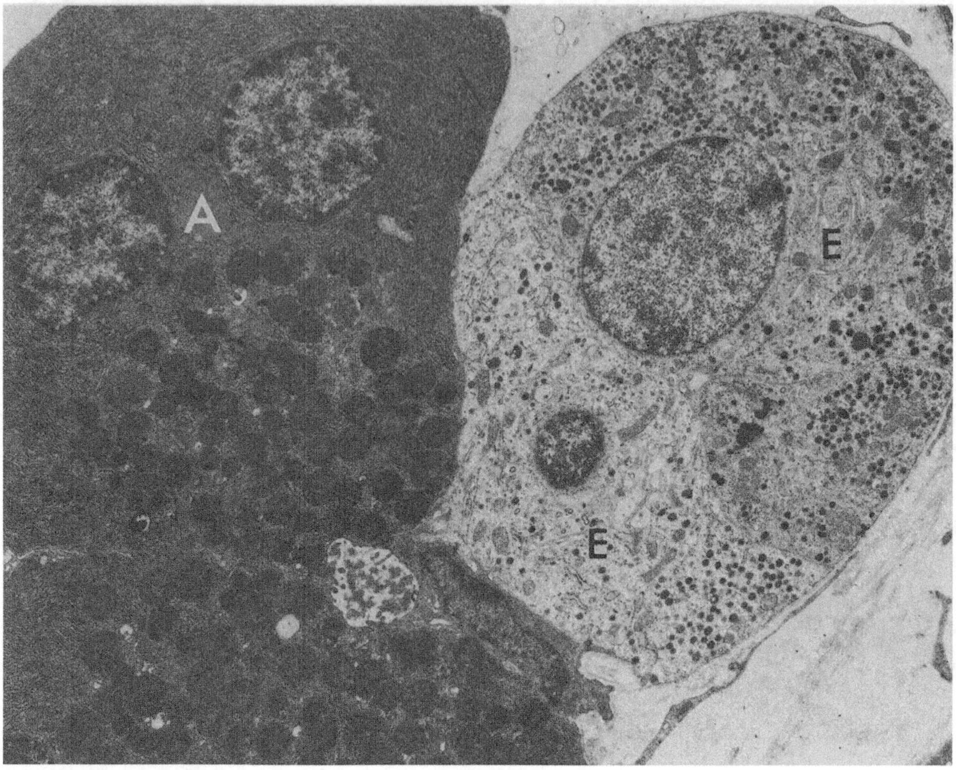

Figure 16. Ultrastructural demonstration of endocrine islet cells, E, possibly derived from acinar cells, A, and beginning to bud off from the acinus. ×6000.

3.3. Islet Cell Distribution and Cytology with Aging

Along with an increase in number and size of pancreatic islets with aging, there were also changes in islet cell distribution. However, reports concerned with an understanding of islet cell distribution as a function of age have been scanty. Hellman (1959a) reported an increase in the percentage volume of A (alpha) and a decrease in volume of B (beta) cells in islets of aged Wistar rats (ages of rats: five days, 3.3 months, and 16 months). The percentage volume contributed by the remaining islet constituents (sinusoids, other islet cells, etc.) in the older rats was double that of younger ones. Remacle *et al.* (1977), on the other hand, showed that the relative volumes of total A and B cells per islet remained unchanged in albino rats at 6 and 24 months. In a more recent study, Reaven *et al.* (1979) determined that the absolute number of B cells increased by 50% in 12-month-old Sprague–Dawley rats compared to rats at two months. In addition, these investigators determined that the volume of mature B cell granules per B cell increased by a factor of 2 with age. These results were reflected in the measurement of immunoreactive insulin content in aged islets which were twice that of younger ones. These results are in contrast to a previously published report by Remacle *et al.* (1977) who showed no significant change in the volume of B cell cytoplasmic granules in 24-month-old rats compared to rats at 6 months although the total islet tissue mass had increased in the older animals.

The most recent report on changes in islet cell populations with aging was reported by McEvoy (1981), who showed by immunohistochemical and morphometric means that B cell volume increased in Sprague–Dawley rats from 10 days to 210 days (7 months) while A and D cell volumes were maximum at 50 days and 75 days, respectively. No data are available on islet cell volume in animals beyond seven months. Currently, experiments are underway in this author's (BGS) laboratory to determine the topographic distribution and volume of A, B, D (delta), and PP (pancreatic polypeptide) cells in rats up to 18 months old. For example, as in the young rat (two months), the aging rat's (>12 months) islets have abundant insulin-containing cells within their cores when observed with immunostaining techniques (Slavin, unpublished observation). The major difference between the two age groups is that the aging islets are generally larger and more numerous. These findings are in accord with those of Reaven *et al.* (1979), who found in 12- and 18-months-old rats that islets not only contain more B cells but also contain greater numbers of insulin-containing granules per B cell and more immunoreactive insulin. In a preliminary study in our laboratory, glucagon-containing cells were found located mostly at the periphery of the islet and occasionally in the center (Fig. 17).

Few modern fine structural studies of aging islet cells have been reported thus far. The first electron microscopic study of islet cells from aged animals was by Kyogoku and Okamoto (1969), who reported cystic changes in the rough endoplasmic reticulum,

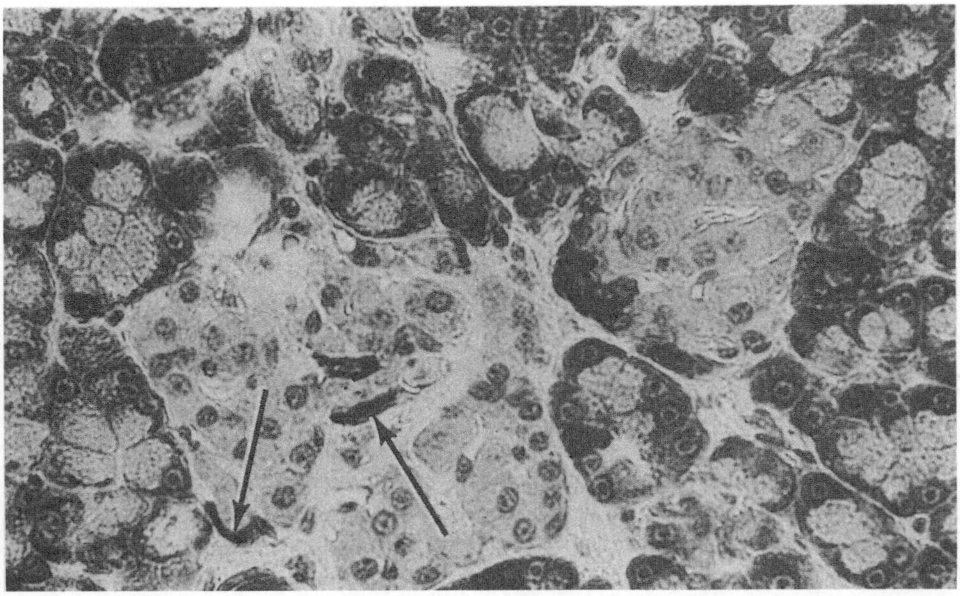

Figure 17. Light micrograph of PAP immunostained section of pancreas from an 18-month-old rat. Stained cells (arrows) indicate immunostaining for glucagon in islet cells. Hematoxylin counterstain. ×503.

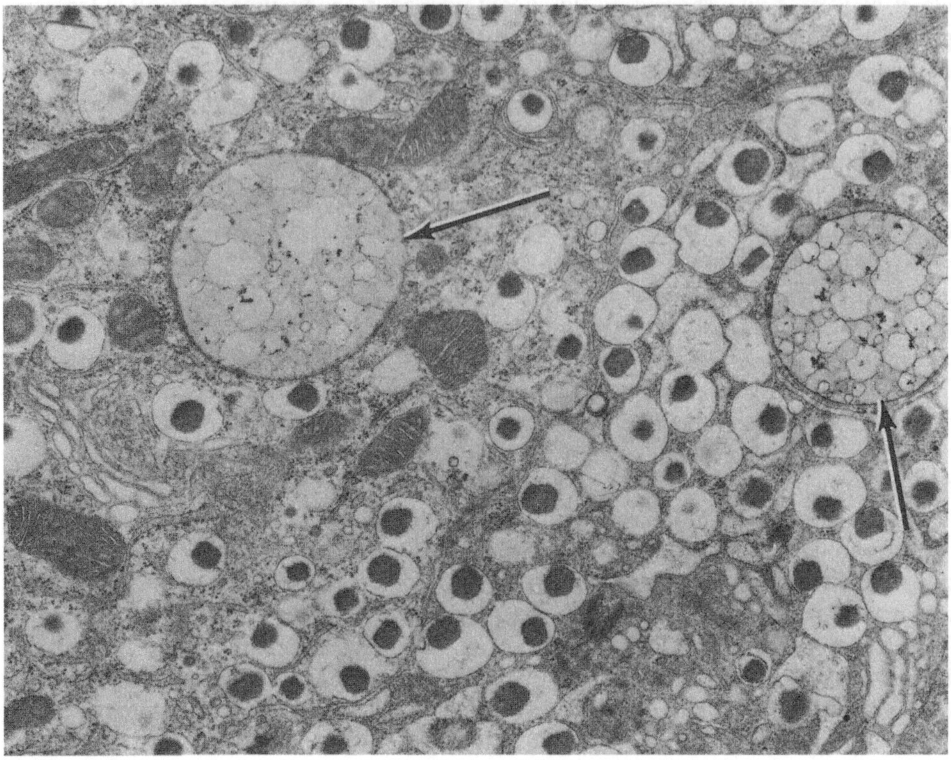

Figure 18. Ceroid or lipofuscin pigment (arrows) contained within B cell of a 23-month-old mouse. ×20,000.

mitochondria, and Golgi regions of some B cells in 18-month-old rats. In our laboratory we found no substantial difference in fine structural features of islet cells in 12-month-old Sprague–Dawley rats compared to those at two or four months. However, in agreement with Saladino and Getty (1972) in dogs, we noted the appearance of ceroid or lipofuscin pigment in pancreatic B cells of C57BL/6J mice and Sprague–Dawley rats usually beyond 18 months of age (Fig. 18).

3.4. Stroma-Vascular Changes

We have had the opportunity to examine the stroma-vasculature of rats from two to 29 months of age. Beginning approximately at 12 months most rat islets showed microscopic evidence of fibrosis, which increased progressively up to 29 months. The fibrosis was usually confined to the pericapillary space and to the periphery of the islet. In the young rat (two to four months), islets were permeated by sinusoid-like capillaries which were closely associated with the surrounding parenchyma (Figs. 19a and 19b). At 12 months, most islets, especially those having undergone hypertrophy, exhibited pericapillary fibrosis (Figs. 20a and 20b). In addition, these capillaries appeared to be

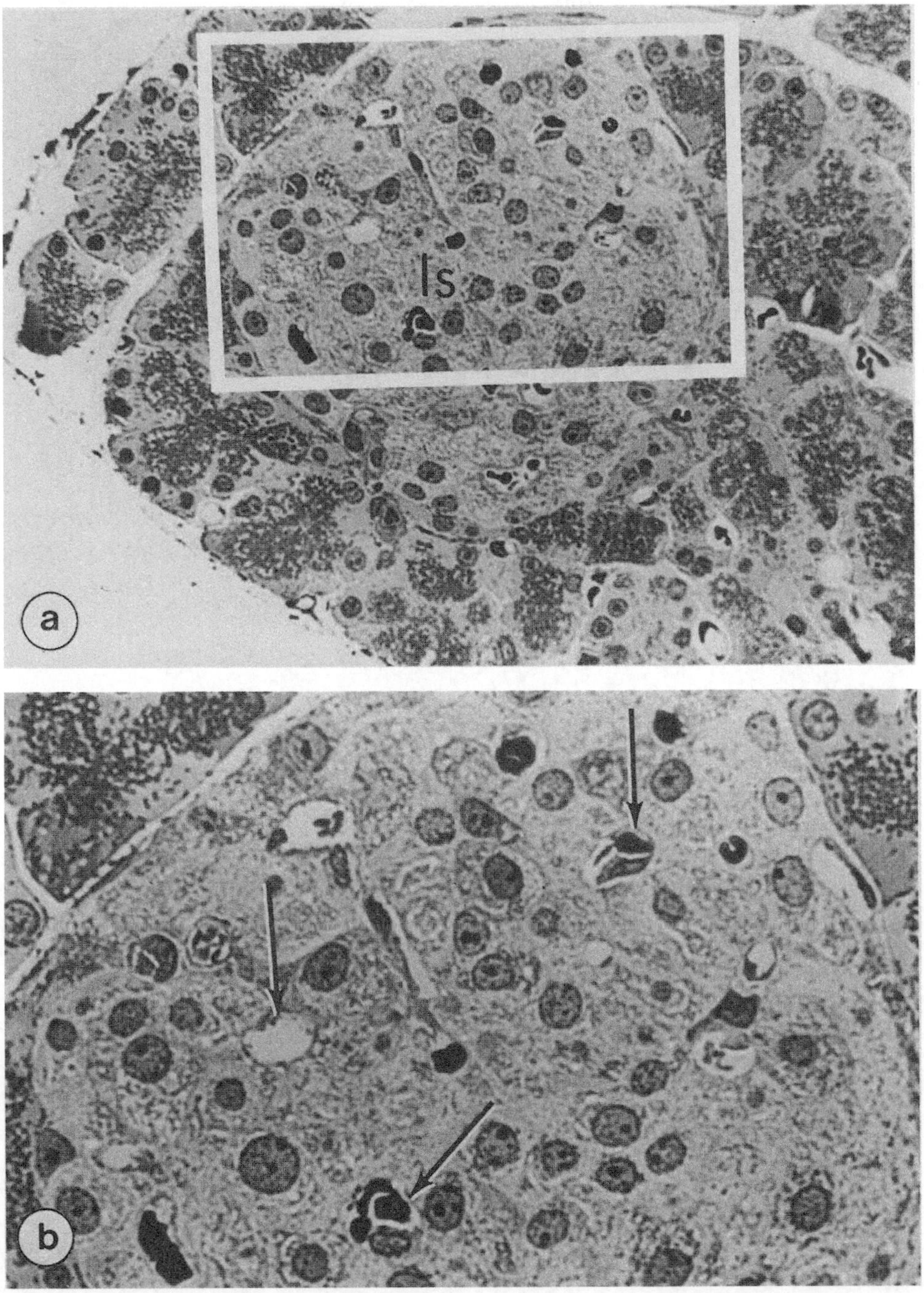

Figure 19. (a) Semithin section of pancreatic islet, Is, from a 3-month-old rat. Note surrounding exocrine tissue. Silver staining. ×553. (b) Enlargement of blocked off area in (a). Note permeation of islet with relatively wide capillaries (arrows). Also, note intimate association of capillaries with surrounding endocrine cells. Very little pericapillary connective tissue can be seen. Silver staining. ×1005, reproduced at 95%.

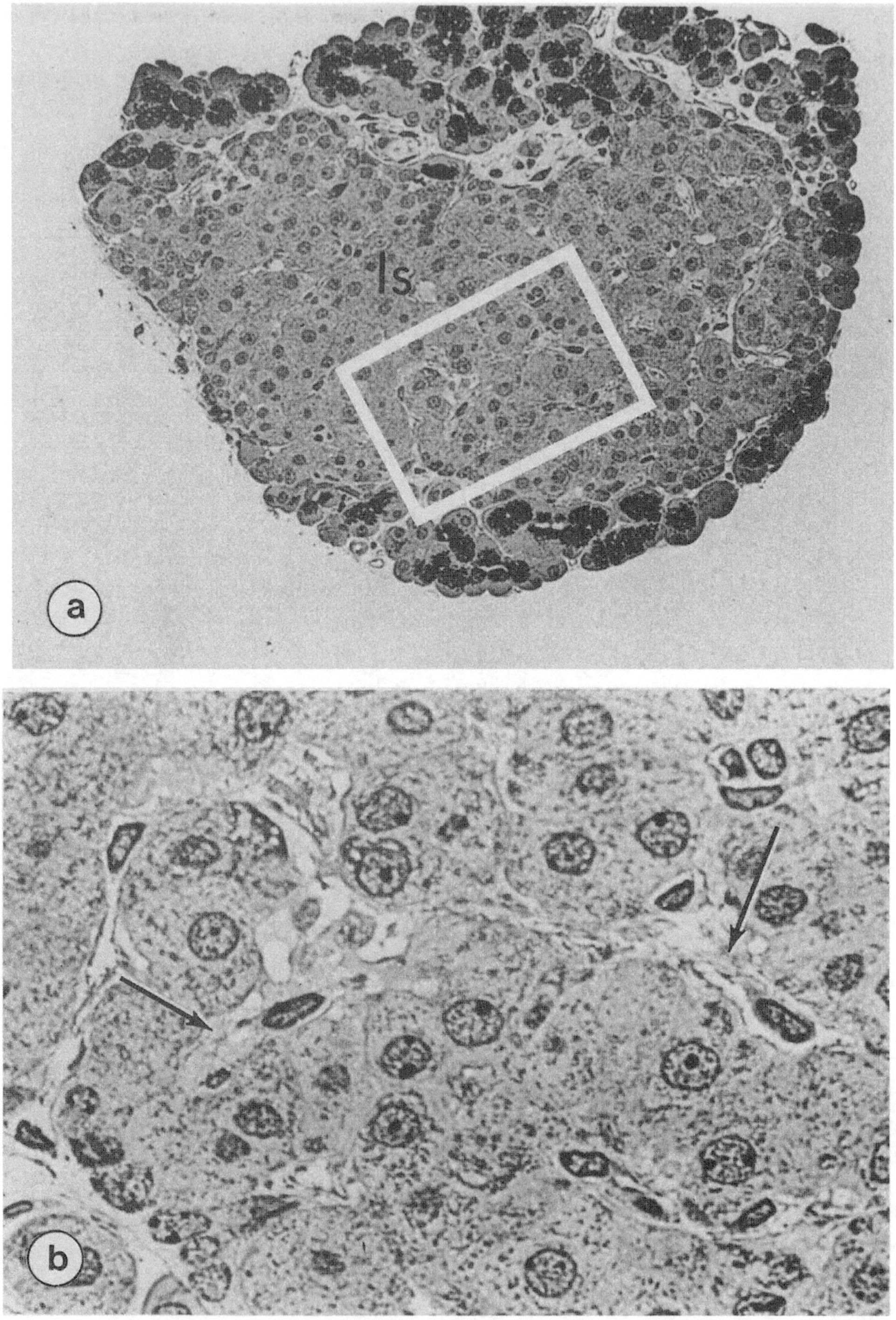

Figure 20. (a) Semithin section of pancreatic islet, Is, from a 12-month-old rat. Note the hypertrophy of this islet. Silver stain. ×300. (b) Enlargement of boxed area above. Arrows demonstrate the extent of pericapillary fibrosis seen in the islets of these rats. Also, capillaries appear compressed. Compare this micrograph to Figure 19b. Silver stain. ×1005, reproduced at 95%.

somewhat flattened or compressed compared to that in young islets. At the ultrastructural level abundant collagen fibers were noted at the islet periphery and surrounding every capillary at 12 months (Figs. 21, 22a and 22b). Furthermore, there appeared to be an increase in the number of endothelial cell folds projecting into the lumen of aged capillaries. Pericapillary and peripheral fibrosis in islets have been reported in aged humans by Zhukov (1964), Houcke *et al.* (1964), Zyss (1968), and in aged rats by Hajdu and Rona (1967), Kyogoku and Okamoto (1969), and Lewis and Wexler (1974). Interestingly, it is well established that pericapillary and peri-islet fibrosis occurs during diabetes (Lazarus and Volk, 1961).

In the present study, using virgin or retired breeder Sprague–Dawley rats, more than 50% at 12 months or older exhibited varying numbers of extensively fibrotic and degenerated islets. The appearance of these highly damaged islets was progressive with increasing age. These islets were characterized by abundant interstitial connective tissue segregating the islet cells into small groups or clusters (Figs. 23a and 23b). A few acinar units were found amongst the islet cells. Also in the vicinity were ductlike structures containing ductal lining cells and endocrine cells. At the fine structural level, islet cells were found in clusters surrounded by abundant collagen fibers and fibroblast cells (Fig. 24). In rats at 19 months, some islets were found which were similar in severity of fibrosis to that at 12 months with the exception that fatty infiltration of the pancreas was more extensive and they contained metaplastic ductlike structures. In addition to the fibrotic changes occurring in aged islets, it was common to find densely stained lipofuscin-containing cells at the islet periphery (Fig. 25).

Pancreatic islets in normal, young animals consisted of a vascular supply characterized by a rich network of capillaries (Fig. 19b). Endothelial cells of islet capillaries were mostly of the fenestrated variety typical of endocrine glands in general, and were surrounded by a fine basal lamina and a thin layer of collagen (Fig. 21). Pericytes or their cellular processes were usually found adjacent to capillary endothelial cells. Remacle *et al.* (1977) found that the relative capillary volume in young rats (6 months) was the same as in 24-month-old rats. We made no quantitative study of islet vascularity, but there appears to be no change in the relative richness of vascular supply to islets between our young and old rats (excluding highly degenerated islets).

3.5. Age-Related Functional Correlates

It is well known that there is a decline in glucose tolerance with aging in humans (Silverstone *et al.*, 1957; Crockford *et al.*, 1966; Williams, 1978; Hayashi, 1982) and rats (Hajdu *et al.*, 1968; Klimas, 1968; Kyogoku and Okamoto, 1969; Gommers and deGasparo, 1972) who show no overt clinical signs of diabetes mellitus. Several explanations have been presented for the abnormal glucose tolerance associated with aging. Silverstone *et al.* (1957), Cerasi and Luft (1972), and Hayashi (1982) have shown that there is a decreased sensitivity of tissues to insulin with aging so that glucose uptake is altered. A second theory states that the insulin-degrading activity (measured in liver, muscle, and kidney) declines with age (Runyan *et al.*, 1979) possibly due to a decrease in receptor sites on cell membranes as suggested by Freeman *et al.* (1972) and Olefsky and Reaven (1975). Evidence to account for age-related decreased glucose tolerance

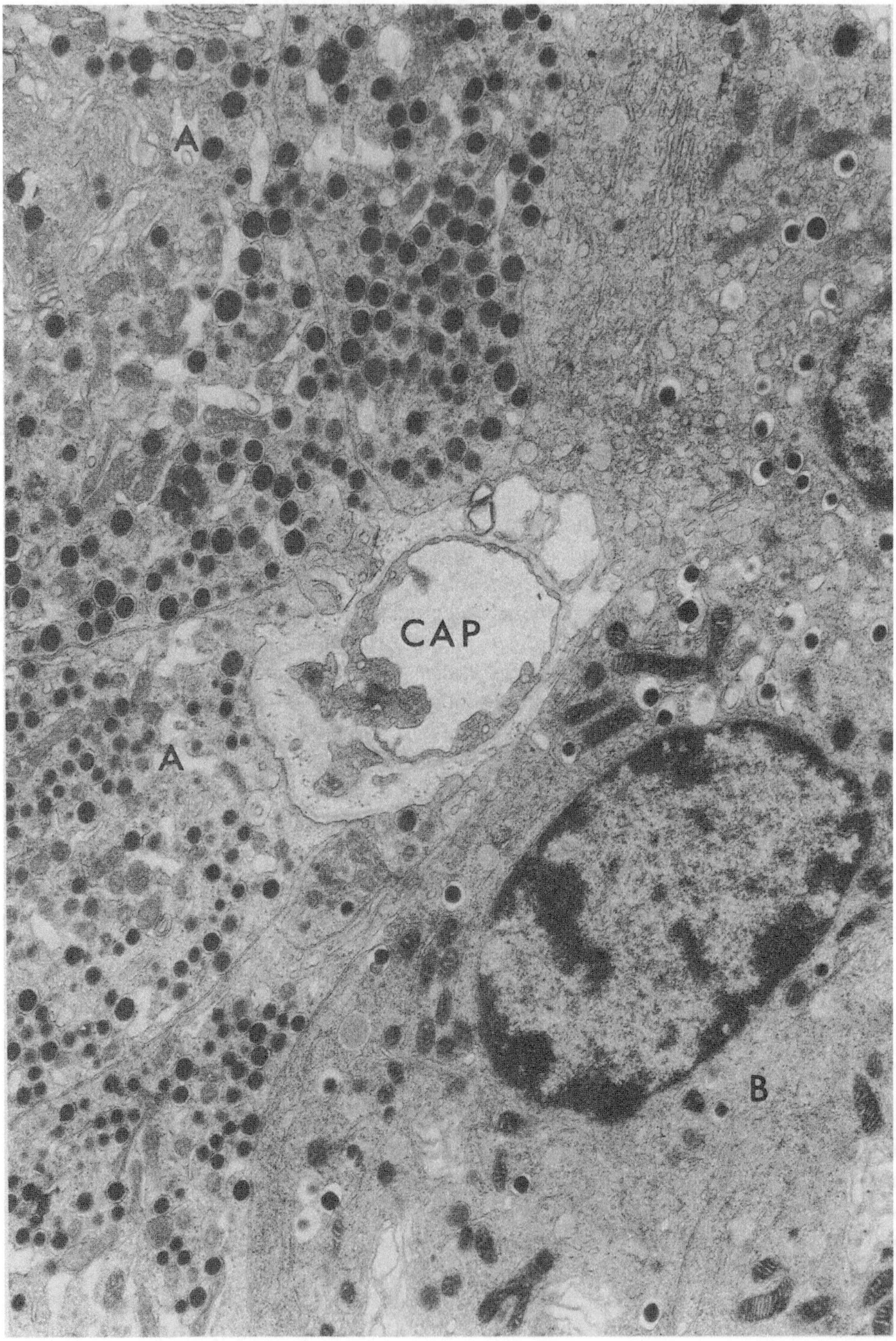

Figure 21. Electron micrograph of a portion of pancreatic islet from a 2-month-old rat. Note typical A and B cells. The capillary, CAP, is typically fenestrated and surrounded by a basal lamina and a small amount of collagen. ×14,400, reproduced at 95%.

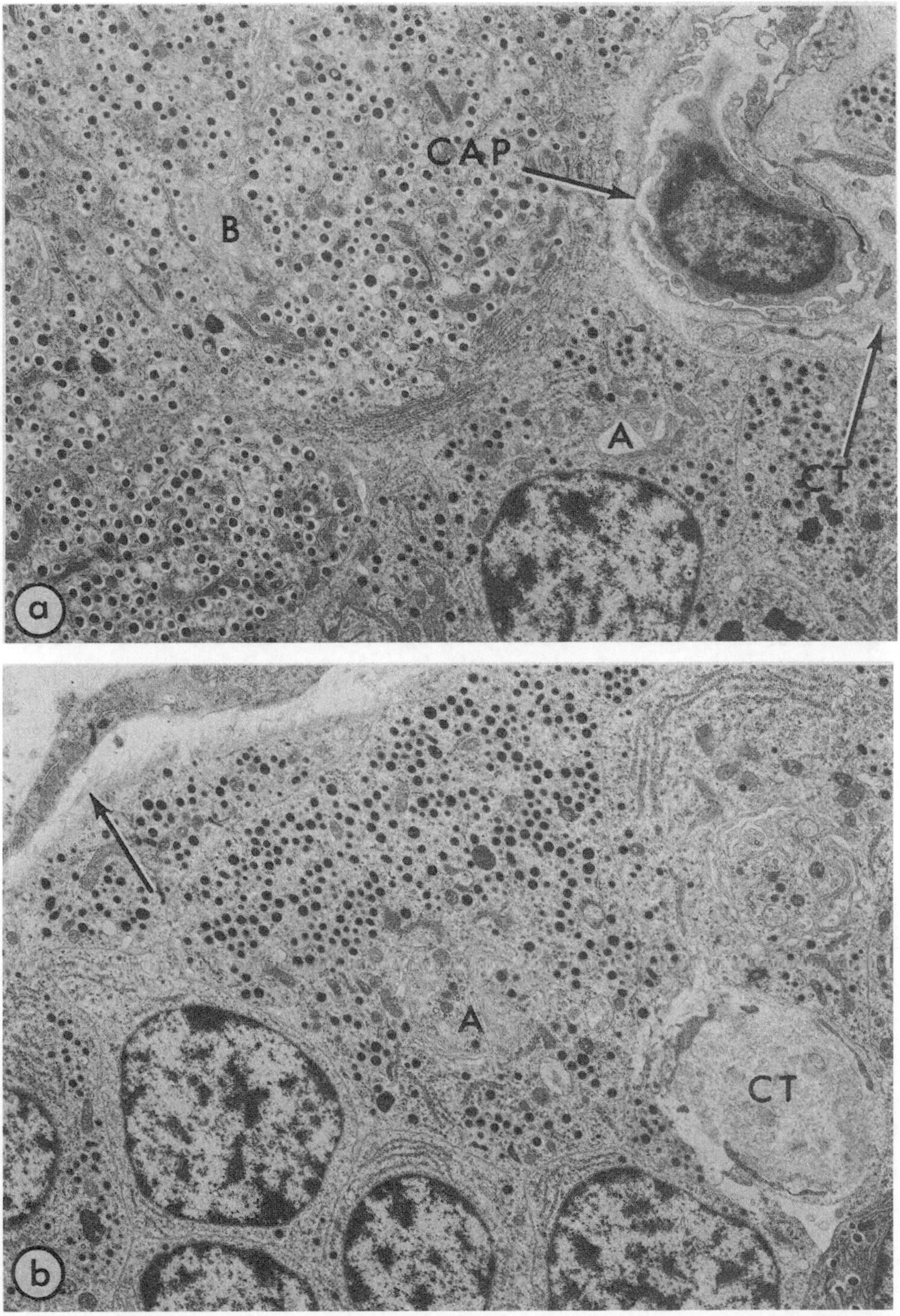

Figure 22. Electron micrographs of pancreatic islet cells from 12-month-old rats. (a) Note A and B cells adjacent to capillary, CAP. Capillary is surrounded by abundant connective tissue, CT. ×7200. (b) At the periphery of an islet from a 12-month-old rat note the extent of peripheral fibrosis (arrow) as well as abundant connective tissue, CT, probably surrounding a capillary as in (a). A, alpha cell. ×7200, reproduced at 95%.

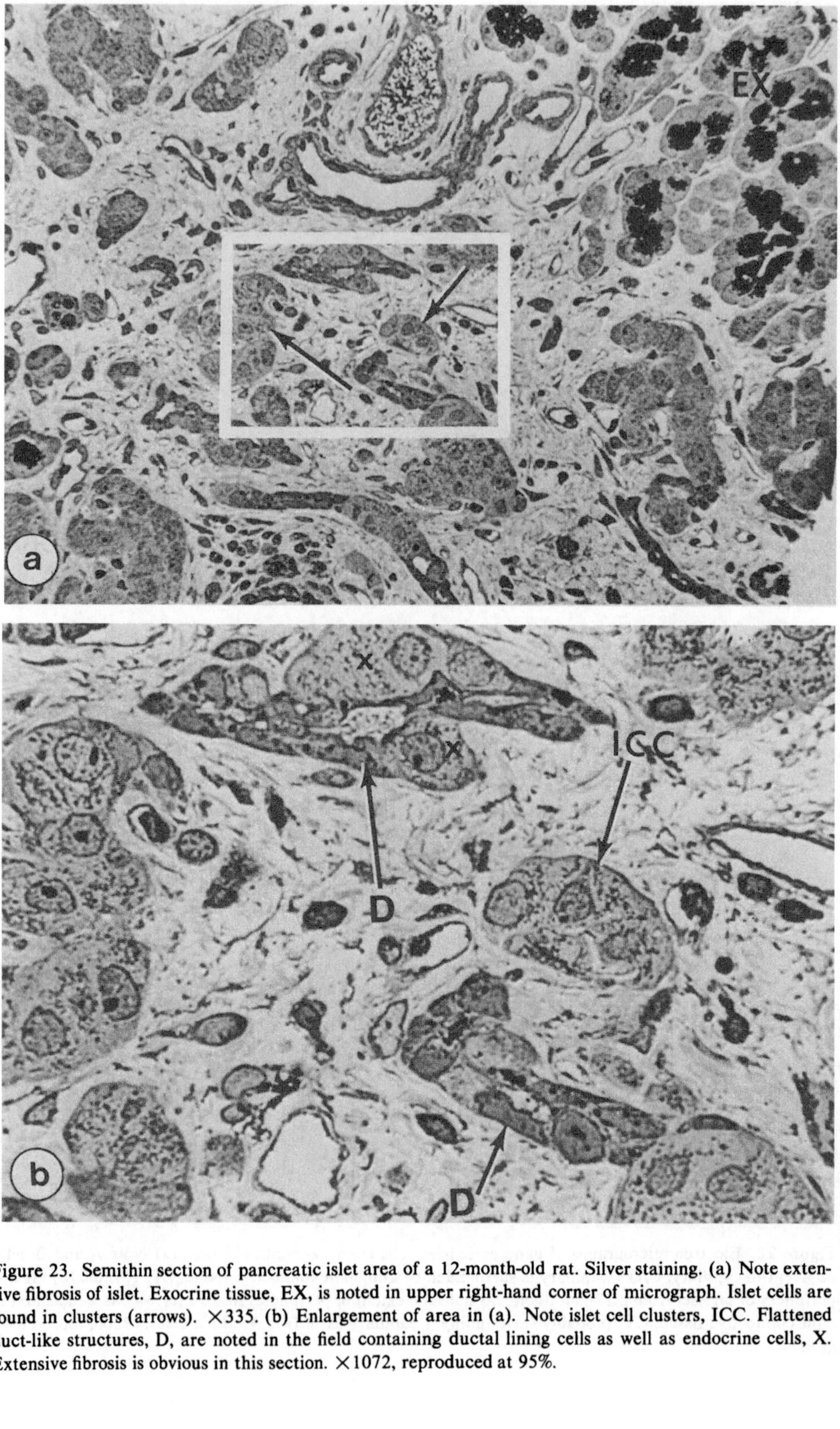

Figure 23. Semithin section of pancreatic islet area of a 12-month-old rat. Silver staining. (a) Note extensive fibrosis of islet. Exocrine tissue, EX, is noted in upper right-hand corner of micrograph. Islet cells are found in clusters (arrows). ×335. (b) Enlargement of area in (a). Note islet cell clusters, ICC. Flattened duct-like structures, D, are noted in the field containing ductal lining cells as well as endocrine cells, X. Extensive fibrosis is obvious in this section. ×1072, reproduced at 95%.

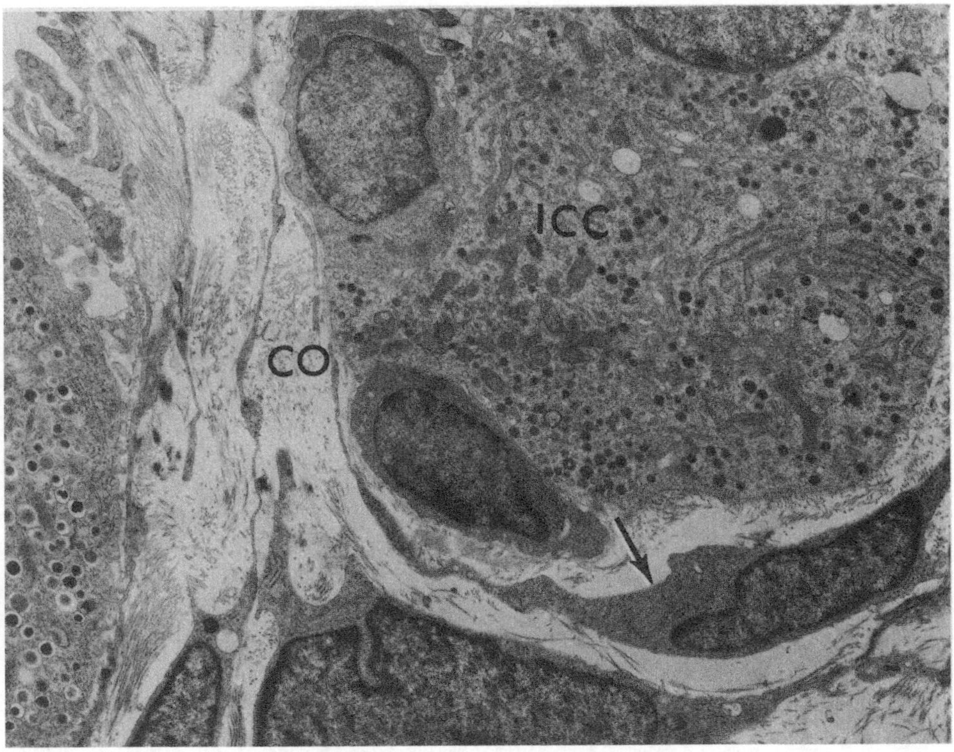

Figure 24. Ultrastructural view of highly degenerated islet similar to that seen in Fig. 23. Note the cluster of islet cells, ICC, surrounded by abundant collagen, CO, and fibroblasts (arrow). ×7200.

has been presented by Duckworth and Kitabchi (1976) and Gold *et al.* (1981) which reveals changes in the ratio of stored and secreted proinsulin to insulin. Another hypothesis suggests that the basic age-related defect is due to a decreased rate of insulin release from B cells of pancreatic islets (Reaven *et al.*, 1979). Obviously, a full description of these factors is beyond the scope of the present report. We shall present those pertinent findings indicating that the cells of the pancreatic islet may lie at the center of the debate.

Aging is associated with a marked decrease in glucose- or leucine-stimulated insulin release from isolated rat pancreatic islets (including islets derived from the ventral-duodenal or splenic regions) (Reaven *et al.*, 1979, 1980, 1982). However, it is known that leucine also stimulates glucagon secretion from islets (Assan *et al.*, 1977), and that glucagon itself can stimulate insulin release (Devrim and Recant, 1966; Grodsky *et al.*, 1967). A reduced oxidation of glucose has been demonstrated in islets from 12-month-old rats compared to 2-month-old rats (Reaven and Reaven, 1980). It has been suggested by the latter authors that the age-related defect in glucose-stimulated insulin secretion may be secondary to the glucose oxidation status in aged islet cells.

The relationship between the different islet cell types is important in this regard in view of the fact that certain islet cell hormones have been shown to inhibit or stimulate the secretion of other islet hormones. The term "paracrine" is currently used to signify

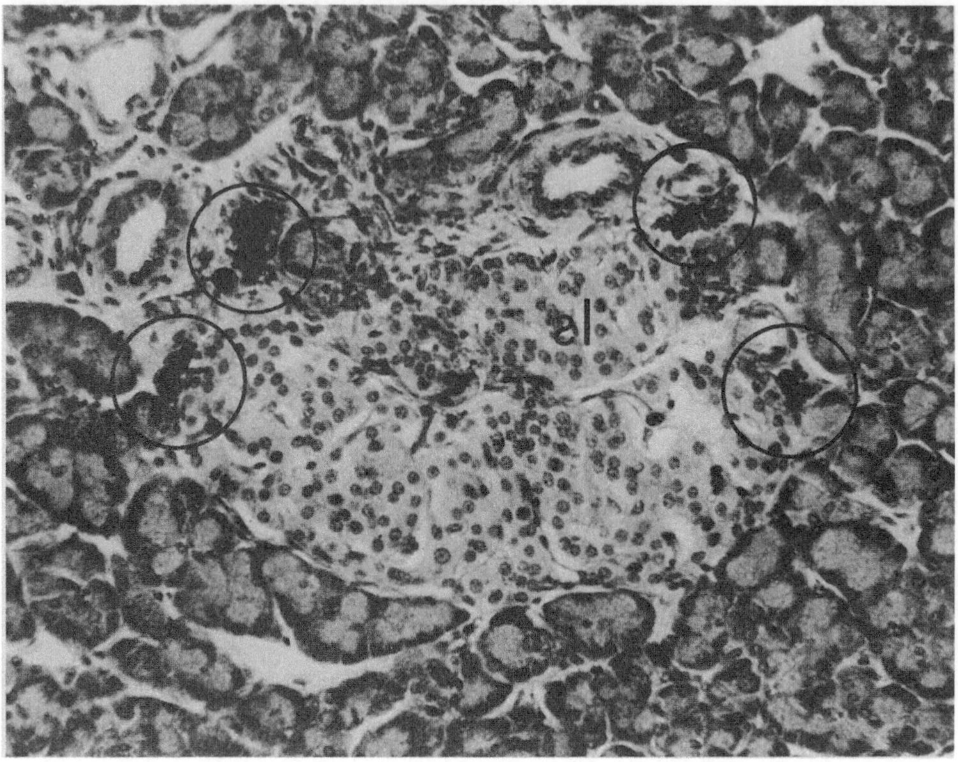

Figure 25. Light micrograph of the pancreas from a 29-month-old rat. Circled areas demonstrate lipofuscin-laden cells at the periphery of the islet, Is. PAS and Hematoxylin. ×302.

an hypothesis which suggests that hormones released by specific endocrine cells can stimulate hormonal secretion by other specific endocrine cells located in the immediate vicinity. This secondary stimulation can occur either by diffusion of the initial hormone into the interstitial space, or by release into a local portal system, or by direct intercellular communication via cell membrane specializations (Unger and Orci, 1977; Forssmann *et al.,* 1978). For example, somatostatin, administered exogenously, has been shown to inhibit the release of a variety of peptide hormones, e.g., growth hormone, thyroid-stimulating hormone, thyroxin, gastrin, insulin, glucagon, and pancreatic polypeptide (Alumets *et al.,* 1979). Somatostatin-containing cells have been found in the stomach, intestine, thyroid gland, brain, and pancreatic islets. In the stomach, it has been found that somatostatin-containing cells have long cellular processes ending in clublike swellings which contact neighboring parietal cells and gastrin cells. This evidence points to the possibility that somatostatin cells inhibit the secretion from gastric cells (Alumets *et al.,* 1979). Also, in the pancreatic islet, somatostatin cells (D cells) occupy a position between insulin (B) and glucagon (A) cells. Thus, they are in the ideal position to affect locally the secretion of insulin and glucagon. Likewise, insulin is known to inhibit glucagon and somatostatin secretion when administered exogenously

(Samols and Harrison, 1976); and exogenous glucagon has been found to stimulate both insulin and somatostatin secretion (Patton *et al.*, 1977; Unger and Orci, 1977). Similarly, synthetic human pancreatic polypeptide has been shown to be a potent stimulator of pancreatic somatostatin secretion in the isolated perfused rat pancreas (Trimble *et al.*, 1982). These effects may also occur by local intraislet cell secretion.

In recent reports it has been shown that certain islet cells undergo a major topographical redistribution in diabetic mice (Baetens *et al.*, 1978; Leiter *et al.*, 1979). For example, D cells appeared to migrate from their normal peripheral position to a more central one with progressive hyperglycemia. Prior to this migration, in the prehyperglycemic state, long cellular processes of D cells penetrated deeply into the islet, which greatly increased the surface contact area between D and B cells. In addition, there was an increase in the number of D cells per islet. Leiter *et al.* (1979) suggested that such topographical alterations may be an attempt by somatostatin, released locally by the D cell, to control the secretion of insulin by the B cell. In fact, Adelman *et al.* (1980), in a preliminary experiment, used antibodies against somatostatin to demonstrate the contribution by somatostatin to controlling insulin secretion in aging.

Experiments are under way in several laboratories to determine the exact age-related changes in islet cell structure and function, which should lead to a clearer understanding of altered glucose metabolism in the aged animal.

Acknowledgments

The authors wish to acknowledge Drs. C. Finch and D. Gee from the Departments of Biological Science and Gerontology, University of Southern California, for contributing C57BL/6J mice to this project. And to Dr. L. Lipson, Department of Medicine, University of Southern California School of Medicine, we wish to acknowledge his contribution of Sprague–Dawley rats. We also wish to thank Mrs. H. Shiroishi for her able and excellent technical assistance. Funds for this project were derived partially from Biomedical Research Support Grants, No. 5S07 RR5356-18 and No. 2 S07 RRDS36-20.

REFERENCES

Adelman, R. C., Obenrader, M. F., Klug, T. L., and Sartin. J. L., 1980, Hormone interactions during aging, in: *Biological Mechanisms in Aging* (R. T. Schimke, ed.), pp. 686–695. National Institute on Aging. U.S. Department of Health and Human Services, Public Health Service, National Institutes of Health.

Alumets, J., 1979, Topography of somatostatin cells in the stomach of the rat: Possible functional significance, *Cell Tiss. Res.* **202**:177–188.

Andrew, W., 1944, Senile changes in the pancreas of Wistar Institute rats and of man with special regard to the similarity of locule and cavity formation, *Am. J. Anat.* **74**:97–127.

Anver, M. R., and Cohen, B. J., 1979, Lesions associated with aging, in: *Biology and Diseases* (H. J. Baker, J. R. Lindsey, and S. H. Weisbroth, eds.), pp. 378–399, Academic Press, New York.

Anver, M. R., Cohen, B. J., Lattvada, C. P., and Foster, S. J., 1982, Age-associated lesions in barrier-reared male Sprague–Dawley rats: A comparison between Hap: (SD) and Crl:COBS$^{(R)}$CD$^{(R)}$ (SD) stocks, *Exp. Aging Res.* **8**:3–24.

Assan, R., Ahali, J. R., Ballerig, G., Boillot, J., and Girard, J., 1977, Glucagon secretion induced by natural and artificial amino acids in the perfused rat pancreas, *Diabetes* **26**:300–307.

Baetens, D., Stefan, Y., Ravazaola, M., Malaisse-Legae, F., Coleman, D. L., and Orci, L., 1978, Alteration of islet cell populations in spontaneously diabetic mice, *Diabetes* **27**:1–7.

Bartos, V., and Groh, J., 1969, The effect of repeated stimulation of the pancreas on the pancreatic secretion in young and aged men, *Gerontol. Clin.* **11**:56–62.

Berg, B. N., 1967, Lonvevity studies in rats: II Pathology of aging rats, in: *Pathology of Laboratory Rats and Mice* (E. Cotchin and R. J. C. Roe, eds.), pp. 749–786, Blackwell Scientific Publications, Oxford and Edinburgh.

Boorman, G. A., and Hollander, C. F., 1973, Spontaneous lesions in the female WAG/Rij (Wistar) rat, *J. Gerontol.* **28**:152–159.

Bunnag, S. C., 1966, Postnatal neogenesis of islets of Langerhans in the mouse, *Diabetes* **15**:480–491.

Cerasi, E., and Luft, R., 1972, Clinical diabetes and theories of pathogenesis, in: *Handbook of Physiology, Endocrinology*, Sec. 7., Vol. 1 (R. O. Greep and E. B. Astwood, eds.), Am Physiol. Soc., Washington, D. C., pp. 627–640.

Chio, K. S., Reiss, G., Fletcher, B., and Tappel, A. L., 1969, Peroxidation of subcellular organelles: Formation of lipofuscin-like fluorescent pigments, *Science* **166**:1535–1536.

Cohen, B. J., Anver, M. R., Ringler, D. H., and Adelman, R. C., 1978, Age-associated pathological changes in male rats, *Fed. Proc.* **37**:2848–2850.

Coleman, G. L., Barthold, S. W., Osbaldiston, G. W., Foster, S. J., and Jonas, A. M., 1977, Pathological changes during aging in barrier-reared Fischer 344 male rats, *J. Gerontol.* **32**:258–278.

Conklin, J. L., 1962, Cytogenesis of the human fetal pancreas, *Am. J. Anat.* **111**:181–193.

Crockford, P. M., Harbeck, R. J., and Williams, R. H., 1966, Influence of age on intravenous glucose tolerance and serum immuno-reactive insulin. *Lancet* **1**:465–576.

Devrim. S., and Recant, L., 1966, Effect of glucagon on insulin release in vitro, *Lancet* **2**:1227–1228.

Duckworth, W. C., and Kitabchi, A. E., 1974, Insulin and glucagon degradation by the same enzyme, *Diabetes* **23**:536–543.

Fikry, M., 1968, Exocrine pancreatic functions in the aged, *J. Am. Geriat. Soc.* **16**:463–467.

Forssman, W. G., Helmstaedter, V., Metz, J., Muhlmann, G., and Feurle, G. E., 1978, Immunohistochemistry and ultrastructure of somatostatin cells with special reference to the gastroenteropancreatic (GEP) system. *Metabolism* **27A**:1179–1191.

Freeman, C., Karoly, K., and Adelman, R. C., 1973, Impairments in availability of insulin in vivo and in binding of insulin to purified hepatic plasma membranes during aging, *Biochem. Biophys. Res. Commun.* **54**:1573–1580.

Gold, G., Reaven, G. M., and Reaven, E. P., 1981, Effect of age on proinsulin and insulin secretory patterns in isolated rat islets, *Diabetes* **30**:77–82.

Gommers, A., deGasparo, M., 1972, Variation de l'insulinemie en fonctions de l'age chez le rat male non traite, *Gerontologia* **18**:176–184.

Grodsky, G. M., Bennett, L. L., Smith, D. F., and Schmid, F. G., 1967, Effect of pulse administration of glucose or glucagon on insulin secretion in vitro, *Metabolism* **16**:223–233.

Hajdu, A., and Rona, G., 1967, Morphological observations on spontaneous pancreatic islet changes in rats, *Diabetes* **16**:108–110.

Hajdu, A., Herr, F., and Rona, G., 1968, The functional significance of a spontaneous pancreatic islet change in aged rats, *Diabetologia* **4**:44–47.

Hard, W. L., 1944, The origin and differentiation of the alpha and beta cells in the pancreatic islet of the rat, *Am. J. Anat.* **75**:369–403.

Hayashi, K., 1982, Insulin insensitivity and hyposuppressibility of glucagon by hyperglycemia in aged Wistar rats, *Gerontology* **28**:10–18.

Hellerstrom, C., 1977, Growth of pancreatic islets in animals, in: *The Diabetic Pancreas* (B. W. Volk and K. F. Wellman, eds.), Plenum Press, New York, pp, 61–97.

Hellman, B., 1959a, The effect of ageing on the total volumes of the A and B cells in the islets of Langerhans of the rat. *Acta Endocrinol.* **32**:92–112.

Hellman, B., 1959b, The effect of ageing on the number of the islets of Langerhans in the rat, *Acta. Endocrinol.* **32**:78–91.

Houcke, E., Houcke, M., and Leblois, J., 1964, Le pancrease du vieillard, *Presse Med.* **72**:1887–1892.

Kendrey, G., and Roe, F. J. C., 1969, Histopathological changes in the pancreas of laboratory rats, *Lab. Anim.* **3**:207–220.

Kim. S. K., Weinhold, P. A., Calkins, D. W., and Hartog, V. W., 1981, Comparative studies of the age-related changes in protein synthesis in the rat pancreas and parotid gland, *Exp. Geront.* **16**:91–99.

Klimas, J. E., 1968, Oral glucose tolerance during the life-span of a colony of rats, *J. Gerontol.* **23**:31–34.

Kyogoku, M., and Okamoto, K., 1969, Experimental studies on the insulin producing activity of the islet of Langerhans in aged rat, *Folia Endocrinol. Jpn.* **45**:794–795, 810–820.

Lazarus, S. S., and Volk, B. W., 1961, Pancreas in mature-onset diabetes, *Arch. Pathol.* **71**:56/44–71/59.

Leiter, E. H., Gapp, D. A., Eppig, J. J., and Coleman, D. L., 1977, Ultrastructural and morphometric studies of delta cells in pancreatic islets from C57BL/KS diabetic mice, *Diabetes* **26**:241–244.

Lewis, B. K., and Wexler, B. C., 1974, Serum insulin changes in male rats associated with age and reproductive activity, *J. Gerontol.* **29**:139–144.

Like, A. A., and Chick, W. L., 1969, Mitotic division in pancreatic beta cells, *Science* **163**:941–943.

Like, A. A., and Chick, W. L., 1974, Pancreatic beta cell replication induced by glucocorticoids in subhuman primates, *Am. J. Pathol.* **75**:329–348.

Like, A. E., and Jones, E. E., 1967, Studies on experimental diabetes in the Wellesley hybrid mouse IV. Morphologic changes in islet tissue, *Diabetologia* **3**:179–187.

Like, A. A., and Miki, E., 1967, Diabetic syndrome in sand rats IV. Morphologic changes in islet tissue, *Diabetologia* **3**:143–166.

Like, A. A., and Orci, L., 1972, Embryogenesis of the human pancreatic islets: A light and electron microscopic study, *Diabetes* **21(Suppl. 2)**:511–534.

Liu, H. M., and Potter, E. L., 1962, Development of the human pancreas, *Arch. Pathol.* **74**:53–66.

Logothetopoulos, J., 1972, Islet cell regeneration and neogenesis, in: *Handbook of Physiology, Endocrinology,* Sec. 7, (D. F. Steiner and N. Freinkel, eds.), Vol. 1, Am. Physiol. Soc., Washington, D.C., pp. 67–76.

Logothetopoulos, J., and Bell, E. G., 1966, Histological and autoradiographic studies of the islets of mice injected with insulin antibody, *Diabetes* **15**:205–211.

McEvoy, R. C., 1981, Changes in the volumes of the A-, B-, and D-cell populations in the pancreatic islets during the postnatal development of the rat, *Diabetes* **30**:813–817.

Meyer, J., and Necheles, H., 1940, Clinical significance of salivary, gastric and pancreatic secretion, *J. Am. Med. Assoc.* **115**:2050–2053.

Necheles, H., Plotke, F., and Meyer, J., 1942, Studies on old age; active pancreatic secretion in aged, *Am. J. Dig. Dis.* **9**:157–159.

Olefsky, J. M., and Reaven, G. M., 1975, The effects of age and obesity on ^{125}I insulin binding to isolated rat adipocytes, *Endocrinology.* **96**:1486–1498.

Patton, G. S., Ipp, E., Dobbs, R. E., Orci, L., Vale, W., and Unger, R. H., 1977, Pancreatic immunoreactive somatostatin release, *Proc. Natl. Acad. Sci.* **74**:2140–2143.

Pictet, R., and Rutter, W. J., 1972, Development of the embryonic endocrine pancreas. in: *Handbook of Physiology, Endocrinology,* Sec. 7 (D. F. Steiner and N. Freinkel, eds.), Vol. 1, Am. Physiol. Soc., Washington, D.C., pp. 25–66.

Pictet, R., Orci, L., Gonet, A. E., Rouiller, C., and Renold, A. E., 1967, Ultrastructural studies of the hyperplastic islets of Langerhans of spiny mice (Acomys cahirinus) before and during the development of hyperglycemia, *Diabetologia* **3**:188–211.

Reaven, E. P., Gold, G., and Reaven, G. M., 1979, Effect of age on glucose-stimulated insulin release by the β-cell of the rat, *J. Clin. Invest.* **64**:591–599.

Reaven, E., Gold, G., and Reaven, G., 1980, Effect of age on leucine-induced insulin secretions by the β-cell, *J. Gerontol.* **35**:324–328.

Reaven, E., Solomon, R., Azhar, S., and Reaven, G., 1982, Functional heterogeneity of pancreatic islets of aging rats, *Metabolism* **31**:859–860.

Reaven, G. M., and Reaven, P. D., 1980, Effect of age on glucose oxidation by isolated rat islets, *Diabetologia* **18**:69–71.

Remacle, G., Hauser, N., Jeanjean, J., and Gommers, A., 1977, Morphometric analysis of endocrine pancreas in old rats, *Exp. Gerontol.* **12**:207–214.

Ritter, V., 1964, Die altersbedingte Funktionseinbusse des Bauchspeicheldruse, *Wien. Med. Wochschr.* **114**:97–98.

Runyan, K., Duckworth, M. C., Kitabchi, A. E., and Huff, G., 1979, The effect of age on insulin-degrading activity in rat tissue, *Diabetes* **28**:324–325.

Rutter, W. J., 1980, The development of the endocrine and exocrine pancreas. in: *The Pancreas* (P. J. Fitzgerald and A. B. Morrison, eds.), Williams and Wilkins, Baltimore, Chap. 2, pp. 30–38.

Saladino, C. F., and Getty, R., 1972, Some historic and electron microscopic observations on the aged beagle pancreas, *Exp. Geront.* **7**:409–411.

Samols, E., and Harrison, J., 1976, Intraislet negative insulin-glucagon feedback, *Metabolism* **25**:1443–1447.

Samorajski, T., and Ordy, J. M., 1967, The histochemistry and ultrastructure of lipid pigment in the adrenal glands of aging mice, *J. Gerontol.* **22**:253–267.

Sanadi, D. R., 1977, Metabolic changes and their significance in aging, in: *Handbook of the Biology of Aging* (C. E. Finch and L. Hayflick, eds.), pp. 73–98, Van Nostrand Reinhold Company, New York.

Silverstone, F. A., Brandfanbrener, M., Shock, N. W., and Tiengst, M. J., 1957, Age differences in the intravenous glucose tolerance tests and the response to insulin, *J. Clin. Invest.* **36**:504–514.

Slavin, B. G., Paule, W. J., and Bernick, S., 1981, Possible mechanisms of pancreatic islet cell hyperplasia in aging rats, *J. Cell Biol.* **91**:219a.

Snook, J. T., 1975, Effect of age and long-term diet on exocrine pancreas of the rat, *Am. J. Physiol.* **228**:262–268.

Strehler, B. L., Mark, D. D., Mildvan, A. S., and Gee, M. V., 1959, Rate and magnitude of age pigment accumulation in the human myocardium, *J. Gerontol.* **14**:430–439.

Tappel, A. L., 1975, Lipid peroxidation and fluorescent molecular damage to membranes, in: *Pathobiology of Cell Membranes,* Vol. 1 (B. F. Trump and A. U. Arstila, eds.), pp. 145–170, Academic Press, New York.

Trimble, E. R., Gerber, P. G., and Renold, A. E., 1982, Effects of synthetic human pancreatic polypeptide, synthetic bovine pancreatic polypeptide and the C-terminal hexapeptide on pancreatic somatostatin and glucagon secretion in the rat, *Diabetes* **31**:178–181.

Unger, R. H., and Orci, L., 1977, Possible roles of pancreatic D-cell in the normal and diabetic states, *Diabetes* **26**:241–244.

Wexler, B. C., and Fischer, C. T., 1963, Hyperplasia of the islets of Langerhans in breeder rats with arteriosclerosis, *Nature* **200**:33–37.

Williams, T. F., 1978, Diabetes mellitus in the aged, in: *Geriatric Endocrinology, Aging,* Vol. 5 (R. B. Greenblatt, ed.), Raven Press, New York, pp. 103–113.

Zhukov, N., 1964, Intrainsular pericapillary fibrosis and its role in functional impairment of the insular apparatus in old age, *Probl. Endokr, Gorm.* **10**:13–17.

Zyss, R., 1968, Morphological counterparts of functional disturbances of pancreatic islets in aged people and in diabetics. Attempts at correlation, *Polish Med. J.* **7**:1448–1456.

INDEX

GPSR Compliance

The European Union's (EU) General Product Safety Regulation (GPSR) is a set of rules that requires consumer products to be safe and our obligations to ensure this.

If you have any concerns about our products, you can contact us on ProductSafety@springernature.com

In case Publisher is established outside the EU, the EU authorized representative is:

Springer Nature Customer Service Center GmbH
Europaplatz 3
69115 Heidelberg, Germany

Batch number: 10692442

Printed by Printforce, the Netherlands